MEAD JOHNSON
RESEARCH CENTER
JAN 22 1979
LIBRARY
PROPERTY OF BRISTOL-MYERS CO.
RESEARCH LIBRARY
WALLINGFORD, CT.

AF578221

Noncatecholic Phenylethylamines

MODERN PHARMACOLOGY-TOXICOLOGY

A Series of Monographs and Textbooks

Volume 1 A Guide to Molecular Pharmacology-Toxicology, Parts I and II
Edited by R. M. Featherstone

Volume 2 Psychopharmacological Treatment: Theory and Practice
Edited by Herman C. B. Denber

Volume 3 Pre- and Postsynaptic Receptors
Edited by Earl Usdin and William E. Bunney, Jr.

Volume 4 Synaptic Receptors: Isolation and Molecular Biology
Eduardo De Robertis

Volume 5 Methods in Narcotics Research
Edited by Seymour Ehrenpreis and Amos Neidle

Volume 6 Horizons in Clinical Pharmacology
Edited by Roger F. Palmer

Volume 7 Synthetic Antidiarrheal Drugs: Synthesis–Preclinical and Clinical Pharmacology
Edited by Willem Van Bever and Harbans Lal

Volume 8 Receptors and Mechanism of Action of Steroid Hormones, Parts I and II
Edited by Jorge R. Pasqualini

Volume 9 Hormone-Receptor Interaction: Molecular Aspects
Edited by Gerald S. Levey

Volume 10 Structure and Function of Monoamine Enzymes
Edited by Earl Usdin, Norman Weiner, and Moussa B. H. Youdim

Volume 11 Receptors in Pharmacology
Edited by John R. Smythies and Ronald J. Bradley

Volume 12 Noncatecholic Phenylethylamines (in two parts)
Edited by Aron D. Mosnaim and Marion E. Wolf

Additional Volumes in Preparation

Noncatecholic Phenylethylamines

PART 1

PHENYLETHYLAMINE: BIOLOGICAL MECHANISMS AND CLINICAL ASPECTS

edited by

Aron D. Mosnaim and Marion E. Wolf

Department of Pharmacology
The Chicago Medical School
University of Health Sciences
Chicago, Illinois

Department of Psychiatry
Rush-Presbyterian-St. Luke's Medical Center
Chicago, Illinois

MARCEL DEKKER, INC. New York and Basel

Library of Congress Cataloging in Publication Data

Noncatecholic phenylethylamines.

(Modern pharmacology-toxicology; v. 12)
Includes indexes.

CONTENTS: pt. 1. Phenylethylamine: biological mechanisms and clinical aspects.
1. Phenylethylamines. 2. Phenylethylamines--Physiological effect. 3. Brain chemistry.
I. Mosnaim, Aron D., 1940- II. Wolf, Marion E., 1948-
QP801.P39N66 599'.01'88 77-26130
ISBN 0-8247-6616-4

MARCEL DEKKER, INC.
270 Madison Avenue, New York, New York 10016

Current printing (last digit):
10 9 8 7 6 5 4 3 2 1

PRINTED IN THE UNITED STATES OF AMERICA

To Giselle and Janet

PREFACE

Studies on the nature of the physiological role of phenylethylamine and its implications in various clinical conditions are of current and widespread interest. Round-table discussions on this field have been held at meetings of the Society for Neurochemistry (1975) and the Society of Biological Psychiatry (1976). In this volume we have included the material presented at these conventions in addition to the contributions of a number of other research groups in an attempt to present an up-to-date balanced view of the knowledge in this field.

The first section of this book deals with the neurochemistry, metabolism, and pharmacology of phenylethylamine. Current knowledge of the mechanisms involved in the regulation of its biosynthesis, storage, metabolism, and excretion are reviewed in detail. The controversial subjects of tissue levels and brain regional and subcellular phenylethylamine distribution are also discussed. The second part examines the changes in animal behavior elicited by phenylethylamine. Different authors discuss the fundamental question of the catecholamine mediation of these effects and also present some evidence for the existence of phenylethylaminergic receptors. The third section reviews recent clinical evidence suggesting a role for phenylethylamine in phenylketonuria and in the modulation of affective and extrapyramidal disorders.

We wish to acknowledge the contributions of all authors to this book. We also appreciate the helpful comments of Drs. Ruven Greenberg, Alfredo Vásquez, Farouk Karoum, Albert Zeller, David Garver, Sabit Gabay, and Maurice Rapport. The generous support of the Chicago Medical School and the School of Graduate and Postdoctoral Studies of the University of Health Sciences throughout this endeavor is also greatly appreciated.

CONTRIBUTORS

FRANCES M. ACHEE, Biochemical Research Laboratory, Veterans Administration Hospital, Brockton, Massachusetts

SEYMOUR M. ANTELMAN, Western Psychiatric Institute and Clinic, Department of Psychiatry, University of Pittsburgh School of Medicine, and Psychobiology Program, Department of Psychology, University of Pittsburgh, Pittsburgh, Pennsylvania

KARL BLAU, Prenatal Biochemistry Unit, Department of Chemical Pathology, Bernhard Baron Memorial Research Laboratories, Queen Charlotte's Maternity Hospital, London, England

RICHARD L. BORISON,* Department of Pharmacology, Chicago Medical School, University of Health Sciences, and Department of Anesthesiology, Mount Sinai Hospital Medical Center, Chicago, Illinois

CLAUS BRAESTRUP, The Research Laboratory, Department E, St. Hans Mental Hospital, Roskilde, Denmark

BRUCE I. DIAMOND, Department of Anesthesiology, Mount Sinai Hospital Medical Center, Chicago, Illinois

DAVID J. EDWARDS, Western Psychiatric Institute and Clinic, Department of Psychiatry, University of Pittsburgh School of Medicine, Pittsburgh, Pennsylvania

ARNOLD J. FRIEDHOFF, Department of Psychiatry, Millhauser Laboratories, New York University School of Medicine, New York, New York

JOSE A. FUENTES,† Laboratory of Preclinical Pharmacology, National Institute of Mental Health, St. Elizabeth's Hospital, Washington, D.C.

---

*Current affiliation: Illinois State Psychiatric Institute, Chicago, Illinois.

†Current affiliation: Section of Pharmacology, Institute of Medicinal Chemistry, C.S.I.C., Madrid, Spain.

SABIT GABAY, Biochemical Research Laboratory, Veterans Administration Hospital, Brockton, Massachusetts

J. CHRISTIAN GILLIN, Division of Special Mental Health Research, Intramural Research Program, National Institute of Mental Health, St. Elizabeth's Hospital, Washington, D.C.

RUVEN GREENBERG, Department of Physiology, College of Medicine, University of Illinois at the Medical Center, Chicago, Illinois

HENRI S. HAVDALA, Department of Anesthesiology, Mount Sinai Hospital Medical Center, Chicago, Illinois

BERNARDO HELLER, Laboratory of Psychopharmacology and Experimental Neuropsychiatry, and School of Psychiatry, Faculty of Medicine, National University of Buenos Aires, Buenos Aires, Argentina

BENG T. HO, Texas Research Institute of Mental Sciences, the University of Texas Health Science Center at Houston, Houston, Texas

MARJORIE G. HORNING, Institute for Lipid Research, Baylor College of Medicine, Houston, Texas

DAVID M. JACKSON, Department of Pharmacology, University of Sydney, Sydney, New South Wales, Australia

GEORGE A. JERVIS, New York State Institute for Basic Research in Mental Retardation, Staten Island, New York

YEN HOONG LOO, New York State Institute for Basic Research in Mental Retardation, Staten Island, New York

JOAN MAY, Department of Pharmacology, Chicago Medical School, University of Health Sciences, and Department of Anesthesiology, Mount Sinai Hospital Medical Center, Chicago, Illinois

EGIDIO A. MOJA,* Division of Special Mental Health Research, Intramural Research Program, National Institute of Mental Health, St. Elizabeth's Hospital, Washington, D.C.

ARON D. MOSNAIM, Department of Pharmacology, Chicago Medical School, University of Health Sciences, Chicago, Illinois

FERNANDO ORREGO, Department of Biochemistry, National Institute of Cardiology, Mexico City, Mexico

*Current affiliation: Istituto di Clinica delle Malattie Nervose e Mentali, University of Cagliari, Italy.

STEPHEN R. PHILIPS, Psychiatric Research Division, University Hospital, Saskatoon, Saskatchewan, Canada

AXEL RANDRUP, The Research Laboratory, Department E, St. Hans Mental Hospital, Roskilde, Denmark

JUAN M. SAAVEDRA, Laboratory of Clinical Science, National Institute of Mental Health, Bethesda, Maryland

HECTOR C. SABELLI,* Department of Pharmacology, Chicago Medical School, University of Health Sciences, Chicago, Illinois

MARÍA CHRISTINA SALDATE, Department of Biochemistry, National Institute of Cardiology, Mexico City, Mexico

JACK W. SCHWEITZER, Department of Psychiatry, Millhauser Laboratories, New York University School of Medicine, New York, New York

S. ROBERT SNODGRASS, Department of Neurology and Neuropathology, Harvard Medical School, and Children's Hospital Medical Center, Boston, Massachusetts

H. SPATZ, Laboratory of Psychopharmacology and Experimental Neuropsychiatry, Hospital Nacional Jose T. Borda, Buenos Aires, Argentina

N. SPATZ, Laboratory of Psychopharmacology and Experimental Neuropsychiatry, Hospital Nacional Jose T. Borda, Buenos Aires, Argentina

DAVID M. STOFF, Division of Special Mental Health Research, Intramural Research Program, National Institute of Mental Health, St. Elizabeth's Hospital, Washington, D.C.

NORMAN J. URETSKY, Department of Neurology and Neuropathology, Harvard Medical School, and Children's Hospital Medical Center, Boston, Massachusetts

WOLFGANG H. VOGEL, Department of Pharmacology, Jefferson Medical College, Thomas Jefferson University, Philadelphia, Pennsylvania

CHRISTOPHER E. WHALLEY, Department of Physiology, College of Medicine, University of Illinois at the Medical Center, Chicago, Illinois

MARION E. WOLF, Department of Psychiatry, Rush-Presbyterian-St. Luke's Medical Center, Chicago, Illinois

---

*Current affiliation: Department of Psychiatry, Rush-Presbyterian-St. Luke's Medical Center, Chicago, Illinois.

RICHARD JED WYATT, Division of Special Mental Health Research, Intramural Research Program, National Institute of Mental Health, St. Elizabeth's Hospital, Washington, D.C.

J. A. YARYURA-TOBIAS, Research Division, North Nassau Mental Health Center, Manhasset, New York

E. ALBERT ZELLER,* Department of Biochemistry, Northwestern University Medical School, Chicago, Illinois

*Current affiliation: National Institute of Environmental Health Sciences, Research Triangle Park, North Carolina.

CONTENTS

Part A

BIOCHEMICAL STUDIES

Chapter 1

BIOCHEMICAL AND PHARMACOLOGICAL CHARACTERIZATION OF THE PHENYLETHYLAMINE METABOLIC PATHWAY

Aron D. Mosnaim

Department of Pharmacology
Chicago Medical School
University of Health Sciences
Chicago, Illinois

Marion E. Wolf

Department of Psychiatry
Rush-Presbyterian-St. Luke's Medical Center
Chicago, Illinois

I. INTRODUCTION

There has been a great interest in the research of the possible physiological role and pharmacological actions of 2-phenylethylamine (PEA), ever since Barger and Dale (1910) described its peripheral sympathomimetic effects (1). Undoubtedly the presence of the 2-phenylethyl radical as a basic component of the catecholamine molecule, as well as that of amphetamine and many other psychotropic drugs, has greatly furthered these studies.

Progress in this field has been slow until recently due to several factors, such as the toxic effects reported after PEA administration to humans (15,46), its short half-life (52,62), and technological problems in the development of a simple, sensitive, and reproducible methodology for the measurement of PEA in biological samples (2,3,12,16,17,18,28,51,55,58,60). The satisfactory resolution of this analytical problem, which also exists for a number of substances in close metabolic relationship with PEA (5,24,35,49, 56,59), strongly deserves further attention.

Studies in our laboratory provided the first conclusive evidence for the identification of PEA in urine (41) and of its presence as an endogenous constituent of mammalian tissues (38,42, 52), including mouse (42,52), rat (2), rabbit (3,38), and human brain (27). We also established the levels for this amine in whole brain (3,38), as well as in selected encephalic areas (39).

Following initial reports describing a correlation between urinary PEA excretion and depression (16; see Schweitzer and Friedhoff, this book), and the striking behavioral effects elicited by the injection of this amine to several animal species (see Jackson, this book), we decided to examine the possible effects of a number of centrally active drugs on the brain PEA levels with the aim of establishing the possible role of this amine in their mechanisms of action. Thus, we found that the brain levels of PEA were altered by a variety of drug treatments. In fact, they were increased by antidepressant drugs such as imipramine (39), L- and D-amphetamine (2,3), monoamine oxidase inhibitors (MAO I) (39), and L-DOPA (52), as well as by other agents such as Δ^9-tetrahydrocannabinol (39) and thyrotropin-releasing hormone (43), and significantly decreased by reserpine (52) and α-methyldopa (4). The integration of these various findings led us to formulate the PEA hypothesis of affective behavior, which in brief states that PEA is a neuromodulator responsible for triggering or sustaining wakefulness, alertness, and excitement (53).

Further studies on the metabolism of brain PEA were then required to achieve a better understanding of its possible role in the pathophysiology of depression and Parkinsonism (see Heller, this book). The findings that either a peripheral and central (α-methyldopa) or only peripheral (α-methyldopa hydrazine) L-aromatic amino acid decarboxylase enzyme inhibitors greatly decreased brain PEA levels (4), in conjunction with previous results showing that the systemic administration of either D- or L-phenylalanine increased these values (39), substantiated the hypothesis that brain PEA originates both from in situ L-phenylalanine decarboxylation and from peripherally formed amine (39,54,56). Further evidence to this effect was provided by the observation that the recovery of radioactive PEA from brain homogenates shortly after the intraventricular injection of labeled L-phenylalanine is dramatically decreased by pretreatment with α-methyldopa but remains unchanged after pretreatment with its hydrazine derivative (56).

Studies conducted to explain the various electrophysiological, behavioral, and clinical effects observed after the administration of PEA itself, its precursor L-phenylalanine, or drugs known to influence its brain levels led to the formulation of a number of possible mechanisms of action for this amine. Indeed, PEA has been demonstrated or postulated to exert its effects by releasing catecholamines (20), by directly acting on PEA receptors (13,34), by playing a neuromodulator and/or co-transmitter role (53), as a dopamine (DA) precursor (56), or through a combination of these (56).

In fact, in discussing the actions of PEA, especially in the CNS, one should consider this amine as an integral part of a complex metabolic pathway which includes a number of other biogenic amines and their corresponding metabolites, some of them known to have physiological actions of their own (23,34). In this paper, we shall describe what we call the "PEA metabolic pathway" based on its pharmacological interactions with a number of drugs, for example, pargyline, probenecid, thyrotropin-releasing hormone (TRH), imipramine, amphetamine, and the various mechanisms for PEA biosynthesis and its conversion to other biogenic amines and their metabolites.

II. MATERIALS AND METHODS

Radioactive {[1-^{14}C]PEA, sp. act. 4.85 mCi/mM and [4-^{3}H(N)]PEA, sp. act. 4.58 Ci/mM}, p-tyramine (TRM) ([1-^{14}C]TRM, sp. act. 12.4 mCi/mM and [^{3}H(G)]TRM , sp. act. 9.75 Ci/mM) and L-[U-^{14}C]phenylalanine (sp. act. 290 mCi/mM) were obtained from New England Nuclear, Boston, Mass. Phenylacetic acid (PAAc) ([1-^{14}C]PAAc, sp. act. 0.13 mCi/mM) was obtained from ICN, Irving, California. Before use, the radiochemical purity of the amines and of PAAc was examined by thin layer chromatography (TLC), followed by radioactivity scanning analysis (in every case >95% purity, Packard Model-7201 radiochromatogram scanner). All other reagents and solvents were of the finest grade commercially available and were used without further purification.

III. BRAIN METABOLIC STUDIES

White male New Zealand rabbits (2.0-2.5 kg; 9-11 weeks) were implanted (under light local anesthesia) with injection cannulae in the right lateral ventricle 2 to 3 days before the experiment. The animals were sacrificed by decapitation 10 min after the intraventricular injection of the labeled precursor. The whole brains were immediately removed, the main surface blood vessels and blood clots carefully dissected out, and the tissue quickly rinsed in chilled saline (0.9% NaCl), patted dry, and weighed. After removal, the injection cannulae corresponding to each group of animals were rinsed thoroughly with saline, and the washings evaporated and counted. Various other drugs were dissolved in saline and injected intraperitoneally as described (Tables 1 and 2). The procedure followed for PEA and TRM determination has been described elsewhere (56).

For PAAc measurements, the pH of the water phase remaining after amine extraction (34) was adjusted to 2.0 (conc. HCl) and the organic acid extracted into chloroform (15 ml x 2). Phenylacetic acid was further purified by its re-extraction into a pH 9.0 solution (NaOH 0.1 N) followed by pH lowering (2.0, HCl) and its final extraction into more chloroform (10 ml x 2). The pooled organic layers were evaporated to dryness (N_2 blowing), and the residue was redissolved in a small volume (100-500 μl) of fresh chloroform. After counting an aliquot (20-100 μl), the rest was spotted on a

TABLE 1

Biosynthesis of PEA, TRM, and PAAc in Rabbit Brain After the Intraventricular Administration of Labeled Precursors—Drug Effects

Pretreatment (time before treatment)	Intraventricular Treatment[a] (time before sacrifice)	$[^{14}C]$PEA Recovered[b]	$[^{14}C]$TRM Recovered[b]	$[^{14}C]$PAAc Recovered[b]
Saline, 2 ml (1 hr, ip)	L-$[^{14}C]$Phenylalanine (U) (10 min)	16.80[c] ± 3.40 (10^{-2})	1.70[c] ± 0.17 (10^{-2})	1.10 ± 0.18
Pargyline, 50 mg/kg (72, 48, 24 hr, ip)	L-$[^{14}C]$Phenylalanine (U) (10 min)	83.27[c] ± 0.52 (10^{-2})	56.33[c] ± 0.29 (10^{-2})	0.08 ± 0.02
Probenecid, 150 mg/kg (2 hr, ip)	L-$[^{14}C]$Phenylalanine (U) (10 min)	21.40 ± 3.80 (10^{-2})	2.60 ± 1.40 (10^{-2})	1.71 ± 0.21
Thyrotropin-releasing hormone, 1 mg/kg (1 hr, ip)	L-$[^{14}C]$Phenylalanine (U) (10 min)	13.40 ± 2.90 (10^{-2})	4.30 ± 2.10 (10^{-2})	1.20 ± 0.24
Saline, 2 ml (1 hr, ip)	p-$[1-^{14}C]$Tyramine (10 min)	1.28[c] ± 0.23	1.29[c] ± 0.39	—
Saline, 2 ml (1 hr, ip)	2-$[1-^{14}C]$Phenylethylamine (10 min)	3.16[c] ± 1.34	0.58[c] ± 0.43	18.41 ± 1.54

[a]Two to four animals per group, two or more groups per experiment; L-$[^{14}C]$phenylalanine (U) (spec. act. 290 mCi/mM, 0.36 μg/10 μl per rabbit); 2-$[1-^{14}C]$phenylethylamine (spec. act. 4.85 mCi/mM, 18.6 μg/10 μl per rabbit); p-$[1-^{14}C]$tyramine (spec. act. 12.4 mCi/mM, 0.34 μg/10 μl per rabbit).

[b]As percent of radioactivity injected.

[c]From Silkaitis and Mosnaim (56).

Quanta/Gram cellulose chromatographic plate along with 5 μg of standard PAAc, and developed unidimensionally using an alkaline solvent system [ethanol (96%) : ammonia (conc.); 20 : 1] for 2 hr at room temperature. After TLC radioscanning (radioactive area at R_F 0.24-0.31), the area containing the standard PAAc was sprayed with a freshly prepared solution (aniline : ethanol : water : n-butanol and glucose; 1 : 10 : 10 : 30 ml and 1 g) and heated in a ventilated oven at 115 to 120°C for 7 to 10 min [orange-brown color over a yellowish background (R_F 0.27-0.31)]. The area containing the labeled substance was scraped off the plate, extracted into chloroform, and counted. For the estimation of PEA, TRM, and PAAc, percent of recovery, ^{3}H-amines and ^{14}C-compounds (amines and PAAc) were used as internal and external standards, respectively.

IV. RESULTS

The data with respect to the recovery of labeled PEA, TRM, and PAAc from brain homogenates after the intraventricular administration of either radioactive L-phenylalanine, PEA, or TRM are listed in Tables 1 and 2. After development and TLC radioscanning, the plates spotted with the brain samples containing the PAAc fraction showed only one radioactive area with R_F similar to that of standard PAAc chromatographed under identical conditions (R_F 0.24-0.31 and 0.27-0.31, respectively). Small but measurable amounts of labeled PEA, TRM, and PAAc were present in brain homogenates 10 min after the intraventricular administration of radioactive L-phenylalanine. In similar experiments, pretreatment with pargyline increased approximately four- and 25-fold, the recoveries of PEA and TRM, respectively, while drastically reducing those of PAAc. When probenecid was used instead, there was no significant change in the amine recoveries, but PAAc values were about 1.5 times higher than in the control experiments. Thyrotropin-releasing hormone did not seem to affect significantly the recoveries of any of these compounds.

As shown in Table 2, psychotropic drugs influenced the absolute as well as relative amounts of PEA, TRM, and PAAc present in brain

TABLE 2

Biosynthesis of TRM and PAAc in Rabbit Brain After the Intraventricular Administration of Labeled PEA—Drug Effects

Pretreatment (time before treatment)	Intraventricular Treatment[a] (time before sacrifice)	$[^{14}C]$PEA Recovered[b]	$[^{14}C]$TRM Recovered[b]	$[^{14}C]$PAAc Recovered[b]
Saline, 2 ml (1 hr, ip)	2-[1-^{14}C]Phenylethylamine (10 min)	3.16[c] ± 1.34	0.58[c] ± 0.43	18.41 ± 1.54
Pargyline, 50 mg/kg (72, 48, 24 hr, ip)	2-[1-^{14}C]Phenylethylamine (10 min)	7.62[c] ± 0.60	1.74[c] ± 0.33	0.61 ± 0.12
Probenecid, 150 mg/kg (2 hr, ip)	2-[1-^{14}C]Phenylethylamine (10 min)	—	—	31.40 ± 5.30
D-Amphetamine•SO_4, 13.5 mg/kg (0.5 hr, ip)	2-[1-^{14}C]Phenylethylamine (10 min)	0.92[d] ± 0.26	—	4.14 ± 0.56
D-Amphetamine•SO_4, 13.5 mg/kg (4 hr, ip)	2-[1-^{14}C]Phenylethylamine (10 min)	1.73[d] ± 0.41	—	8.45 ± 1.38
Thyrotropin-releasing hormone, 1 mg/kg (1 hr, ip)	2-[1-^{14}C]Phenylethylamine (10 min)	6.70 ± 0.43	0.47 ± 0.11	19.41 ± 2.38

[a]Two to four animals per group, two or more groups per experiment; 2-[1-^{14}C]phenylethylamine (spec. act. 4.85 mCi/mM, 18.6 μg/10 μl per rabbit).

[b]As percent of radioactivity injected.

[c]From Silkaitis and Mosnaim (56).

[d]From Borison et al. (3).

10 min after PEA administration. Thus, pargyline increased approximately two- and sixfold the recoveries of PEA and TRM, respectively, while drastically decreasing those of PAAc (18.41 to 0.61%). Amphetamine (0.5 or 4 hr prior to sacrifice) decreased the brain content of labeled PEA and PAAc while not affecting TRM.

The i.p. injection of TRH, prior to the intraventricular administration of labeled PEA, produced a small but significant increase in the recoveries of both PEA and TRM while not affecting PAAc. As expected, probenecid markedly increased the values obtained for labeled PAAc.

The results recorded in Tables 1 and 2 have been corrected for percent recovery. In the experiments in which either L-phenylalanine or TRM was used as labeled precursor, the percent recovery ranged from 54 to 92% for PEA ([^{3}H]PEA internal standard), 49 to 67% for TRM and 62 to 81% for PAAc. When PEA was injected intraventricularly, the percent recovery ranged from 15 to 31% for TRM ([^{3}H]TRM internal standard), 43 to 66% for PEA and 57 to 83% for PAAc.

V. DISCUSSION

Techniques involving the injection of small amounts of labeled substances into the lateral ventricles or the basal cisterns of the brain, thus by-passing the blood-brain barrier, have facilitated many studies of biochemical pathways occurring in the brain. The characteristics of this technique impose, however, some severe limitations as to the extent of using the data thus obtained to discuss the nature and mechanisms of physiologically occurring processes. The destruction of tissue around the inserted cannula and the possibility of formation of artifactual pools of injected substance and its possible different distribution in both brain regions and subcellular structures as compared with its endogenous counterpart are some of the factors to be considered in discussing the results obtained when using this model (21).

As could be expected from its rapid exit from brain tissue and metabolization, mainly to PAAc (34), only a small fraction of the originally injected PEA is present in brain homogenates either as unchanged PEA or aminated PEA metabolites shortly after its intraventricular administration. This poses a serious analytical problem in measuring the concentration of these substances, making critical the amount and the specific activity of the injected PEA and the time interval between injection and animal sacrifice. These problems are greatly increased when estimating the even smaller recoveries of labeled PEA and metabolites after the intraventricular administration of radioactive PEA precursors, for example L-phenylalanine and TRM (56).

Our studies of TRM and PEA recoveries after the intraventricular administration of labeled PEA indicate 10 min as a useful time interval to use in our experimental model since when injecting the specified amounts of labeled PEA (Tables 1 and 2), this timing allowed the formation and accumulation of measurable amounts of TRM and PAAc, whereas affording the maximal recovery values for unchanged PEA. Similarly, 10 min had also proved to be a suitable time interval in studying the formation of PEA and TRM and of PEA from intraventricularly administered L-phenylalanine or TRM, respectively (56).

In attempting to explain the different biochemical, electrophysiological, behavioral, and clinical effects observed after the administration of PEA or attributed to this amine after the administration of its amino acid precursor L-phenylalanine (39) (see Yaryura-Tobias, this book), one should consider that PEA could be acting by itself, for example, on phenylethylaminergic receptors and/or via its conversion to other substances involved in the PEA pathway. In Fig. 1, we present a diagram which is useful in the understanding of the overall mechanism of actions of PEA as presently known.

Recent work has substantiated the participation of a number of biochemical reactions in the disposition of PEA. Besides its main metabolic route, conversion to PAAc via phenylacetaldehyde (34), brain PEA has been shown to proceed also to other physiologically active amines such as hydroxyphenylethylamine (OHPEA) (36), recently

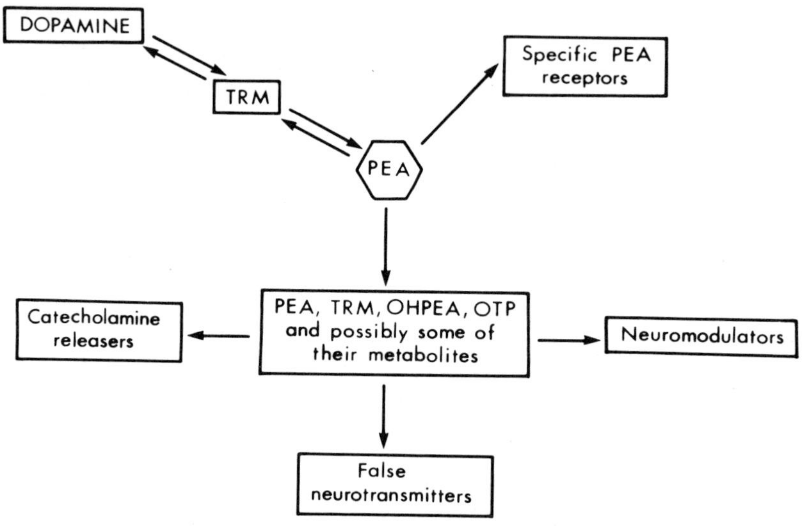

FIG. 1. Possible mechanisms of action of brain PEA.

identified as an endogenous component of mammalian brain (27), and to TRM (31,56). From this amine, it could also be converted to octopamine (7,25) and DA (31,56). Furthermore, the formation of other aminated PEA metabolites, for example, m-TRM (49), N-methyl (53) and N-acetyl PEA (26), and pyridoxylidene-2-PEA (32) should also be considered. Moreover, there are strong possibilities that some deaminated metabolites of these substances could have physiological effects of their own (23,34). The relative physiological and quantitative importance of any one of these processes would depend, among other factors, on the concentration and localization, regional (5,39,51,60) as well as subcellular (5, see elsewhere this book), of available PEA and the enzymes and receptor sites with which it interacts.

The fact that PEA readily crosses the blood-brain barrier (45, 47,49,52), actually peripherally administered amine has been shown to accumulate in the brain (45), makes apparent that net variations on its peripheral availability resulting from noncompensated changes in processes regulating its biosynthesis and/or metabolism, among others the levels of L-phenylalanine (8,9), L-aromatic aminoacid decarboxylase (3,33), and/or type B monoamine oxidase (MAO) activity (63), could in turn result in significant variations on brain PEA levels, and be ultimately reflected in the delicate balance of the substances involved in brain phenylethylaminergic and/or adrenergic mechanisms (Fig. 1).

Although under physiological conditions the peripheral decarboxylation to PEA could be expected to be a relative minor metabolic route for L-phenylalanine [Km 10^{-2} M (S), whereas its plasma levels are in the order of 10^{-4} M], recent evidence indicates that when the circulating levels of this aminoacid are substantially increased, for example, loading L-phenylalanine and ingesting a large protein-rich meal, its conversion to PEA becomes a major metabolic route (8). This enhanced PEA biosynthesis, which takes place not only in the liver, kidney, and other tissues, but perhaps also in erythrocytes (red cells contain L-phenylalanine and L-aromatic amino acid decarboxylase), could be expected to increase brain PEA levels. The duration, extent, and physiological importance of these increases are the overall result of the different reactions involved in PEA biosynthesis and metabolism.

It is generally accepted that the quantitatively most important route for PEA metabolization is its deaminative oxidation by type B MAO (63). The characteristics of this PEA-MAO interaction and the possible clinical implications of the strong dependency of PEA concentration on MAO activity deserves further study. In fact, changes on platelets MAO activity, which have been shown to occur in several disease states, including juvenile diabetes mellitus (61) as well as during different stages of some physiological cycles (30), will produce variations on the PEA availability which in turn could influence a number of peripheral as well as brain processes. The recent evidence for decreased platelet MAO in chronic schizophrenic patients (11) and bipolar affective disorders (44) gives support to the so-called PEA hypothesis of affective behavior (53). It would be expected that these patients will have higher base levels of blood PEA and therefore brain PEA levels, which in turn will result in sustained high brain levels of a number of other biogenic amines, particularly important in this case DA, and some of their metabolites. Free blood PEA levels, in dynamic equilibrium with brain PEA concentration and, therefore, with a number of substances involved in the PEA metabolic pathway, could thus represent a peripheral mechanism for the regulation of some brain processes.

The results in Tables 1 and 2 corroborates previous studies on brain PEA biosynthesis (54), and the role of this amine as a precursor for brain TRM and PAAc (56). They also provide with further evidence to support the suggested involvement of an active transport system in the disposition of brain PAAc [(48), see probenecid pretreatment], and for the in vivo and in vitro occurrence of a brain dehydroxylation reaction converting TRM to PEA (37,56). This biochemical pathway, together with the already reported brain conversion of DA to TRM (31,53), could in certain conditions, for example, excess brain DA, represent an important route in catecholamines metabolism. For example, this pathway could be useful in explaining the reported increase in mouse brain PEA levels after the intraperitoneal administration of L-DOPA (52).

The large increase in TRM and decrease in PAAc recoveries obtained from animals pretreated with pargyline (after either L-phenylalanine or PEA intraventricularly injection) are most probably due to the inhibition of TRM and PEA deaminative oxidation. p-Tyramine values also reflect its increased biosynthesis due to extra PEA availability.

Phenylacetic acid recovered after the intraventricular administration of L-phenylalanine (pargyline pretreatment) probably represents not only an incomplete MAO I but also the contribution of the L-phenylalanine → phenylpyruvic → PAAc pathway, a metabolic route clearly deserving further study (34, see Loo et al. and Blau, this book).

The variable decrease in radioactive PEA and PAAc recoveries obtained from brain homogenates of animals pretreated with D-amphetamine (after the intraventricular injection of labeled PEA) are probably the overall result of a complex interaction of the drug with one or more processes related to the PEA biochemical pathway. Thus, amphetamine is known to release PEA and to have weak MAO I properties. It may also compete with PEA for uptake by nerve terminals for β-hydroxylation and for p-hydroxylation (2-4). Indeed, pretreatment with amphetamine increases the brain levels of endogenous PEA (2,3), as well as the recoveries of labeled amine from brain homogenates of animals intraventricularly injected with radioactive L-phenylalanine (4). The results showing an increase in both endogenous brain PEA (43) and PEA recoveries (after intraventricular injection of this amine) in animals pretreated with TRH could prove of particular importance in explaining some of the alleged effects of this hormone (10). Our preliminary results also show that TRH does potentiate the excitatory effects of PEA when using a protocol similar to the DOPA-potentiation test (43).

In summary, we have analyzed in this paper the various biosynthetic and degradative pathways for PEA, the influence of certain drugs on its metabolism, and the different mechanisms of action through which this amine exerts its physiological and pharmacological effects.

ACKNOWLEDGMENTS

Part of the research reported in this review was carried out in collaboration with Mr. R. Silkaitis, Mr. Madubuike, Mr. C. Whalley, Mr. M. Mason, Dr. R. Borison, and Dr. E. Inwang. We also acknowledge the skilled technical assistance of Mr. S. Myles, Ms. D. Edstrand, and Ms. L. Jones in the various aspects of this work. This project was supported in part by grants from the General Research Support

Branch, Division of Research Facilities and Resources, NIH General Research Support Grant FR-5366. Drugs were kindly donated by Merck, Sharp and Dohme and by Abbott Laboratories.

REFERENCES

1. A. Barger and H. H. Dale (1910). Chemical structure and sympathomimetic action of amines. *J. Physiol. (Lond.)*, 41:19-59.

2. R. L. Borison (1975). Endogenous brain 2-phenylethylamine: Biochemical and pharmacological studies on its origin and possible role as a major mediator of the central actions of amphetamine and other psychotropic drugs. Ph.D. Thesis in Pharmacology, The University of Health Sciences/The Chicago Medical School.

3. R. L. Borison, A. D. Mosnaim, and H. C. Sabelli (1975). Brain 2-phenylethylamine as a mediator for the central actions of amphetamine and methylphenidate. *Life Sci.*, 17:1331-1344.

4. R. L. Borison, A. D. Mosnaim, and H. C. Sabelli (1974). Biosynthesis of brain 2-phenylethylamine: Influence of decarboxylase inhibitors and D-amphetamine. *Life Sci.*, 15:1837-1848.

5. A. A. Boulton and G. B. Baker (1975). The subcellular distribution of β-phenylethylamine, p-tyramine and tryptamine in rat brain. *J. Neurochem.*, 25:477-481.

6. A. A. Boulton and L. Milward (1971). Separation, detection and quantitative analysis of urinary β-phenylethylamine. *J. Chromatogr.*, 57:287-296.

7. A. A. Boulton and P. H. Wu (1972). Biosynthesis of cerebral phenolic amines. I. In vivo formation of p-tyramine, octopamine, and synephrine. *Can. J. Biochem.*, 50:261-267.

8. J-C. David, W. Dairman, and S. Udenfriend (1974). Decarboxylation to tyramine: A major route of tyrosine metabolism in mammals. *Proc. Nat. Acad. Sci. USA*, 71:1771-1775.

9. J-C. David, W. Dairman, and S. Udenfriend (1974). On the importance of decarboxylation in the metabolism of phenylethylamine, tyrosine, and tryptophan. *Arch. Biochem. Biophys.*, 160:561-568.

10. K. L. Davis, L. E. Hollister, and P. A. Berger (1975). Thyrotropin-releasing hormone in schizophrenia. *Am. J. Psychiat.*, 132(9):951-953.

11. E. F. Domino and S. S. Khanna (1976). Decreased blood platelet MAO activity in unmedicated chronic schizophrenic patients. *Am. J. Psychiat.*, 133(3):323-326.

12. D. A. Durden, S. R. Philips, and A. A. Boulton (1973). Identification and distribution of β-phenylethylamine in the rat. *Can. J. Biochem.*, 51:995-1002.

13. J. M. Feldman and H. E. Lebovitz (1971). The nature of the interaction of amines with the pancreatic beta cell to influence insulin secretion. *J. Pharmacol. Exp. Ther.*, 179:56-65.

14. L. R. Fernandez, C. E. Rodriguez, H. Spatz, and N. Spatz (1974). Influence of anti-depressive therapy on phenethylaminuria. *An. Psiq. Bio.*, 3:237-238.

15. E. Fischer. Personal communication.

16. E. Fischer, B. Heller, and A. H. Miro (1968). β-Phenylethylamine in human urine. *Arzneim.-Forsch.*, 18:1486.

17. E. Fischer, H. Spatz, R. S. Fernandez Labriola, E. M. Rodriguez Casanova, and N. Spatz (1973). Quantitative gas-chromatographic determination and infrared spectrographic identification of urinary phenethylamine. *Biol. Psychiat.*, 7:161-165.

18. E. Fischer, H. Spatz, B. Heller, and H. Reggiani (1972). Phenylethylamine content of human urine and rat brain, its alterations in pathological conditions and after drug administration. *Experientia*, 28:397-398.

19. R. W. Fuller and B. W. Roush (1972). Substrate-selective and tissue-selective inhibition of monoamine oxidase. *Arch. Int. Pharmacodyn. Ther.*, 198:270-276.

20. L. Fuxe, H. Grobecker, and J. Johnson (1967). The effect of β-phenylethylamine on central and peripheral monoamine-containing neurons. *Eur. J. Pharmacol.*, 2:202-207.

21. D. W. Gallager, E. Sanders-Bush, G. K. Aghajanian, and F. Sulser (1975). An evaluation of the use of intraventricularly administered [^{3}H]5-hydroxytryptamine as a marker for endogenous brain 5-hydroxytryptamine. *Brain Res.*, 93:111-122.

22. W. J. Giardina, W. A. Pedemonte, and H. C. Sabelli (1973). Iontophoretic study of the effects of norepinephrine and 2-phenylethylamine on single cortical neurons. *Life Sci.*, 12:153-161.

23. W. J. Giardina and H. C. Sabelli (1974). Evidence for the biological activity of the deaminated metabolites of the adrenergic amines. *Biol. Psychiat.*, 9(1):11-23.

24. L-M. Gunne and J. Jonsson (1965). On the occurrence of tyramine in the rabbit brain. *Acta. Physiol. Scand.*, 64:434-438.

25. A. J. Harmar and A. S. Horn (1976). Octopamine in mammalian brain: rapid post-mortem increase and effects of drugs. *J. Neurochem.*, 26:987-994.

26. L. L. Hsu, M. A. Geyer, and A. J. Mandell (1976). Extrapineal amine N-acetylation in rat brain. Regional and subcellular distribution and enzyme kinetics. *Biochem. Pharmacol.*, 25: 815-819.

27. E. E. Inwang, A. D. Mosnaim, and H. C. Sabelli (1973). Isolation and characterization of phenylethylamine and phenylethanolamine from human brain. *J. Neurochem.*, 20:1469-1473.

28. E. E. Inwang, B. J. Primm, F. de L. Jones, H. Dekirmenjian, J. M. Davis, and E. T. Henderson (1975/76). Metabolic disposition of 2-phenylethylamine and the role of depression in methadone-dependent and detoxified patients. *Drug and Alcohol Dependence,* 1:295-303.

29. J. P. Johnston (1968). Some observations upon a new inhibitor of monoamine oxidase in brain tissue. *Biochem. Pharmacol.,* 17:1285-1297.

30. E. L. Klaiber, D. M. Broverman, W. Vogel, Y. Kobayashi, and D. Moriarty (1972). Effects of estrogen therapy on plasma MAO activity and EEG driving responses of depressed women. *Am. J. Psychiat.,* 128:1492.

31. L. Lemberger, A. Klutch, and R. Kuntzman (1966). The metabolism of tyramine in rabbits. *J. Pharmacol. Exp. Ther.,* 153(2):183-190.

32. Y. H. Loo and W. M. Cort (1972). Assay of pyridoxal and its derivatives. In: *Methods of Neurochemistry, Vol. 2,* edited by R. Fried. Marcel Dekker, New York, pp. 169-204.

33. W. Lovenberg, H. Weissback, and S. Udenfriend (1962). Aromatic L-amino acid decarboxylase. *J. Biol. Chem.,* 237:89-93.

34. U. P. Madubuike (1975). Phenylacetic acid (a new neuroacid) isolation, quantification and pharmacological studies. M.S. Thesis in Pharmacology, The University of Health Sciences/The Chicago Medical School.

35. J. D. Majer and A. A. Boulton (1970). Absolute, unambiguous ultramicroanalysis of metabolites present in complex biological extracts. *Nature,* 225:658-660.

36. B. P. Molinoff, L. Landsberg, and J. Axelrod (1969). An enzymatic assay for octopamine and other β-hydroxylated phenylethylamines. *J. Pharmacol. Exp. Ther.,* 170:253.

37. A. D. Mosnaim, D. L. Edstrand, M. E. Wolf, and R. P. Silkaitis (1977). Evidence for an enzymatic p-tyramine-dehydroxylating system in rabbit brain preparations in vitro. *Biochem. Pharmacol.,* in press.

38. A. D. Mosnaim and E. E. Inwang (1973). A spectrophotometric method for the quantification of 2-phenylethylamine in biological specimens. *Anal. Biochem.,* 54:561-577.

39. A. D. Mosnaim, E. E. Inwang, and H. C. Sabelli (1974). The influence of psychotropic drugs on the levels of endogenous 2-phenylethylamine in rabbit brain. *Biol. Psychiat.,* 8(2): 277-234.

40. A. D. Mosnaim, E. E. Inwang, J. H. Sugerman, W. J. DeMartini, and H. C. Sabelli (1973). Ultraviolet spectrophotometric determination of 2-phenylethylamine in biological samples and its possible correlation with depression. *Biol. Psychiat.,* 6:235-257.

41. A. D. Mosnaim, E. E. Inwang, J. H. Sugerman, and H. C. Sabelli (1973). Identification of 2-phenylethylamine in human urine by infrared and mass spectroscopy and its quantification in normal subjects and cardiovascular patients. *Clin. Chim. Acta,* 46: 407-413.

42. A. D. Mosnaim and H. C. Sabelli (1971). Quantitative determination of the brain levels of β-phenylethylamine-like substance in control and drug-treated mice. *Pharmacologist,* 13:283.

43. A. D. Mosnaim, M. E. Wolf, and P. Plotnikoff (1976). Possible involvement of the 2-phenylethylamine pathway in the central effects of thyrotropin-releasing hormone. *Soc. Biol. Psychiat.* (Abstract), California.

44. D. L. Murphy and R. Weiss (1972). Reduced monoamine oxidase activity in blood platelets from bipolar depressed patients. *Am. J. Psychiat.,* 128:1351-1357.

45. T. Nakajima, Y. Kakimoto, and I. Sano (1964). Formation of phenylethylamine in mammalian tissues and its effect on motor activity in the mouse. *J. Pharmacol. Exp. Ther.,* 143:319.

46. T. Nakanishi (1952). The therapeutic effect of β-phenylethylamine on the mental disorders. *Biological Amines* (in Japanese), 148.

47. W. H. Oldendorf (1971). Brain uptake of radiolabeled amino acids, amines, and hexoses after arterial injection. *Am. J. Physiol.,* 221:1629-1639.

48. W. A. Pedemonte, A. D. Mosnaim, and M. Bulat (1976). Penetration of phenylacetic acid across the blood-cerebrospinal fluid barrier. *Res. Commun. Chem. Pathol. Pharmacol.,* 14:111-116.

49. S. R. Philips, B. A. Davis, D. A. Durden, and A. A. Boulton (1975). Identification and distribution of m-tyramine in the rat. *Can. J. Biochem.,* 53:65-69.

50. S. R. Philips, D. A. Durden, and A. A. Boulton (1974). Identification and distribution of p-tyramine in the rat. *Can. J. Biochem.,* 52:366-373.

51. J. M. Saavedra (1974). Enzymatic isotopic assay for and presence of β-phenylethylamine in brain. *J. Neurochem.,* 22:211-216.

52. H. C. Sabelli, W. J. Giardina, A. D. Mosnaim, and N. H. Sabelli (1973). A comparison of the functional roles of norepinephrine, dopamine and phenylethylamine in the central nervous system. Proc. Satellite Symposium "Central and Peripheral Adrenergic Systems," XXV int. Congress of Physiological Sciences, Warsaw, August, 3-6, 1971/*Acta Physiologica Polonica,* 24:33-40.

53. H. C. Sabelli and A. D. Mosnaim (1974). Phenylethylamine hypothesis of affective behavior. *Am. J. Psychiat.,* 131: 695-699.

54. H. C. Sabelli, W. A. Pedemonte, C. Whalley, A. D. Mosnaim, and A. J. Vazquez (1974). Further evidence for a role of 2-phenylethylamine in the mode of action of Δ^9-tetrahydrocannabinol. *Life Sci.*, 14:149-156.

55. J. W. Schweitzer, A. J. Friedhoff, and R. Schwartz (1975). Phenethylamine in normal urine: failure to verify high values. *Biol. Psychiat.*, 10:277-285.

56. R. P. Silkaitis and A. D. Mosnaim (1976). Pathways linking L-phenylalanine and 2-phenylethylamine with p-tyramine in rabbit brain. *Brain Res.*, 114:105-115.

57. H. Spatz, B. Heller, M. Nachon, and E. Fischer (1975). Effects of D-phenylalanine on clinical picture and phenethylaminuria in depression. *Biol. Psychiat.*, 10(2):253-239

58. H. Spatz and N. Spatz (1972). Spectrophotophlorometric determination of Beta-phenylethylamine in blood and urine. *Biochem. Med.*, 6:1-6.

59. S. Spector, K. Melmon, W. Lovenberg, and A. Sjoerdsma (1963). The presence and distribution of tyramine in mammalian tissues. *J. Pharmacol. Exp. Ther.*, 140:229-235.

60. J. Willner, H. F. LeFevre, and E. Costa (1974). Assay by multiple ion detection of phenylethylamine and phenylethanolamine in rat brain. *J. Neurochem.*, 23:857-859.

61. M. E. Wolf, A. D. Mosnaim, and E. A. Zeller (1977). Platelet monoamine oxidase: Reduced activity in patients with juvenile diabetes mellitus. *Fed. Proc.* (Abstract).

62. P. H. Wu and A. A. Boulton (1975). Distribution, metabolism, and disappearance of intraventricularly injected p-tyramine in the rat. *Can. J. Biochem.*, 52:374-381.

63. H. Y.-T. Yang and N. H. Neff (1973). β-Phenylethylamine: A specific substrate for type B monoamine oxidase of brain. *J. Pharmacol.*, 187:365-371.

64. J. A. Yaryura-Tobias, B. Heller, H. Spatz, and E. Fischer (1974). Phenylalanine for endogenous depression. *J. Ortho. Psychiat.* 3(2):80-81.

Chapter 2

PHENYLETHYLAMINE IN BRAIN FUNCTION AND DYSFUNCTION: NEUROCHEMICAL AND PHARMACOLOGICAL STUDIES

David J. Edwards

Western Psychiatric Institute and Clinic
Department of Psychiatry
University of Pittsburgh School of Medicine
Pittsburgh, Pennsylvania

Seymour M. Antelman

Western Psychiatric Institute and Clinic
Department of Psychiatry
University of Pittsburgh School of Medicine
and
Psychobiology Program
Department of Psychology
University of Pittsburgh
Pittsburgh, Pennsylvania

I. INTRODUCTION

Interest in the actions of phenylethylamine (PEA) in the CNS has centered around neurochemical, pathological, and pharmacological questions. Briefly, these areas of investigation have sought to determine: 1) what, if any, is the role of PEA in normal brain function, 2) whether alterations in either the content or metabolism of PEA might be related in some manner (whether as a causative factor or as a secondary effect) to certain disorders, and finally, 3) what the mechanisms are by which exogenously administered doses of PEA produce their behavioral and neurochemical effects.

Impetus to study the latter two questions were largely derived from two unrelated findings made in 1963. In that year, Mantegazza and Riva (1) reported that PEA produced an amphetaminelike action when injected into rats and mice. In that same year, Oates et al. (2) observed that PEA was excreted in amounts up to 3 mg/day by uncontrolled phenylketonuric patients receiving a monoamine oxidase inhibitor (MAOI). Moreover, the MAOI also produced an increase in tremor and ataxia in these patients, suggesting that PEA might be responsible for some of the clinical symptoms which are associated with this disorder. Indeed, since hyperactivity is often seen in phenylketonuria, this symptom might have a neurochemical basis in common with the increased locomotor activity observed in PEA-injected experimental animals.

The question as to a possible role of PEA in normal brain function is much more difficult to answer. The potent behavioral effects of pharmacological doses of PEA and the high turnover rate of endogenous PEA have given rise to speculations that this amine might also have a role in normal brain function, either as a chemical transmitter or as a synaptic modulator (3).

The first part of this chapter reviews some of the present evidence for the possible actions of PEA in normal brain function and dysfunction. Then in Section II we present new findings resulting from other studies carried out, in order to elucidate the neurochemical mechanisms by which pharmacological doses of PEA produce their behavioral effects.

II. PEA IN NORMAL BRAIN FUNCTION AND PATHOLOGICAL CONDITIONS

A. Possible Role of Endogenous PEA in Brain Function

1. Endogenous Content of PEA in Brain

Elucidation of the possible role of endogenous PEA in brain function has been hampered by the lack of sufficiently sensitive and specific techniques. Nakajima et al. (4) failed to detect PEA either in brain or in other tissues of untreated rabbits, using ninhydrin to visualize the amines which had been extracted from 20 g of tissue and separated by paper electrophoresis. Since their procedures were able to detect as little as 0.03 μg of PEA, they estimated that the amount of PEA in rabbit tissues, if present, must be less than 1 ng/g.

Similar results were reported for rats by Edwards and Blau (5) using gas chromatography combined with electron-capture detection. With this procedure (6), PEA was extracted from brain tissue by the use of organic solvents and subsequently reacted with dinitrobenzene sulfonic acid to form the dinitrophenyl derivative. As little as 0.1 ng of this derivative could be measured by the electron-capture detector. Although this meant that as little as 20 ng/g of PEA in the original tissue could be detected (6), PEA could still only be measured in the tissues of rats pretreated with an MAOI (60 ng/g) but not of untreated rats (5).

Recently, highly sensitive and specific techniques such as solid probe mass spectrometry (7), enzymatic radiochemical techniques (8), and gas chromatography/mass spectrometry (9) have been utilized for the measurement of PEA in rat brain. The results of these three independent techniques appear to show remarkable agreement and indicate that the endogenous concentration of PEA in whole rat brain is, indeed, very low (i.e., 1-2 ng/g).

2. Distribution of PEA in Brain Regions and in Subcellular Fractions

The extremely low concentrations of PEA which are apparently present in brain of untreated animals need not exclude the possibility that this compound is important in normal brain function. Compartmentation of PEA at particular cellular locations (e.g., synaptic endings) could serve to increase the effective concentration and, hence, the physiological action of this amine. Moreover, the rate of formation of PEA, particularly if it is formed at these key cellular sites, may be more important than the total tissue content in determining its neurochemical action. Indeed, the rapid turnover rate of PEA is apparent from the large accumulation of this amine in tissues after the administration of an MAOI (5,8,10).

It has been established by several laboratories that PEA is unevenly distributed among different regions of the brain, regardless of whether the accumulation of the endogenous amine following pargyline treatment (4,5), the accumulation of exogenously administered PEA (11-13), or the endogenous concentration in untreated animals (9) is measured. Each of these studies found the highest concentration of PEA to be in the striatum.

The addition of PEA either in vitro or in vivo has been found to result in an accumulation of PEA in synaptosomes. For example, when high concentrations (.07 mM) of [^{14}C]PEA were incubated with brain minces obtained from pargyline-treated rats, some of the radioactive label sedimented on a sucrose density gradient with the synaptosomal fraction, although no attempt was made to characterize the labeled compound (14). Edwards and Blau (5) were able, however, to directly measure PEA in synaptosomes isolated from rats which had been pretreated with pargyline and phenylalanine, suggesting that PEA can, indeed, be detected in synaptosomes, at least under these conditions used to produce an experimental model of phenylketonuria (PKU). Boulton and Baker (15) recently have been able to demonstrate the presence of PEA in synaptosomal fractions of untreated rats.

3. Significance of Type B Monoamine Oxidase (MAO) to PEA Metabolism in the Brain

Studies originally carried out by Johnston in 1968 (16) demonstrated that two forms of MAO (types A and B) could be distinguished by their substrate specificity and sensitivity to inhibition by the drug, clorgyline. The clorgyline-sensitive activity was designated as type A MAO, and appears to be responsible for the oxidative deamination of serotonin (5-HT), norepinephrine (NE), and probably dopamine (DA). Phenylethylamine deamination is carried out preferentially by type B MAO and is blocked only by 10^2 to 10^3 higher concentrations of clorgyline (17). On the other hand, type B MAO was found to be selectively blocked by the drug deprenyl which is structurally similar to clorgyline but which, like PEA, contains an unsubstituted aromatic ring (18). These drugs also selectively block the oxidative deamination of amines in vivo. For example, clorgyline, but not deprenyl (each 1 mg/kg, i.v.), causes an increase in the brain concentrations of NE and 5-HT (19). In contrast, deprenyl, but not clorgyline, resulted in a decrease in the disappearance of [^{14}C]PEA, which had been intraventricularly injected (19).

Studies carried out in our laboratory have indicated that the enzymatic activities for types A and B MAO [assayed with the substrates 5-HT and PEA, respectively (20)] were not uniformly distributed in the brain (21). Moreover, a comparison of these enzymatic activities in developing rats of different ages revealed that types A and B MAO also show different patterns of development (21). The type B enzymatic activity of suckling rats was especially low and could possibly result in high concentrations of PEA in the brain of very young rats (and presumably also of fetal rats). Although the brain concentrations of PEA in fetal rats have not yet been determined, the maximum concentrations of phenylethanolamine and octopamine were found to occur at 16 to 17 days of gestation, when they actually exceeded that of NE (22).

The activities of types A and B MAO determined in subcellular fractions of rat brain also indicated that these two enzymatic

activities were differently distributed (23, and manuscript in press). For example, the ratio of the activities of types A and B MAO was only one-half as large in the synaptosomal fraction as in the free (nonsynaptosomal) mitochondria. Furthermore, the PEA : 5-HT ratio exhibited a fourfold variation among the synaptosomal fractions obtained from different brain regions (23). These results suggest the possibility that certain populations of synaptic endings contain a relatively high activity of type B MAO, whereas other populations of synaptic endings contain little or no type B MAO. This notion may have important implications for the metabolism and possible physiological functions of PEA in the brain. For example, PEA might be expected to accumulate in nerve endings (such as noradrenergic ones) which have the decarboxylase (and, thus, the ability to synthesize PEA) but which perhaps lack a significant activity of type B MAO. On the other hand, a relatively high activity of type B MAO in certain nerve endings would raise the question of the significance of this enzyme to the function of these neurons. For example, does the type B MAO activity in these neurons function simply to protect them from toxic effects due to an accumulation of PEA or does this enzymatic activity act to terminate (or regulate) specific functions carried out by PEA in these neurons? The key to our understanding of the possible role of PEA in brain function may, indeed, lie in the answer to this question.

B. Alterations of PEA Metabolism in Pathological Conditions or in Drug Therapy

1. Phenylketonuria

Since phenylketonuria (PKU) is dealt with elsewhere in this monograph, this topic will not be discussed in detail in this chapter. At present, PKU is the only disease in which PEA is known unequivocally to be altered, so that an understanding of the behavioral consequences would be important in comparison with other disorders (e.g., schizophrenia) in which PEA may be involved.

However, the importance of PEA to the production of the clinical features of PKU is not yet certain, since increased PEA production is only one of the biochemical abnormalities resulting from PKU. Considering the multiplicity of symptoms associated with PKU [e.g., mental retardation, agitated behavior, EEG abnormalities, hyperactivity, tremors and seizures (24)], it seems unlikely that a single biochemical aberration will be found to account for them all. Nevertheless, that PEA (or a related amine) may be responsible for certain of these symptoms is suggested by the increase in tremor and ataxia observed after the administration of MAOI to uncontrolled PKUs (2). In view of the known behavioral effects of PEA (see discussion that follows), it seems likely that certain psychotic symptoms often seen in untreated PKU (25) could also be related to the overproduction of this amine.

2. Depression

Sabelli and Mosnaim (26) have proposed a PEA hypothesis of affective disorders, stating that certain forms of endogenous depression are characterized by a decrease in the concentration or turnover of PEA in the brain. Although the urinary excretion of PEA was reported to be decreased in certain groups of depressed patients (7,26), there is still some disagreement concerning the levels of PEA in urine, since lower concentrations have been found using other procedures (27; Edwards, unpublished observation). The validity of these results obtained for depressed patients cannot be judged until they are replicated using more specific procedures for the determination of PEA.

If depression is, indeed, related to a deficiency of PEA in the brain, the possible mechanism for that deficiency should be examined. Although such a deficiency of PEA could result from either a decrease in its rate of synthesis or an increase in its rate of degradation, present evidence points to an increase in the rate of degradation (via MAO) as the probable cause of reduced PEA levels. One proposition suggests that an increase in the accessibility of monoamines to MAO due to an increased permeability of cellular membranes may be one possible cause of depression (28). Alternatively, an increase in the activity of MAO due to hormonal influences, aging processes, or other physiological factors could also play a role in controlling the tissue content of PEA. There is already some evidence for an increased MAO activity in depressed patients, suggesting that a reduced PEA content in depression would probably occur by this mechanism. Nies et al. (29) reported that the activity of MAO in platelets from depressed patients was significantly higher than from normal individuals. Klaiber et al. (30) also found that plasma MAO activity was higher in depressed women than in nondepressed women, although the procedures which these investigators used to determine plasma MAO activity appear to measure a combination of plasma amine oxidase and platelet MAO activities (31). Since human blood platelets contain only the type B MAO (20,31), alterations in platelet MAO activity do, indeed, suggest that the metabolism of PEA, but not necessarily NE and 5-HT (which are substrates for type A MAO), is abnormal in certain depressed patients.

An important question raised by the notion that PEA metabolism is altered in depression is whether the deficit of PEA is a causative factor in the production of this illness or whether it is simply a secondary result of other metabolic changes. If the behavioral symptoms are caused by a deficiency of PEA, this would imply that PEA plays an essential role in the normal brain function. Moreover, the depressive illness should be ameliorated by treatment with L-phenylalanine or with specific inhibitors of type B MAO. On the other hand, if the deficit of PEA is simply a secondary result of other metabolic changes, the determination of this deficiency may, nevertheless, aid in diagnosing subtypes of depression and in predicting therapeutic responses to various psychopharmacological treatments.

3. Antidepressant Drug Action

The effectiveness of antidepressant agents in the treatment of certain types of depressive disorders has generally been attributed to alterations in NE metabolism which overcome a functional deficit of this neurotransmitter amine (32). Alternative mechanisms of action suggest that these agents act via 5-HT metabolism. Tricyclic antidepressants have been thought to potentiate the actions of either

NE or 5-HT by blocking the neuronal re-uptake of these amines (33). However, there is now reason to believe that the antidepressant effects of these two classes of drugs may not be entirely explained by their actions on either of these amines. For example, the pharmacological properties of iprindole offer a serious challenge to the contention that either NE or 5-HT is involved in the antidepressant effects of tricyclic drugs, since this drug, although approximately as effective as imipramine as an antidepressant agent (34), does not inhibit the brain uptake of either [^{3}H]NE (35) or [^{3}H]5-HT (36).

An alternative mechanism for the clinical effectiveness of these drugs is suggested by the recent finding that all tricyclic antidepressant drugs tested (including iprindole) selectively inhibit type B MAO (37,38). Thus, an interference in the degradation of PEA might account for the antidepressant effects of these drugs. Although studies carried out so far have indicated, at least in the rat, that relatively high doses of these drugs are required in order to block the in vivo deamination of [^{14}C]PEA, which has been administered by i.p. (39), i.v. (40), or intraventricular injection (Edwards and Antelman, unpublished observation), it is conceivable that, since MAO activity is widely distributed among different organs, an inhibition of PEA metabolism at critical cellular locations may not be reflected in the measurement of the deamination of exogenously administered PEA. A more definitive answer as to the role of type B MAO inhibition in the antidepressant action of tricyclics may be obtained by study of the effects of these agents upon the metabolism of the endogenous PEA content of specific brain areas.

The antidepressant action of MAOI such as Parnate could also be due to their effects on PEA metabolism. Although these agents are generally considered to act by blocking NE and 5-HT deamination, the MAOI used clinically are nonspecific in their actions upon types A and B MAO. Since these MAOI actually produce much greater increases in PEA concentrations [40-fold (5,8)] than in NE or 5-HT, the possible role of PEA in producing the antidepressant effects of these drugs should also be considered.

4. Schizophrenia

Just as depression has been proposed to be due to a deficiency in PEA, schizophrenia may be related to an excess or overproduction of this amine (41). This hypothesis is supported by the fact that PEA has an amphetaminelike action in animals. Since amphetamine psychosis in humans is strikingly similar to and often confused with schizophrenia, an excess of PEA could act as an endogenously produced amphetaminelike compound. Furthermore, the effects of PEA upon DA-containing neurons (see below), and the supposed involvement of these neurons in schizophrenia (42) suggest a possible link between PEA and schizophrenia.

A mechanism by which an increase in PEA concentrations could occur in schizophrenia is suggested by reports that the activity of platelet MAO activity is reduced in certain chronic schizophrenics (43). Although this finding has been replicated in several laboratories (44-47), it is not clear what significance a reduced platelet MAO activity may have in the production of the behavioral features of schizophrenia. First, no differences were found in either types A or B MAO activity in brain between schizophrenics and normals, although a selective loss in certain brain regions cannot be ruled out (48). Moreover, if PEA were responsible for producing schizophrenialike symptoms, these symptoms should be even more evident in PKUs or in patients treated with MAOI. Although this does not appear to be generally true, occasional psychotic symptoms are observed in PKU (25). It is still conceivable, however, that an increase in PEA in certain critical brain areas might, indeed, be related to certain symptoms associated with schizophrenia.

III. EXPERIMENTAL STUDIES

A. Introduction

The structure of PEA is basic to the sympathomimetic amines, for example, amphetamine (AM), NE, and DA, and a number of similar behavioral effects have been reported. For instance, in common with AM,

PEA has been shown to increase locomotor activity (1), potentiate self-stimulation (49), and induce stereotypy (50). These effects have been attributed to an indirect, AM-like release of the catecholamines (CA) from a presynaptic, extragranular pool (50). In the studies discussed below, we examined the effects of PEA on indices of nigrostriatal DA function and now report both behavioral and biochemical data which suggest that, in addition to its indirect, presynaptic-releasing action, PEA also appears to exert a direct, stimulating effect on postsynaptic DA receptors.

B. Behavioral Data

The site of action of PEA in the nigrostriatal DA (NSDA) pathway was studied behaviorally using the rotation paradigm developed by Ungerstedt (51). In this model, the presynaptic NSDA neurons are lesioned unilaterally by 6-hydroxydopamine (6-OHDA) and rotational behavior is then observed following the administration of either a DA-releasing agent such as amphetamine or a DA-receptor stimulating drug such as apomorphine. Since the paradigm calls for a unilaterally denervated preparation, AM-like compounds increase release of DA only from the unlesioned side of the brain and produce vigorous circling behavior toward the side of the lesion (i.e., ipsilateral rotations). Conversely, since receptors on the denervated striatum appear to become more responsive following denervation (51), agents, such as apomorphine, which act to stimulate postsynaptic DA receptors directly, preferentially act on the denervated striatum and result in circling movements away from the lesioned side (i.e., contralateral rotation) (51). By observing the directionality of circling to a given drug then, it may be possible to infer whether it acts pre- or postsynaptically within the NSDA system (51,52).

C. Procedures and Results

Male albino rats of the Sprague-Dawley strain (140-160 g) were used in these experiments (Supplier: Zivic-Miller, Pittsburgh). All animals received unilateral 6-OHDA lesions (8 μg/4 μl, free base) of the nigrostriatal bundle (NSB) at the point where the bundle emerges from the substantia nigra (51,53). Testing for rotational behavior was begun approximately 1 week after 6-OHDA treatment. Animals were initially tested for rotational responses to AM and apomorphine (APO). Only those animals showing clear-cut ipsilateral rotation to AM and contralateral rotation to APO were tested with PEA. The results of the first experiment are indicated in Table 1. As can be seen, PEA treatment (100 mg/kg) produced contralateral, APO-like rotation in all nine animals tested (median contralateral rotations = 161, range = 92-472; median ipsilateral rotations = 0, range = 0-14). Rotation behavior typically began within 2-3 min after PEA and lasted for 10-15 min. Since rotation induced by direct DA receptor agonists is known to be unaffected by inhibition of DA synthesis (52), we tested the effects of the tyrosine hydroxylase inhibitor, α-methyl tyrosine (AMT), on PEA-induced rotation. The AMT (2 x 100 mg/kg) (N = 4) failed to prevent the contralateral rotational response to PEA (median contralateral rotations = 207, range = 151-421; median ipsilateral rotations = 0, range = 0). These data support the notion that PEA may stimulate striatal DA receptors directly without acting through presynaptic neurons.

In order to verify the effectiveness of our 6-OHDA treatment, animals were sacrificed and their caudates rapidly dissected out and assayed for DA (54). The results (Table 2) indicate a 98% depletion of DA in the caudate ipsilateral to our 6-OHDA injection ($P < 0.001$, t-comparison with untreated control). In addition, however, we also obtained a substantial, highly significant, 67% depletion of caudate DA on the uninjected side ($P < 0.001$, t-comparison relative to untreated controls). In our efforts to trace the source of this unexpected finding, we discovered that 6-OHDA had adventitiously been injected at the rate of 3 μl/min rather than at 1 μl/3 min as we had intended. Since unilateral electrolytic lesions of the NSB failed to produce any depletion of DA in the contralateral caudate (Antelman and Rowland, unpublished observation), this suggested that the bilateral (although unequal) depletion obtained in the present

TABLE 1

Rotational Behavior of Rats After Unilateral Injection of 6-OHDA into the NSB—Direction of Rotation[a]

6-OHDA (8 μl/4 μl) Injection Side	AM (2 mg/kg)	APO (0.5 mg/kg)	PEA (100 mg/kg)	PEA (100 mg/kg) + AMT (2x100 mg/kg)
		Experiment 1		
Left	Left (5/5)	Right (5/5)	Right (5/5)	Right (3/3)
Right	Right (4/4)	Left (4/4)	Left (4/4)	Left (2/2)
		Experiment 2		
Left	Left (6/6)	Right (6/6)	Left (4/6)[b]	Right (6/6)
Right	Right (3/3)	Left (3/3)	Right (3/3)	Left (3/3)

[a]Numbers in parentheses indicate number of animals tested under each drug condition. All drugs were injected intraperitoneally. With the exception of AMT, drugs were injected immediately prior to testing. α-Methyl tyrosine was administered at 6 and 2 hr before testing.

[b]Two of the animals in this group failed to show unidirectional rotation.

TABLE 2

Caudate DA Concentrations After
Unilateral Injection of 6-OHDA into the NSB[a]

	Caudate DA (μg/gm)		
	Untreated Controls	6-OHDA-Treated Animals (8 μg/4 μl)	
		Injected Side	Uninjected Side
Experiment 1	6.77 ± 0.24 (6)	0.13 ± 0.07 (6)	2.26 ± 0.48 (6)
Experiment 2	8.01 ± 0.80 (5)	0.19 ± 0.05 (8)	5.23 ± 0.33 (8)

[a]Number of samples in each group is indicated in parentheses. Dopamine values are not corrected for recovery, which was 81% in Experiment 1 and 87% in Experiment 2.

experiment was, in all likelihood, due to diffusion of the 6-OHDA to the uninjected nigral area following its rapid injection. Due to the possibility that the partial, though considerable, destruction of the contralateral side may have been a determining factor in the APO-like rotation seen with PEA, we repeated our initial experiment, this time using a slower rate of injection for 6-OHDA (1 μl/3 min).

In stark contrast to the results of Experiment 1, all animals displaying unidirectional rotation to PEA following "slow" 6-OHDA injections (7 out of 9) showed AM-like, ipsilateral rotation (Table 1) (one animal displayed bidirectional rotation, whereas the other failed to rotate at all after PEA); median ipsilateral rotations = 150, range = 74-406; median contralateral rotations = 0.5, range = 0-24. Rotation behavior to PEA in the present experiment had a slower onset (10-15 min) and was of a longer duration than that observed in the first experiment, often lasting up to 1 hr.

Since AM-like turning implies presynaptic release of DA from the nondenervated NSB (51,53), such turning by PEA should be blocked by AMT (52). Moreover, and more importantly, AMT should produce a temporary state in the NSB, which is functionally analagous to the unequal bilateral denervation which resulted from our "fast" (1 μl/20 sec) 6-OHDA injections, and, therefore, PEA rotation might be expected to change from ipsilateral AM-like rotation to contralateral APO-like rotation. Indeed, this prediction was dramatically borne out by the results reported in Table 1. As can be seen, *all* animals tested with PEA and AMT pretreatment (including the two mentioned above which had previously failed to show unidirectional rotation) reversed their directional preference and now showed contralateral, APO-like turning (median contralateral rotations = 112, range = 34-330; median ipsilateral rotations = 4, range = 0-19). Furthermore, when these same animals were retested 1 week later with PEA alone, seven out of nine reverted to AM-like rotation. The two remaining animals which had initially failed to display unidirectional rotation to PEA alone, also reverted to their previous behavior patterns.

Caudate assay revealed that whereas DA depletion on the injected side was identical to that obtained in the first experiment (98% depletion, $P < 0.001$), the caudate on the uninjected side showed only a 34% depletion of DA ($P < 0.05$) in contrast to the 67% seen in Experiment 1 (Table 2). This finding of DA depletion in the uninjected NSB in spite of our slow injection procedure suggests an obvious shortcoming in experiments which use this side as a control for unilateral 6-OHDA lesions. Caution would seem to be especially necessary when placements close to the midline are involved.

Taken collectively, the results of our investigation strongly suggest that PEA can act to stimulate caudate DA receptors directly. This action of PEA is most readily demonstrated in the denervated animal and is easily obscured in the intact organism, where the indirect, DA-releasing action of PEA predominates.

D. Neurochemical Effects of PEA in the Striatum

Almost all studies involving the neurochemical effects of PEA administration have been concerned with changes in the content of various biogenic amines in the whole brain. Thus, PEA has been reported to cause a depletion in the whole brain content of NE (55,56), DA (56), and 5-HT (57). These and other observations have been interpreted to suggest that PEA acts by a release of neurotransmitter amines from their presynaptic stores. This explanation, however, fails to account for the increase (+24%) in brain DA concentrations which has been reported after a single injection of 100 mg/kg PEA, in contrast to depletion (32%) after eight injections of 50 mg/kg each (56). Instead, these data suggest that PEA may have at least two separate effects on neurotransmitter mechanisms.

In view of the apparent dual action of PEA upon rotational behavior of rats after unilateral 6-OHDA lesions in the nigrostriatal pathway (see previous discussion), we investigated the biochemical effects of PEA administration upon DA metabolism in the striatum of unlesioned rats.

Sprague-Dawley male rats (150-200 g) were used in all of these studies. PEA was injected i.p. at a dose of 100 mg/kg, and the rats were killed by decapitation at various intervals after injection. The brains were rapidly removed, and the striata were then homogenized in 1 N HCl. After centrifugation of the homogenates, aliquots were taken for the analysis of PEA, dopamine, and homovanillic acid (HVA) concentrations. PEA was determined as the dinitrophenyl derivative by gas chromatography and electron-capture detection using the method of Edwards and Blau (6), except that the silylation step for derivatizing hydroxylated amines was omitted. Dopamine was isolated on alumina columns and quantitated by an automated fluorometric procedure as the trihydroxyindoles (54). Striatal HVA was determined as the pentafluoropropionyl derivative by gas chromatography combined with electron-capture detection, according to the method of Watson et al. (58).

As shown in Fig. 1, maximum concentrations of PEA were reached 20 min after injection, although the concentrations in the caudate were higher than in the rest of the brain at all times examined. After the maxima, PEA concentrations showed a rapid decline. However, even 60 min after injection, PEA could still be measured in the caudate (1.5 μg/g) as well as in the rest of the brain (0.37 μg/g). In contrast, previous investigators were unable to detect PEA in the

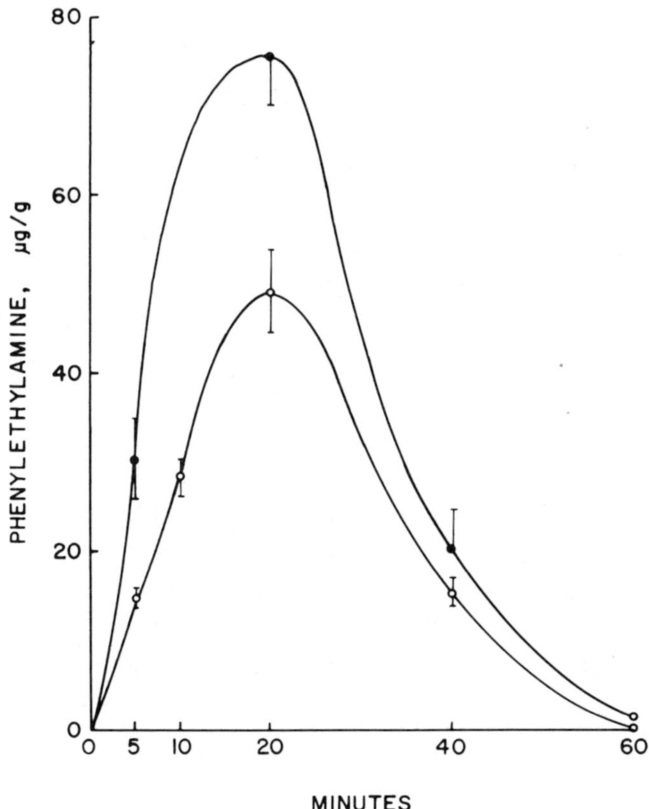

FIG. 1. Concentration of PEA in striatum (closed circle), rest of brain (open circle) after an injection of 100 mg/kg i.p.

brain of experimental animals after more than 20-30 min post-injection (12,59), leading these and other investigators to conclude that behavioral and pharmacological effects occurring after this period of time must be due to the formation of a metabolite of PEA. Since our results demonstrate that the levels of PEA are still far in excess of endogenous levels even 60 min after injection, it would seem plausible that all the behavioral effects observed during the hour after an injection of 100 mg/kg PEA might be due to actions caused by PEA itself. Indeed, the AM-like rotational behavior described previously typically lasted until approximately 60 min after the injection of PEA. This time course, therefore, closely resembles that observed in Fig. 1 for the elevation of PEA concentrations.

Striatal HVA concentrations, which provide an index of the activity of DA-containing neurons in the nigrostriatal pathway, were significantly reduced 5 min after an injection of PEA to 78% of control levels ($P < 0.02$). (See Fig. 2.) After this decline, HVA concentrations showed a gradual increase, returning to control levels by 10-20 min and then dramatically continuing to rise until they were significantly elevated to 147 and 157%, respectively (for each, $P < 0.001$) by 40-60 min.

The initial decline in the striatal HVA concentration suggests that the short term effects of PEA are reflected in a reduction of the activity of DA-containing neurons in the striatum. Since the decrease in HVA was accompanied by an increase in striatal DA (+13%), it cannot be explained merely as a consequence of reduced transmitter levels. Rather, coupled with the behavioral data presented above, it suggests that the immediate effect of PEA may be as a directly acting DA agonist. Thus, apomorphine, which is thought to be a directly acting agent, also decreases striatal HVA concentration (60). Moreover, we have already pointed out that the two agents produce similar effects on the rotational model. Finally, the apomorphinelike rotation to PEA begins almost immediately and is typically over within 10-15 min, a time course corresponding to its early effects on HVA.

The subsequent increase in striatal HVA concentrations, which is accompanied by a return of DA to control levels, may reflect an increase in DA turnover, presumably due to a release in DA from presynaptic storage sites. This later effect of PE on HVA also appears to have its behavioral concomitants in that (as mentioned above) the amphetaminelike effects of PEA on rotation tended to have both a longer latency (about 20 min) and an increased duration, lasting up to 60 min. Thus, these results apparently demonstrate neurochemically the dual action of PEA which was also demonstrated

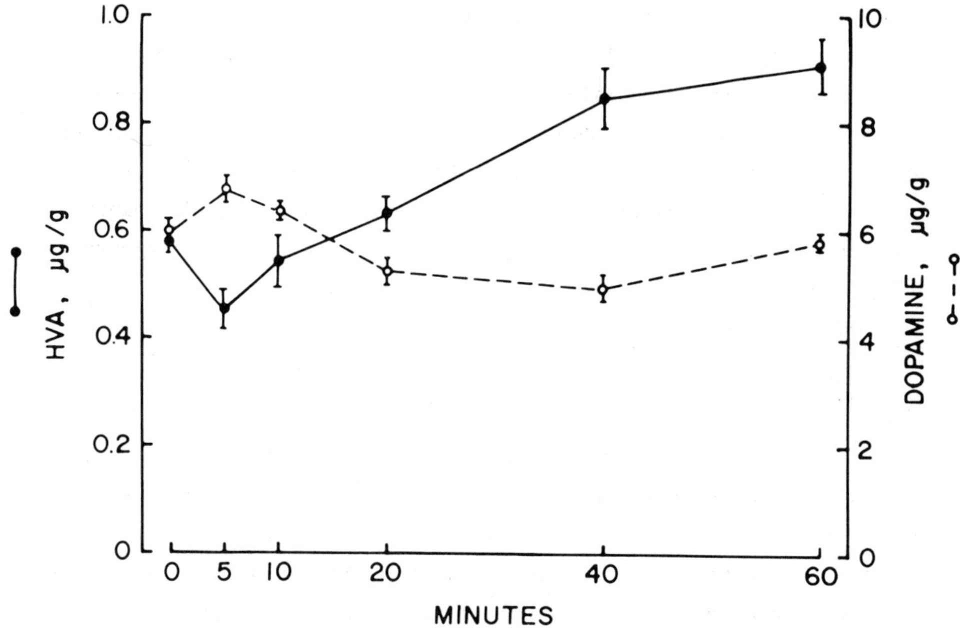

FIG. 2. Effect of PEA on striatal HVA (closed circle) and DA (open circle) concentrations. N = 5-10 animals.

on a behavioral level by the use of the rotational model. To recapitulate, both the neurochemical and the behavioral approaches indicate that PEA can have an apomorphinelike, direct postsynaptic receptor stimulating action in addition to an AM-like, indirect presynaptic releasing action. Moreover, the time course of the neurochemical changes following PEA correlate well with corresponding effects on rotational behavior.

Further evidence for a direct, postsynaptic action of PEA is suggested by preliminary data obtained for the effects of PEA in vitro upon the DA-sensitive adenylate cyclase in the striatum (Edwards and Antelman, unpublished observation). In these studies, PEA has been found to produce a small, though significant, stimulation of cyclic AMP formation in rat striatal homogenates. A concentration of PEA equal to 100 µM produced a 10% stimulation of adenylate cyclase activity. This stimulation was completely blocked by haloperidol, a selective blocker of the DA-sensitive adenylate cyclase (61). For comparison, an equal concentration of DA resulted in a 108% increase in activity. Although the degree of stimulation produced by PEA was small, relative to that produced by DA, taken together with the behavioral and neurochemical data reported above, these results, nevertheless, suggest that PEA may play an important role in DA-receptor function, particularly under conditions of *reduced* DA release.

IV. CONCLUSION

The role of the endogenous PEA in brain remains obscure. However, several unanswered questions suggest directions for future research which might have a significant bearing upon this problem. For example, we do not yet understand: 1) why PEA appears to be more greatly concentrated in the striatum as compared with other brain regions, 2) what the function of type B MAO is, and 3) whether specific receptors for PEA exist in the brain. If PEA does, indeed, have an important role in brain function, one could postulate that mental disorders could arise either from an excess or from a deficiency of the normal content of PEA.

Regardless of the possible role of PEA in normal brain function, it may still be able to produce a behavioral abnormality when formed in excess. There is, for example, considerable reason to believe that PEA may mediate some of the behavioral consequences produced in PKU or by the action of MAOI. Unfortunately, few behavioral studies to date have been carried out using the selective inhibitors of types A and B MAO. One recent study, however, indicated that deprenyl (a selective inhibitor of type B MAO) produces a marked and sustained increase in motor activity in rats (62). Although type B MAO has previously been thought to utilize *both* PEA and DA as substrates (19), additional data obtained by Waldmeier et al. (62) suggest that DA is, in fact, almost entirely deaminated in vivo by type A MAO. Thus, only the deamination of PEA is completely blocked by deprenyl, raising the possibility that the behavioral effects produced by deprenyl may be mediated via PEA.

In this regard, our finding that PEA has a direct as well as an indirect stimulating action on striatal DA receptors may be of particular significance. It has been recently demonstrated, for instance, that deprenyl is efficacious in treating the symptoms of Parkinson's disease (63), which is characterized by a degeneration of DA-containing neurons in the striatum. Since PEA can be formed extraneuronally (in addition to intraneuronally) and is able to act at DA receptors directly, its action may become more important relative to DA under conditions in which degeneration of NSDA-containing neurons occurs (e.g., in Parkinson's disease), thus suggesting a possible mechanism for the therapeutic effect of deprenyl in treating this disease.

ACKNOWLEDGMENTS

We thank Marina Lin, Peter Chin, and Ling Ling Kung for excellent technical assistance. Supported in part by USPHS Grant 24114 to S.M.A.

REFERENCES

1. P. Mantegazza and M. Riva (1963). Amphetamine-like activity of β-phenylethylamine after monoamine oxidase inhibitor in vivo. *J. Pharm. Pharmacol.*, 15:472-478.
2. J. A. Oates, P. Z. Nirenberg, J. B. Jepson, A. Sjoerdsma, and S. Udenfriend (1963). Conversion of phenylalanine to phenylethylamine in patients with phenylketonuria. *Proc. Soc. Exp. Biol. Med.*, 112:1078-1081.
3. H. C. Sabelli, A. D. Mosnaim, and A. J. Vazquez (1974). Phenylethylamine: Possible role in depression and antidepressive drug action. In: *Neurohumoral Coding of Brain Function*, edited by R. O. Meyers and R. R. Drucker-Colin. Plenum Press, New York, pp. 331-357.
4. T. Nakajima, Y. Kakimoto, and I. Sano (1964). Formation of β-phenylethylamine in mammalian tissue and its effect on motor activity in the mouse. *J. Pharmacol.*, 143:319-325.

5. D. J. Edwards and K. Blau (1973). Phenylethylamines in brain and liver of rats with experimentally induced phenylketonuria-like characteristics. *Biochem. J.*, 132:95-100.

6. D. J. Edwards and K. Blau (1972). Analysis of phenylethylamines in biological tissues by gas-liquid chromatography with electron-capture detection. *Anal. Biochem.*, 45:387-402.

7. D. A. Durden, S. R. Philips, and A. A. Boulton (1973). Identification and distribution of β-phenylethylamine in the rat. *Can. J. Biochem.*, 51:995-1002.

8. J. M. Saavedra (1974). Enzymatic isotopic assay for and presence of β-phenylethylamine in brain. *J. Neurochem.*, 22:211-216.

9. J. Willner, H. F. LeFevre, and E. Costa (1974). Assay by multiple ion detection of phenylethylamine and phenylethanolamine in rat brain. *J. Neurochem.*, 23:857-859.

10. A. A. Boulton, A. V. Jourio, S. R. Philips, and P. H. Wu (1975). Some arylalkylamines in rabbit brain. *Brain Res.*, 96:212-216.

11. D. J. Edwards and S. M. Antelman (1975). Phenylethylamine: Interaction with dopaminergic neurons. Roundtable Discussion on Phenylethylamine, American Society for Neurochemistry, Mexico City, March.

12. D. M. Jackson and D. B. Smythe (1973). The distribution of β-phenylethylamine in discrete regions of the rat brain and its effect on brain noradrenaline, dopamine and 5-hydroxytryptamine levels. *Neuropharmacology*, 12:663-668.

13. P. H. Wu and A. A. Boulton (1975). Metabolism, distribution, and disappearance of injected β-phenylethylamine in the rat. *Can. J. Biochem.*, 53:42-50.

14. R. J. Baldessarini and M. Vogt (1971). The uptake and subcellular distribution of aromatic amines in the brain of the rat. *J. Neurochem.*, 18:2519-2533.

15. A. A. Boulton and G. B. Baker (1975). The subcellular distribution of β-phenylethylamine, p-tyramine and tryptamine in rat brain. *J. Neurochem.*, 25:477-481.

16. J. P. Johnston (1968). Some observations upon a new inhibitor of monoamine oxidase in brain tissue. *Biochem. Pharmac.*, 17:1285-1297.

17. H.-Y. T. Yang and N. H. Neff (1973). β-Phenylethylamine: A specific substrate for type B monoamine oxidase of brain. *J. Pharmacol. Exp. Ther.*, 187:365-371.

18. J. Knoll and K. Magyar (1972). Some puzzling pharmacological effects of monoamine oxidase inhibitors. *Adv. Biochem. Psychopharmacol.*, 5:393-407.

19. H.-Y. T. Yang and N. H. Neff (1974). The monoamine oxidases of brain: Selective inhibition with drugs and the consequences for the metabolism of the biogenic amines. *J. Pharmacol. Exp. Ther.,* 189:733-740.

20. D. J. Edwards and S.-S. Chang (1975). Multiple forms of monoamine oxidase in rabbit platelets. *Life Sci.,* 17:1127-1134.

21. D. J. Edwards (1976). Monoamine oxidases in brain and platelets: Implications for the role of trace amines and drug action. In: *Trace Amines and the Brain,* edited by E. Usdin and M. Sandler, Marcel Dekker, New York, pp. 59-81.

22. J. M. Saavedra, J. T. Coyle, and J. Axelrod (1974). Developmental characteristics of phenylethanolamine and octopamine in the rat brain. *J. Neurochem.,* 23:511-515.

23. A. K. Student and D. J. Edwards (1975). Subcellular location of types A & B monoamine oxidase in rat brain. *Neurosci. Abs.,* 1:360.

24. W. E. Knox (1972). Phenylketonuria. In: *The Metabolic Basis of Inherited Disease,* 3rd ed., edited by J. B. Stanbury, J. B. Wyngaarden, and D. S. Frederickson. McGraw-Hill, New York, pp. 266-295.

25. T. L. Perry, S. Hansen, B. Tischler, F. M. Richards, and M. Sokol (1973). Unrecognized adult phenylketonuria. Implications for obstetrics and psychiatry. *New Eng. J. Med.,* 289: 395-398.

26. H. C. Sabelli and A. D. Mosnaim (1974). Phenylethylamine hypothesis of affective behavior. *Am. J. Psychiat.,* 131: 695-699.

27. J. W. Schweitzer, A. J. Friedhoff, and R. Schwartz (1975). Phenylethylamine in normal urine: Failure to verify high values. *Biol. Psychiat.,* 10:277-285.

28. M. Sandler, S. B. Carter, M. F. Cuthbert, and C. M. B. Pare (1975). Is there an increase in monoamine oxidase activity in depressive illness? *Lancet,* i:1045-1049.

29. A. Nies, D. S. Robinson, C. L. Ravaris, and J. M. Davis (1971). Amines and monoamine oxidase in relation to aging and depression in man. *Psychosomat. Med.,* 33:470.

30. E. L. Klaiber, D. M. Broverman, W. Vogel, Y. Kobayashi, and D. Moriarty (1972). Effects of estrogen therapy on plasma MAO activity and EEG driving responses of depressed women. *Am. J. Psychiat.,* 128:1492-1498.

31. D. L. Murphy and C. H. Donnelly (1974). Monoamine oxidase in man: Enzyme characteristics in platelets, plasma and other human tissues. In: *Neuropsychopharmacology of Monoamines and Their Regulatory Enzymes,* edited by E. Usdin, Raven Press, New York, pp. 71-85.

32. J. J. Schildkraut (1973). Neuropharmacology of the affective disorders. *Ann. Rev. Pharmacol.*, 13:427-454.

33. J. A. Biel and B. Bopp (1974). Antidepressant drugs. In: *Psychopharmacological Agents*, Vol. 3, edited by M. Gordon. Academic Press, New York, pp. 283-341.

34. W. E. Fann, J. M. Davis, D. S. Janowsky, J. S. Kaufmann, J. H. Cavanaugh, and J. A. Oates (1974). Effect of antidepressant and antimanic drugs on amine uptake in man. *J. Nerv. Ment. Dis.*, 158:361-368.

35. M. I. Gluckman and T. Baum (1969). The pharmacology of iprindole, a new antidepressant. *Psychopharmacologia (Berl.)*, 15: 169-185.

36. A. S. Horn and R. C. A. M. Trace (1974). Structure-activity relations for the inhibition of 5-hydroxytryptamine uptake by tricyclic antidepressants into synaptosomes from serotoninergic neurones in rat brain homogenates. *Br. J. Pharmacol.*, 51: 399-403.

37. D. J. Edwards and M. O. Burns (1974). Effect of tricyclic antidepressants upon human platelet monoamine oxidase. *Life Sci.*, 15:2045-2058.

38. J. A. Roth and C. N. Gillis (1974). Deamination of β-phenylethylamine by monoamine oxidase inhibition by imipramine. *Biochem. Pharmacol.*, 23:2537-2545.

39. D. J. Edwards (1974). Inhibition of human platelet monoamine oxidase (MAO) by tricyclic antidepressants. *Pharmacologist*, 16:265.

40. P. F. Von Voigtlander and E. G. Losey (1976). Inhibition of phenylethylamine metabolism in vivo—Effect of antidepressants. *Biochem. Pharmacol.*, 25:217-218.

41. E. Fischer, H. Spatz, J. M. Saavedra, H. Reggiani, A. H. Miro, and B. Heller (1972). Urinary elimination of phenethylamine. *Biol. Psychiat.*, 5:139-147.

42. S. H. Snyder, S. P. Banerjee, I. H. Yamamura, and D. Greenberg (1974). Drugs, neurotransmitters and schizophrenia. *Science*, 184:1243-1253.

43. D. L. Murphy and R. J. Wyatt (1972). Reduced monoamine oxidase activity in blood platelets from schizophrenic patients. *Nature*, 238:225-226.

44. H. Meltzer and S. M. Stahl (1974). Platelet monoamine oxidase activity and substrate preferences in schizophrenic patients. *Res. Comm. Chem. Pathol. Pharmacol.*, 7:419-431.

45. E. A. Zeller, B. Boshes, J. M. Davis, and M. Thorner (1975). Molecular aberration in platelet monoamine oxidase in schizophrenia. *Lancet*, i:1385.

46. E. F. Domino and S. S. Khanna (1976). Decreased blood platelet MAO activity in unmedicated chronic schizophrenic patients. *Am. J. Psychiat.*, 133:323-326.

47. J. J. Schildkraut, J. M. Herzog, P. J. Orsulak, S. E. Edelman, H. M. Shein, and S. H. Frazier (1976). Reduced platelet monoamine oxidase activity in a subgroup of schizophrenic patients. *Am. J. Psychiat.*, 133:438-440.

48. M. A. Schwartz, R. J. Wyatt, H.-Y. T. Yang, and N. H. Neff (1974). Multiple forms of brain monoamine oxidase in schizophrenic and normal individuals. *Arch. Gen. Psychiat.* 31: 557-560.

49. C. D. Wise and L. Stein (1970). Amphetamine: Facilitation of behavior by augmented release of norepinephrine from the medial forebrain bundle. In: *International Symposium on Amphetamines and Related Compounds*, edited by E. Costa and S. Garattini. Raven Press, New York, pp. 463-485.

50. K. Fuxe and U. Ungerstedt (1970). Histochemical, biochemical and functional studies on central monoamine neurons after acute and chronic amphetamine administration. In: *International Symposium on Amphetamines and Related Compounds*, edited by E. Costa and S. Garattini. Raven Press, New York, pp. 257-288.

51. U. Ungerstedt (1974). Brain dopamine neurons and behavior. In: *The Neuroscience Third Study Program*, edited by F. O. Schmitt, F. G. Schmitt and F. G. Worden. MIT Press, Cambridge, pp. 695-711.

52. P. Von Voigtlander and K. E. Moore (1973). Turning behavior of mice with unilateral 6-hydroxydopamine lesions in the striatum: Effects of apomorphine, L-DOPA, amatadine, amphetamine and other psychomotor stimulants. *Neuropharmacology*, 12: 451-462.

53. K. E. Moore (1974). Behavioral effects of direct- and indirect-acting dopaminergic agonists. In: *Neuropsychopharmacology of Monoamines and Their Related Enzymes*, edited by E. Usdin, Raven Press, New York, pp. 403-414.

54. P. Waldmeier and L. Maitre (1973). An automated fluorometric method for the estimation of dopamine in brain tissue extracts. *Anal. Biochem.*, 51:474-481.

55. J. Jonsson, H. Grobecker, and P. Holtz (1966). Effect of β-phenylethylamine on content and subcellular distribution of norepinephrine in rat heart and brain. *Life Sci.*, 5:2235-2246.

56. K. Fuxe, H. Grobecker, and J. Jonsson (1967). The effect of β-phenylethylamine on central and peripheral monoamine-containing neurons. *Eur. J. Pharmacol.*, 2:202-207.

57. Y. H. Loo (1974). Serotonin deficiency in experimental hyperphenylalaninemia. *J. Neurochem.*, 23:139-147.

58. E. Watson, B. Travis, and S. Wilk (1974). Simultaneous determination of 3,4-dihydroxyphenylacetic acid and homovanillic acid in milligram amounts of rat striatal tissue by gas-liquid chromatography. *Life Sci.*, 15:2167-2178.

59. I. Cohen, J. F. Fischer, and W. H. Vogel (1974). Physiological disposition of β-phenylethylamine, 2,4,5-trimethoxyphenylethylamine, 2,3,4,5,6-pentamethoxyphenylethylamine and β-hydroxymescaline in rat brain, liver and plasma. *Psychopharmacologia*, 36:77-84.

60. B.-E. Roos (1969). Decrease in homovanillic acid as evidence for dopamine receptor stimulation by apomorphine in the neostriatum of the rat. *J. Pharm. Pharmacol.*, 21:263-264.

61. J. W. Kebabian, G. L. Petzold, and P. Greengard (1972). Dopamine-sensitive adenylate cyclase in caudate nucleus of rat brain, and its similarity to the "dopamine receptor." *Proc. Nat. Acad. Sci.*, 69:2145-2149.

62. P. C. Waldmeier, A. Delini-Stula, and L. Maitre (1976). Preferential deamination of dopamine by an A type monoamine oxidase in rat brain. *Naunyn-Schmiedeberg's Arch. Pharmacol.*, 292:9-14.

63. W. Birkmayer, P. Riederer, M. B. H. Youdim, and W. Linauer (1975). The potentiation of the anti-akinetic effect after L-Dopa treatment by an inhibitor of MAO-B, deprenyl. *J. Neural Transm.*, 36:303-326.

Chapter 3

INHIBITION OF 2-PHENYLETHYLAMINE METABOLISM IN BRAIN BY TYPE B MONOAMINE OXIDASE BLOCKERS

Jose A. Fuentes*

Laboratory of Preclinical Pharmacology
National Institute of Mental Health
St. Elizabeth's Hospital
Washington, D.C.

*Current affiliation: Section of Pharmacology, Institute of Medicinal Chemistry, C.S.I.C., Madrid, Spain.

I. MONOAMINE OXIDASE AND THE METABOLISM OF 2-PHENYLETHYLAMINE

In the cardiovascular system (Fuentes and Neff, unpublished results) 2-phenylethylamine (PEA) is apparently metabolized by two different monoamine oxidases (MAO) [monoamine : O_2 oxireductase (deaminating); EC.1.4.3.4]. One enzyme is sensitive to inhibition by semicarbazide and cuprizone, while the other enzyme is blocked by high concentration of clorgyline or pargyline and not by semicarbazide or cuprizone. The enzyme that is blocked by clorgyline and pargyline is present in brain; however, the enzyme that is blocked by semicarbazide and cuprizone is not (24). Phenylethylamine may follow several metabolic pathways in the CNS. It appears to be a substrate for dopamine-β-hydroxylase (EC.1.14.2.1) (18) and for N-acetyltransferase (EC.2.3.1.5) (26). In the periphery as well as in the CNS, however, oxidative deamination is the major route for the catabolism of PEA. In fact, PEA increases in the CNS by 50-fold after MAO blockade by pargyline (3,19).

II. MULTIPLE FORMS OF MONOAMINE OXIDASE

In 1963, Gorkin (7) found two different enzymes associated with rat liver mitochondria that were capable of deaminating p-nitrophenylethylamine and m-nitro-p-hydroxybenzylamine, respectively. Several years later Youdim and Sandler (27), using a solubilized preparation from rat liver and human placenta as the enzyme sources, found several bands of MAO activity by polyacrylamide gel electrophoresis. Rat brain was also shown to contain several forms of the enzyme by cellulose acetate electrophoresis (13). Since these observations, multiple forms of MAO have repeatedly been identified in a variety of tissues from different animal species (21).

Houslay and Tipton (11) suggested that the multiple forms found by the electrophoretic techniques might be artifactual. They suggested that phospholipids attached to the enzyme protein may determine enzyme mobility during electrophoresis. Thus, the different forms of "solubilized" MAO might represent a single enzyme with varying amounts of phospholipid material bound to it. To counter their notion, Sandler and Youdim (21) speculated that the same membrane fragmentation pattern would not result so consistently in many tissues if the multiple forms of the enzyme were a consequence of the enzyme solubilization process. In line with this argument, Ekstedt and Oreland (1) removed the phospholipids from a rat liver MAO preparation by extraction with aqueous methyl ethyl ketone, and lost one form of enzyme activity but did not produce a parallel increase in the other form of activity. These results do not support the hypothesis that the multiple forms of MAO are the consequences of phospholipid binding to a single enzyme species.

Hartman and Udenfriend (8), by studying antigen-antibody cross-reactivity, examined the degree of structural relatedness of MAO derived from bovine liver and brain. While only one MAO form was immunologically demonstrated in the liver, 20% of the brain mitochondrial MAO activity did not cross-react with the antibody to liver MAO. Therefore, in the bovine brain two MAO proteins with different structural characteristics seem to be present. These data have been extended by McCauley and Racker (14) who also reported the existence of two immunologically distinct deaminating enzymes in the bovine brain.

III. SPECIFIC MONOAMINE OXIDASE SUBSTRATES AND INHIBITORS

The MAOs can be differentiated with drugs that preferentially block the various forms of the enzyme. In 1968, Johnston (12) reported that increasing concentrations of the inhibitor clorgyline produced a stepwise inhibition of rat brain MAO activity when tyramine was used as the substrate. He interpreted this observation as evidence for the existence of two forms of the enzyme: a form resistant to the inhibitor and a form relatively sensitive to the inhibitor. These two forms have been designated as type A MAO and type B MAO, as there is evidence that they may represent classes of enzymes with similar characteristics rather than single enzymes (17). In the

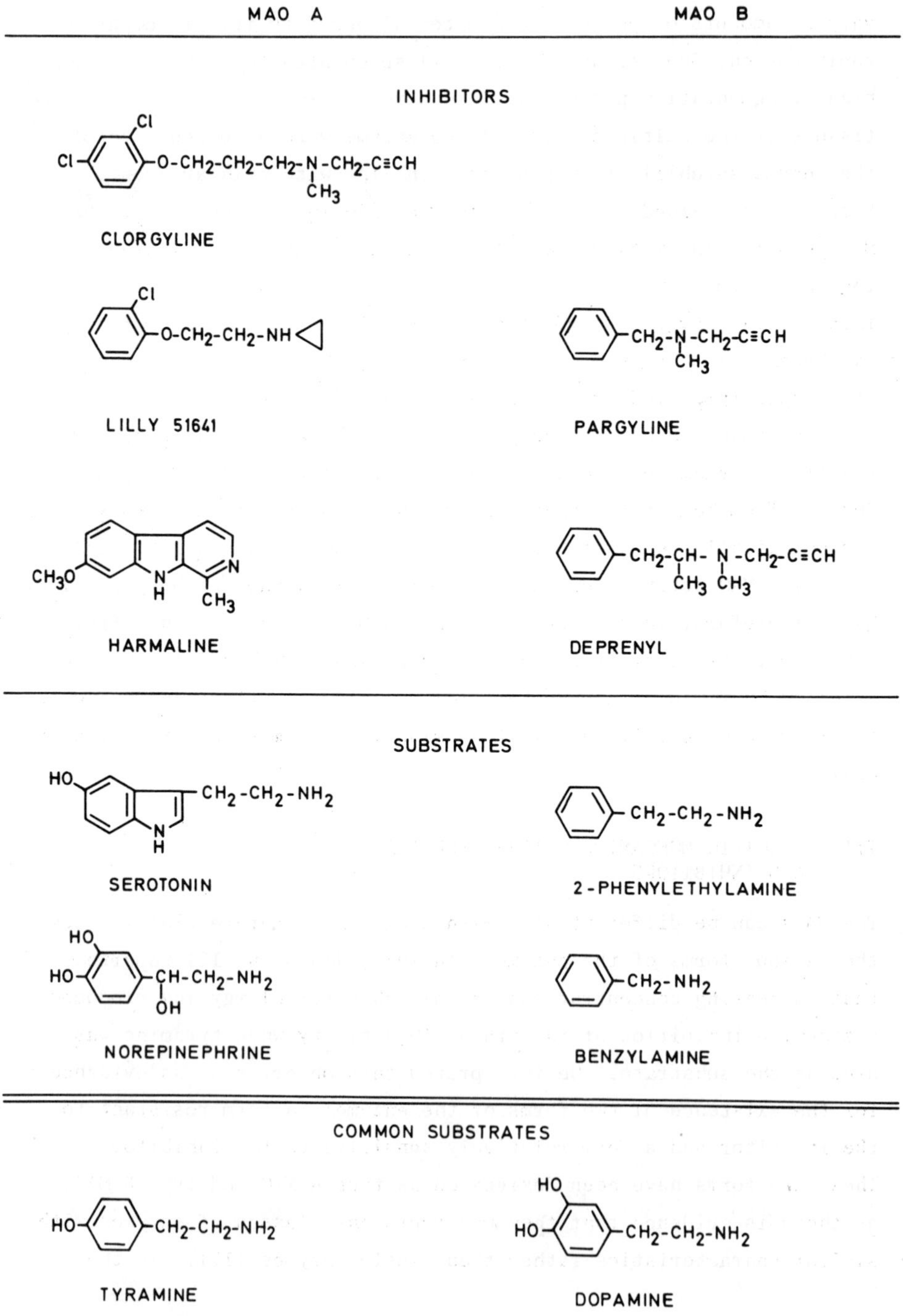

MAO A	MAO B

INHIBITORS

Cl, Cl, O-CH$_2$-CH$_2$-CH$_2$-N-CH$_2$-C≡CH, CH$_3$

CLORGYLINE

Cl, O-CH$_2$-CH$_2$-NH

LILLY 51641

CH$_2$-N-CH$_2$-C≡CH, CH$_3$

PARGYLINE

CH$_3$O, N, H, N, CH$_3$

HARMALINE

CH$_2$-CH- N -CH$_2$-C≡CH, CH$_3$ CH$_3$

DEPRENYL

SUBSTRATES

HO, CH$_2$-CH$_2$-NH$_2$, N, H

SEROTONIN

CH$_2$-CH$_2$-NH$_2$

2-PHENYLETHYLAMINE

HO, HO, CH-CH$_2$-NH$_2$, OH

NOREPINEPHRINE

CH$_2$-NH$_2$

BENZYLAMINE

COMMON SUBSTRATES

HO, CH$_2$-CH$_2$-NH$_2$

TYRAMINE

HO, HO, CH$_2$-CH$_2$-NH$_2$

DOPAMINE

FIG. 1. The structure and characteristics of some MAO I and substrates.

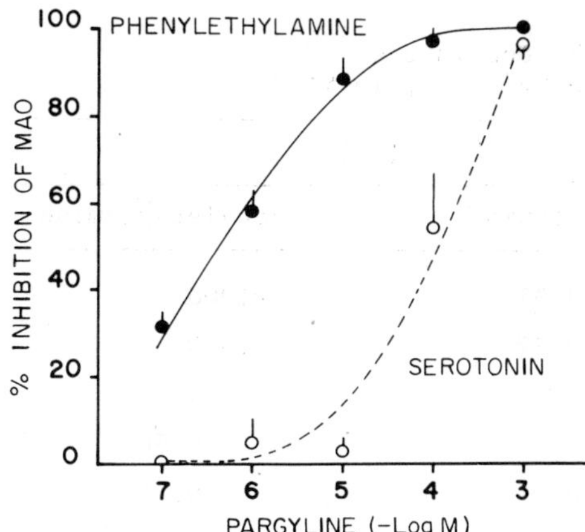

FIG. 2. Inhibition of rat brain MAO by increasing concentrations of pargyline using serotonin and 2-phenylethylamine as substrates.

past few years, relatively specific substrates for each one of the units of this binary system were discovered (Fig. 1); for example, serotonin (5-HT) and norepinephrine (NE) are deaminated by the enzyme type most sensitive to clorgyline (type A MAO), whereas PEA and benzylamine are metabolized by the enzyme type that is least sensitive to clorgyline (type B MAO). Dopamine (DA), tyramine, and tryptamine are apparently deaminated by both enzymes (25).

The use of specific substrates may provide a means to classify new chemical structures as selective MAO inhibitors. For instance, a 10 μM concentration of pargyline preincubated with homogenates of rat brain inhibits PEA deamination by about 90% (Fig. 2), while a concentration 100 times higher is required to block 5-HT deamination. Thus, pargyline can be classified as a type B MAO blocker. In Fig. 1, the chemical structures of several selective MAO inhibitors are shown.

When concentrations required to inhibit each enzyme type by 50% (EC_{50}) are compared (Table 1) clorgyline and (-)-deprenyl are revealed as the most specific MAO blockers among the drugs listed. The isomers of tranylcypromine, in contrast to the other drugs listed, appear the least specific in their ability to differentially block the deamination of PEA and 5-HT. There is a trend, however, toward a lower EC_{50}

TABLE 1

In Vitro Inhibition by Drugs of the Deamination of Serotonin or 2-Phenylethylamine by Rat Brain MAO[a]

Drug	EC_{50} (μM)	
	Serotonin	2-Phenylethylamine
Clorgyline	0.07	>1000
Harmaline	0.57	1000
Lilly 51641	0.63	47
(-)-Deprenyl	560	0.50
Pargyline	100	0.75
(-)-Tranylcypromine	11	120
(+)-Tranylcypromine	0.42	1.8

[a]Data (EC_{50}) are expressed as effective concentration (μM) of drug which inhibited enzyme activity by 50%. Values of EC_{50} were estimated by probit analysis using at least four concentrations of inhibitor drug.

TABLE 2

Apparent in Vivo Inhibition of the Oxidative Deamination of Serotonin and 2-Phenylethylamine[a]

Drug	Dose (μmol/kg)	Percent Inhibition of MAO Activity ± SEM	
		Serotonin	2-Phenylethylamine
Clorgyline	3.2	99 ± 1	10 ± 4
Lilly 51641	0.5	91 ± 3	27 ± 5
Harmaline	3.9	64	20
(-)-Deprenyl	4.4	10 ± 6	83 ± 2
Pargyline	2.5	21 ± 13	82 ± 10

[a]Animals were sacrified 2 hr after the i.v. administration of the drugs (harmaline, 15 min) and brain MAO activity measured. Values for the reversible inhibitor harmaline were obtained by extrapolating to zero volume from a graph in which volumes of phosphate buffer used to homogenate the tissue versus percent inhibition of MAO were plotted.

value for PEA deamination. This is not surprising as tranylcypromine is similar in structure to the type B inhibitor deprenyl.

Perhaps the most promising property of the selective monoamine oxidase inhibitors (MAOI) in biochemical pharmacology is that their administration (Table 2) to animals results in a specific MAO blockade in vivo. As a consequence of in vivo administration, the metabolism of some amines and not others is curtailed. As an example, DA concentrations in brain are increased after injection of deprenyl while the concentration of 5-HT and NE are not. Clorgyline diminishes the metabolism of the catecholamines and 5-HT, but, contrary to the effect of treatment with deprenyl, injected radioactive PEA is not protected from attack by type B MAO (25). Furthermore, pretreatment with pargyline or tranylcypromine has recently been reported (23) to produce a clear-cut delay in the metabolism of radioactive PEA in brain. That DA is a common substrate for both forms of MAO has been confirmed in vivo by measuring DA levels in brain after treatment with selective enzyme inhibitors. Deprenyl administered together with clorgyline increases the concentration of DA in brain even more than following either drug alone (25).

IV. DISTRIBUTION OF TYPES A AND B MONOAMINE OXIDASE IN THE BRAIN

When homogenates of different rat brain regions are used as a source of enzyme, there is not a striking difference between 5-HT and PEA deaminating activities (Table 3). The distribution of types A and B MAO does not appear to correlate with the regional levels of the endogenous biogenic amines. The greatest differences between specific activities were found in the pineal and pituitary glands. The pineal gland shows a greater type B MAO activity than any brain region, while 5-HT is deaminated by homogenates of pituitary gland at a quicker rate than PEA. Saavedra et al. (20) have reported a higher type A MAO activity than type B MAO activity when homogenates of the anterior, intermediate, and posterior pituitary were used as an enzyme source.

TABLE 3

Comparison of Monoamine Oxidase Activity in Rat Brain Regions Using 2-Phenylethylamine or Serotonin as Substrate[a]

Brain Region	2-Phenylethylamine	Serotonin	Ratio of Activity
	(nmol/mg of protein/min ± SEM)		Serotonin/2-Phenylethylamine
Cerebellum	1.7 ± 0.2	2.3 ± 0.2	1.4
Medulla oblongata	1.3 ± 0.1	2.2 ± 0.0	1.8
Hypothalamus	2.4 ± 0.4	3.0 ± 0.3	1.3
Striatum	1.3 ± 0.2	2.0 ± 0.3	1.6
Midbrain	1.3 ± 0.2	2.1 ± 0.3	1.7
Hippocampus	1.8 ± 0.2	2.6 ± 0.3	1.4
Cortex	1.4 ± 0.2	2.6 ± 0.3	1.9
Pineal gland	3.8 ± 0.8	1.5 ± 0.3	0.4
Pituitary gland	1.2 ± 0.2	2.9 ± 0.1	2.7

[a]Brain tissue was dissected as described by Glowinski and Iversen (6), homogenized in 67 mM buffer phosphate (pH 7.2), and incubated with either 0.2 mM 2-phenylethylamine or 1.2 mM serotonin. The reaction products were isolated by column chromatography.

Human brain contains types A and B MAO activities. These activities were identified with the drugs clorgyline and deprenyl and with the substrates 5-HT and PEA. As in rats (Table 3), enzyme activity was found to be rather uniformly distributed when evaluated in 15 regions of the human brain (22).

V. THE USE OF SELECTIVE MONOAMINE OXIDASE INHIBITORS TO EVALUATE FUNCTIONS MODULATED BY THE VARIOUS ENZYMES

Specific MAO blockers can be used as aids to determine the enzyme form which is associated with a particular physiological process or pharmacological effect (16). Once the modulating enzyme is ascertained, the search for the involved neurotransmitter will be narrowed. Drugs that block MAO are used to treat depression. Selective MAO blockade may thus be used to determine the enzyme form and the monoamines involved in the antidepressant mechanisms.

A. Antidepressant Effect of the Monoamine Oxidase Inhibitors: Blockade of Monoamine Oxidase or Amine Re-uptake?

The interaction of a neurotransmitter and its receptor may be terminated by metabolism by either catechol-O-methyl transferase or MAO, or the amine may be recaptured by the nerve endings.

The clinical effects of MAOI drugs as well as tricyclic antidepressants are generally attributed to a greater availability of neurotransmitter at the receptor. Tricyclic antidepressants are usually thought to act through inhibition of amine re-uptake, while the mechanism of action of the MAOI appears to be controversial since some produce blockade of re-uptake as well as inhibit MAO. Hendley and Snyder (9) have suggested that there is a correlation between clinical efficacy of the MAOI drugs and their ability to inhibit re-uptake.

Reserpine treatment produces a model depression in rats which can be prevented by prior treatment with MAOI or tricyclic antidepressants. Tranylcypromine, a MAOI antidepressant drug, used in the clinic is a mixture of (+)- and (-)-tranylcypromine. (+)-Tranylcypromine has been described as the more potent MAO I, while (-)-tranylcypromine is the more potent inhibitor of amine re-uptake (10). These two isomers administered separately can then be used to define whether the antidepressant mechanisms of the racemic mixture of tranylcypromine are associated with MAO inhibition or with blockade of amine re-uptake. An enzymatic isotopic method to measure brain levels of (+)- or (-)-tranylcypromine has recently been developed (4). Therefore, a correlation can be made between MAO inhibition in brain and the concentration of tranylcypromine that blocks amine re-uptake with the isomer's ability to prevent reserpine depression (Table 4). After an i.v. injection of 2.5 mg/kg of either drug, (-)-tranylcypromine reaches higher levels in the rat brain than (+)-tranylcypromine. According to the data published by Horn and Snyder (10), this concentration of (-)-tranylcypromine is about six to 20 times greater than required to block catecholamine re-uptake; however, it produces only about a 30% blockade of MAO. Furthermore, (-)-tranylcypromine did not prevent the depression of spontaneous motor activity observed after reserpine. Interestingly, even though (+)-tranylcypromine reached only about half the level of (-)-tranylcypromine in brain, (+)-tranylcypromine almost completely blocked MAO, and it antagonized reserpine depression. Therefore, in this laboratory test and for the case of tranylcypromine, antipressant activity seems to correlate better with MAO inhibition than with blockade of re-uptake (5).

B. Effect of Specific Monoamine Oxidase Inhibitors on Reserpine Sedation in the Rat

In 1972, Fischer et al. reported that urinary PEA was increased in schizophrenic patients and reduced in depressed patients (2). Murphy and Wyatt (15) showed diminished MAO activity in the platelets of chronic schizophrenics, thus suggesting that diminished type B MAO activity in tissues might be responsible for some forms of mental illness.

TABLE 4

Correlation Between Biochemical and Behavioral Effects Induced by Tranylcypromine (TCP) Isomers[a]

Treatment (mg/kg)	TCP Concentration in Brain [μM ± SEM (5)]	% MAO Inhibition (tyramine as substrate)	Catecholamine Uptake Blockade [ID_{50}(μM)][b]	Prevention of Reserpine Sedation
(+)-TCP 2.5	4.9 ± 0.2	95	1.4 - 4.0	yes
(-)-TCP 2.5	8.1 ± 0.9	30	0.4 - 1.2	no

[a]Animals were sacrified 15 min after the i.p. administration of the drug and either TCP level or MAO activity was measured in the brain. For the behavioral studies reserpine (2 mg/kg) was intravenously injected 15 min after TCP and 30 later motor activity was recorded for a 15 min period (5).

[b]From Horn and Snyder (10). ID_{50} : TCP concentration range that inhibits catecholamine uptake by 50% into the hypothalamus and corpus striatum.

Presently, there is no suitable animal model of depression. The motor depression induced by reserpine administration, however, is often used as a model to test antidepressants. Apparently, this is because depressive symptoms occur in man after high doses of reserpine; moreover, drugs used in the treatment of human depression are capable of preventing reserpine depression when administered to animals.

Selective MAOI might be used to search out the specific enzyme substrate involved in preventing a depressive state. With this line of reasoning, rats were treated with either type A or type B MAOI and 2 hr later, reserpine (2 mg/kg, i.p.) was administered and motor activity recorded. Type A MAOI were found to prevent the reserpine sedation, while type B MAOI did not (Table 5). However, high doses of pargyline that block not only type B but also type A MAO, prevented

TABLE 5

Spontaneous Motor Activity After Reserpine Alone or After MAOI Plus Reserpine

Treatment	Dose (μmol/kg)	Motor Activity ± SEM
Saline	—	129 ± 13
Reserpine (R)	3.3	10 ± 3
Clorgyline + R	3.2	149 ± 21
Lilly 51641 + R	0.5	110 ± 9
Harmaline + R	3.9	75 ± 9
Pargyline + R	2.5	7 ± 2*
(-)-Deprenyl + R	4.4	13 ± 5*
Pargyline + R	25.0	147 ± 26

[a]Rats were injected with MAOI drugs 2 hr (harmaline, 15 min) prior to reserpine. Motor activity was recorded 30 min later for 15 min. Drugs were injected intravenously.

*$P > 0.05$ when compared to the reserpine treated (+R) group.

the reserpine depression (3). In conclusion, the delay of the metabolism of PEA after type B MAO blockade does not seem to produce any effect on the animal model of depression.

VI. SUMMARY

Oxidative deamination is the major metabolic pathway for the degradation of PEA. Multiple functional forms of MAO have been demonstrated in the CNS and in the peripheral nervous system as well. The MAO sensitive to clorgyline is type A MAO, and that which is less sensitive to it is type B MAO. These two enzyme types can be selectively blocked by inhibitors, and they show substrate specificity. The substrates for type A MAO are 5-HT and NE, while PEA and benzylamine are deaminated by type B MAO. Types A and B MAO activities are rather uniformly distributed in the rat and in human brain. In contrast, the pineal gland shows a higher type B than type A activity, while a higher type A than type B activity is found in the pituitary gland.

MAOI rather than blockade of uptake seems to be involved in the antidepressant effect of tranylcypromine. Specific MAOI may be used to determine the MAO form associated with a given pharmacological or physiological process. For example, type A and not type B MAOI are associated with prevention of reserpine depression. Thus, delayed PEA metabolism does not appear to be a requirement for the reversal of depressed motor activity in the rat after treatment with reserpine.

REFERENCES

1. B. Ekstedt and L. Oreland (1976). Effect of lipid-depletion on the different forms of monoamine oxidase in rat liver mitochondria. *Biochem. Pharmacol.*, 25:119-124.

2. E. Fischer, H. Spatz, B. Heller, and H. Reggiani (1972). Phenylethylamine content of human urine and rat brain, its alterations in pathological conditions and after drug administration. *Experientia (Basel)*, 28:307-308.

3. J. A. Fuentes and N. H. Neff (1975). Selective monoamine oxidase inhibitor drugs as aids in evaluating the role of type A and B enzymes. *Neuropharmacology,* 14:819-825.

4. J. A. Fuentes, M. A. Oleshansky, and N. H. Neff (1975). A sensitive enzymatic assay for dextro- or levo-tranylcypromine in brain. *Biochem. Pharmacol.,* 24:1971-1973.

5. J. A. Fuentes, M. A. Oleshansky, and N. H. Neff (1976). A comparison of the antidepressant activity of (-) and (+) tranylcypromine in an animal model. *Biochem. Pharmacol.,* 25:801-804.

6. J. Glowinski and L. L. Iversen (1966). Regional studies of catecholamines in the rat brain—I. *J. Neurochem.,* 13:655-6

7. V. Z. Gorkin (1963). Partial separation of rat liver mitochondrial amine oxidase. *Nature,* 200:77-78.

8. B. K. Hartman and S. Udenfriend (1972). The use of immunological techniques for the characterization of bovine monoamine oxidase from liver and brain. In: *Advances in Biochemical Psychopharmacology, Vol. 5,* edited by E. Costa and P. Greengard. Raven Press, New York.

9. E. D. Hendley and S. H. Snyder (1968). Relationship between the action of monoamine oxidase inhibitors on the noradrenaline uptake system and their antidepressant efficacy. *Nature,* 220: 1330-1331.

10. A. S. Horn and S. H. Snyder (1972). Steric requirements for catecholamine uptake by rat brain synaptosomes: studies with rigid analogs of amphetamine. *J. Pharmacol. Exp. Ther.,* 180: 523-530.

11. M. D. Houslay and K. F. Tipton (1973). The nature of the electrophoretically separable multiple forms of rat liver monoamine oxidase. *Biochem. J.,* 135:173-186.

12. J. P. Johnston (1968). Some observations upon a new inhibitor of monoamine oxidase in brain tissue. *Biochem. Pharmacol.,* 17:1285-1297.

13. H. C. Kim and A. D'Iorio (1968). Possible isoenzymes of monoamine oxidase in rat tissues. *Can. J. Biochem.,* 46: 295-297.

14. R. McCauley and E. Racker (1973). Separation of two monoamine oxidases from bovine brain. *Fed. Proc.,* 32:797 Abst.

15. D. L. Murphy and R. J. Wyatt (1972). Reduced MAO activity in blood platelets from schizophrenic patients. *Nature,* 238: 225-226.

16. N. H. Neff and J. A. Fuentes (1976). The use of selective monoamine oxidase inhibitor drugs for evaluating pharmacological and physiological mechanisms. In: *Monamine Oxidase and Its Inhibition,* edited by Ciba Foundation. Elsevier/Excerpta Medica/North-Holland, Amsterdam.

17. N. H. Neff and C. Goridis (1972). Neuronal monoamine oxidase specific enzyme types and their rates of formation. In: *Advances in Biochemical Psychopharmacology, Vol. 5,* edited by E. Costa and P. Greengard. Raven, New York.

18. J. M. Saavedra and J. Axelrod (1973). Demonstration and distribution of phenylethanolamine in brain and other tissues. *Proc. Nat. Acad. Sci. USA,* 70:769-772.

19. J. M. Saavedra (1974). Enzymatic isotopic assay for and presence of p-phenylethylamine in brain. *J. Neurochem.,* 22:211-216.

20. J. M. Saavedra, M. Palkovits, J. S. Kizer, M. Brownstein, and J. A. Zivin (1975). Distribution of biogenic amines and related enzymes in the rat pituitary gland. *J. Neurochem.,* 25:257-260.

21. M. Sandler and M. B. H. Youdim (1972). Multiple forms of monoamine oxidase: Functional significance. *Pharmacol. Rev.,* 24:331-348.

22. M. A. Schwartz, R. J. Wyatt, H.-Y. T. Yang, and N. H. Neff (1974). Multiple forms of brain monoamine oxidase in schizophrenic and normal individuals. *Arch. Gen. Psychiat.,* 31: 557-560.

23. P. F. Von Voigtlander and E. G. Losey (1976). Inhibition of phenylethylamine metabolism in vivo—Effects of antidepressants. *Biochem. Pharmacol.,* 25:217-218.

24. H.-Y. T. Yang and N. H. Neff (1973): β-Phenylethylamine: A specific substrate for type B monoamine oxidase of brain. *J. Pharmacol. Exp. Ther.,* 187:365-371.

25. H.-Y. T. Yang and N. H. Neff (1974). The monoamine oxidases of brain: Selective inhibition with drugs and the consequences for the metabolism of the biogenic amines. *J. Pharmacol. Exp. Ther.,* 189:733-740.

26. H.-Y. T. Yang and N. H. Neff (1976). N-acetyltransferase of brain: Some properties of the enzyme and the identification of carboline inhibitor compounds. *Mol. Pharmacol.,* 12:69-72.

27. M. B. H. Youdim and M. Sandler (1967). Isoenzymes of soluble monoamine oxidase from human placenta and rat liver mitochondria. *Biochem. J.,* 105:43P.

[illegible] N. H. Neff and [illegible] (19[illegible]). [illegible] monoamine oxidase specific enzymes [illegible] and their rates of formation. In: Advances in Biochemical Psychopharmacology, [illegible] edited by [illegible] Costa and [illegible], New York.

[illegible] M. [illegible] (197[illegible]). [illegible] and dis[illegible] tribution of [illegible]

[illegible] M. [illegible] (197[illegible]) [illegible] presence of [illegible] in [illegible] 2[illegible]-21[illegible]

[illegible]

[illegible]

[illegible]

[illegible]

[illegible]

[illegible]

[illegible] brain [illegible] of [illegible] 12:[illegible]

[illegible]

Chapter 4

SUBSTRATE- AND INHIBITOR-RELATED CHARACTERISTICS OF MONOAMINE OXIDASE IN PURIFIED BRAIN MITOCHONDRIA WITH REFERENCE TO PHENYLETHYLAMINE

Sabit Gabay and Frances M. Achee

Biochemical Research Laboratory
Veterans Administration Hospital
Brockton, Massachusetts

I. INTRODUCTION

This review, which is more representative than comprehensive, concentrates primarily on studies carried out in the author's laboratories and on an appraisal of those selected investigations which appear relevant to the interpretations of recently published articles concerning β-phenylethylamine (PEA) as a substrate of monoamine oxidase (MAO) and its reactivity with specific inhibitors. It specifically discusses the current criteria with which PEA and serotonin (5-HT) metabolism by MAO have been selectively differentiated by (1) their substrate variability and specific affinity,

(2) parameters affecting the enzyme molecularity, (3) thermostability properties of MAO, and (4) inhibitory characteristics of specific agents, be they monoamine oxidase inhibitors (MAOI) or the so-called tricyclics. Consideration is given to the complexities of a membrane-bound enzyme, such as MAO, and the microenvironment that influences its activity (13). Thus, this paper describes primarily work carried out on purified brain mitochondrial preparation because it may be a useful model with which to initiate studies at a subcellular level. These organelles, and the membranes derived therefrom, may be isolated as relatively biochemically homogeneous (as monitored by enzyme markers) and structurally intact (assessed by electron microscopy) for carrying out a number of systematic studies of this crucial problem (1). In the interest of brevity, only references related to both brain MAO and PEA have been eclectically undertaken.

Phenylethylamine has long been known to be a substrate for MAO. In 1916, Guggenheim and Loffler (14) gave PEA hydrochloride orally to rabbits and recovered phenylacetic acid in the urine. Since that time, other investigators have shown that PEA is oxidatively deaminated by MAO in animal tissues (3,28,36,42). However, the metabolism of this amine has taken on considerable significance only within the past decade with the awareness that PEA may be a neuromodulator substance and there may be a specific form of MAO for its deamination.

In 1963, Mantegazza and Riva (19) reported that treatment of animals with iproniazid, an MAOI, and PEA resulted in amphetamine-like behavior. Fischer (7) found that not only PEA, but also its precursor, phenylalanine, in combination with iproniazid prevented and antagonized the psychomotor depression produced by adrenaline and DOPA. The possibility of ergotropic action of brain PEA was thus suggested for the first time. Subsequently, Fischer et al. (8) reported that the PEA content in human urine was decreased in cases of endogenous depression and was elevated in schizophrenia and mania. Working with rat brain, these investigators also found that PEA concentrations were increased after treatment with iproniazid, imipramine, and reserpine. More recently, Boulton (4) suggested that further study of not only PEA but also the less prominent amines, tyramine, and tryptamine might be justifiable in light of their properties, which could be significant in the etiology of psychiatric disorders. As an alternative to the catecholamine theory (34), a "phenylethylamine hypothesis of affective disorders" has been proposed recently, whereby depression relates to a deficiency of PEA in the brain (23,32,33,37). Thus, the mood lifting effects of such drugs as MAO I and imipramine could be due in part to their role in increasing levels of brain PEA.

Within the past few years, there have been an ever increasing number of reports dealing with the multiple nature of MAO. In 1968, Johnston reported studies with a new MAOI, clorgyline, which showed an abnormal double sigmoid dose response curve. To explain these results, it was proposed that MAO consisted of a binary system of enzymes with differing sensitivities to the inhibitor under study. The clorgyline-sensitive enzyme was designated as Enzyme A and the clorgyline-resistant enzyme as Enzyme B. Serotonin appeared to be preferred as a substrate for the A enzyme, while tyramine appeared to be deaminated by both forms, nonpreferentially. Later studies by Yang and Neff (39,40) revealed that a natural substrate for the type B enzyme was PEA. The deamination of PEA was also found to be preferentially inhibited by deprenyl (39), which had previously been shown to be selective for benzylamine (18). From these studies and others have evolved a generalized concept of the binary nature of MAO as differentiated by such properties as substrate specificity, inhibitor sensitivity, and thermostability. However, the nature of MAO is by no means clear cut, and there is still considerable uncertainty as to whether variations in properties represent distinct enzyme forms, different active sites, or modulation due to the microenvironment. Thus, the characterization of MAO remains an important problem in gaining some insight into the biochemical determinants of affective disorders and a better understanding of the action of antidepressant drugs in terms of enzyme-substrate interaction and inhibitor-related characteristics.

II. STUDIES ON MAO CHARACTERISTICS IN BRAIN MITOCHONDRIA RELATIVE TO METABOLISM OF PEA

Herein is reported our studies on the characterization of MAO in a purified, intact mitochondrial fraction from beef brain, prepared by a process employing mechanical and proteolytic homogenization of cortical gray matter, followed by differential centrifugation as described by Achee et al. (1). The mitochondrial fraction obtained, rather than the solubilized enzyme, was chosen for study because of our contention that the natural cellular environment may exert a profound influence on the mode of action of an enzyme, such as MAO, which is known to be strongly attached to the outer membrane of the mitochondrion. There are many complexities associated with biological membrane systems which could modify enzyme activity by cooperative and regulatory effects. This may be particularly true for MAO, which has such a broad substrate specificity. Monoamine oxidase plays a significant role in the metabolism of not only PEA, but also many other important biogenic amines. Of these amines, 5-HT, dopamine (DA), tryptamine, and tyramine, in addition to PEA, were also used in our studies, along with the nonphysiological substrates, kynuramine and m-iodobenzylamine.

A. Parameters Affecting Enzyme Activity

Several properties of MAO activity were found to vary with substrate. While these substrate variations may be a matter of the possible existence of multiple forms of MAO, it appeared that there were similarities observed for some properties which could be related to similarities in substrate structure. Thus, for PEA, as for the other phenylalkylamines, tyramine, and DA, the optimal pH was found to be about pH 7.4, whereas for the indolealkylamines, 5-HT, and tryptamine, the pH optimum was around 8.2. The nonbiogenic amines, kynuramine and m-iodobenzylamine, were deaminated optimally at pH 9.1 and 8.6, respectively (1,13).

Monoamine oxidase activity is affected to some extent by the presence of anions (13). In particular, phosphate and chloride ions were found to cause a slight decrease in activity (Table 1), with

TABLE 1

Effect of Anions on MAO Activity in Bovine Brain Mitochondria[a]

Anion	mM Conc.	Ionic Strength	Phenyl-ethylamine	Tyramine	Dopamine	Tryptamine	Serotonin	Kynuramine
K-Phosphate	50	0.135	97 ± 2	93 ± 0	92 ± 3	95 ± 1	98 ± 1	81 ± 2
	100	0.270	88 ± 4	88 ± 3	82 ± 3	94 ± 0	95 ± 2	67 ± 1
	200	0.540	82 ± 1	86 ± 0	74 ± 1	92 ± 2	87 ± 2	56 ± 1
K-Chloride	50	0.077	91 ± 0	98 ± 2	95 ± 2	91 ± 2	95 ± 2	86 ± 3
	100	0.127	85 ± 1	90 ± 1	91 ± 2	85 ± 1	89 ± 3	64 ± 2
	200	0.227	76 ± 0	88 ± 1	84 ± 2	80 ± 4	80 ± 2	46 ± 2
	400	0.427	—	78 ± 2	74 ± 2	71 ± 1	64 ± 1	32 ± 2

[a]All assays were carried out in 10 mM potassium phosphate buffer, pH 7.4 at 37°C using standard assay procedures, so that salt concentrations represent concentrations in addition to 10 mM K-phosphate. The exception to this is the phosphate salts which represent the total concentration in the assay. The ionic strength is the estimated total ionic strength of the assay. Activities are expressed relative to the activity in 10 mM phosphate which was taken to be 100. Values are the means ± SEM. (Sodium salts yielded similar results; thus, cationic species are ruled out.)

the deamination of kynuramine being more affected than the other substrates. This inhibition by ions was not just due to the increase of ionic strength since pyrophosphate had little or no effect on MAO activity in brain mitochondria at ionic strengths of up to 0.50. It should be noted that both sodium and potassium salts gave similar results indicating that the effect was not due to the cationic species. While there seemingly was little variation among the biogenic amine substrates with respect to the anion effect on MAO, the type of inhibition found with chloride was noncompetitive with constant K_m values and decreasing V_{max} values at increasing chloride concentrations. In contrast, the inhibition was mixed competitive for PEA and tyramine with increasing K_m values and decreasing V_{max} values (Table 2). Here again, it appeared that one could distinguish the action of MAO on phenylalkylamine substrates from indoleamines, similar to the pH optima properties.

B. Thermostability Properties

Thermostability studies are often used to differentiate between different molecular forms of an enzyme. In our studies on the thermal inactivation of MAO, we have observed only slight differences in activity with respect to the seven substrates used. Mitochondrial suspensions were heated at 50°C in 0.27 M sucrose, and the activity remaining after heating for various times up to 60 min was determined. Under these conditions, a loss of 50% of the activity with all substrates occurred after 40 min of heating (Table 3). From Fig. 1, in which the log of the percent remaining activity is plotted against heating time, it appeared that the PEA-deaminating activity was slightly more stable than that of 5-HT, the variation being about 15%. It was thus evident that the differences in thermal inactivation of MAO when measured with all the substrates were not of sufficient magnitude to earmark definitely different molecular forms in the brain mitochondria. This is in marked contrast to what has been reported by Yang and Neff (39) for rat brain mitochondrial MAO, which quickly lost its PEA-deaminating

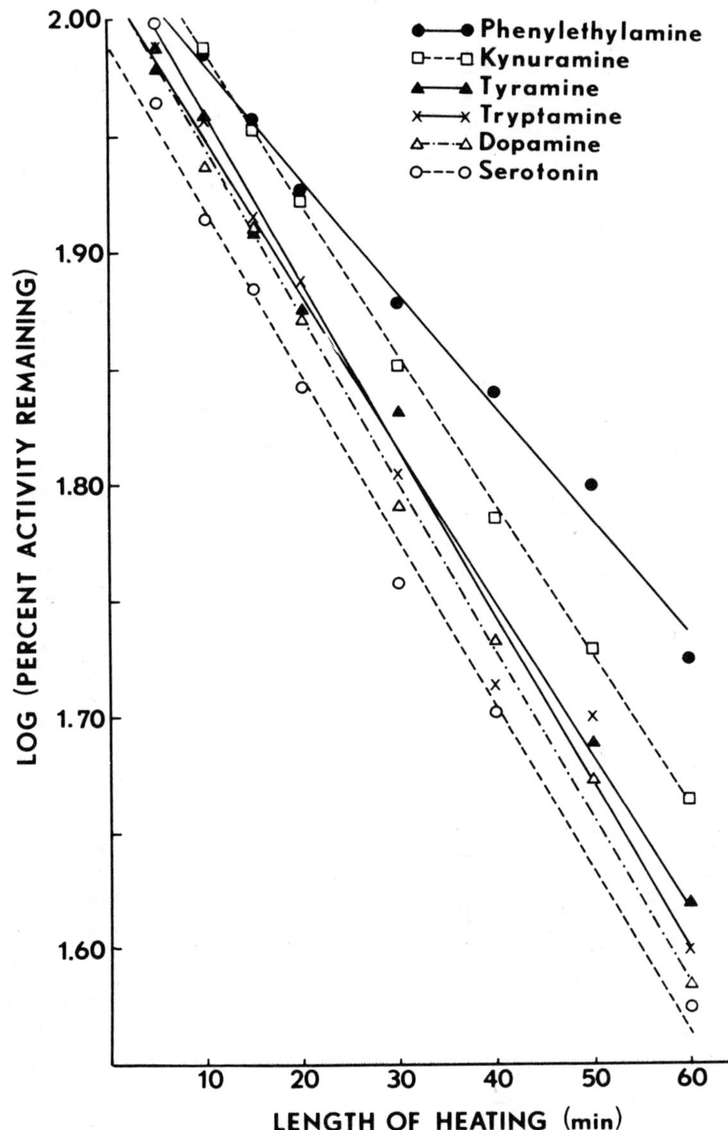

FIG. 1. Thermostability of MAO in beef brain mitochondria at 50°C. Mitochondrial fractions in 0.27 M sucrose were heated at 50°C for the time intervals indicated. The percent activity remaining after heating was determined by assaying with the various substrates in comparison to an unheated control sample. Data points are the means of two to four determinations.

TABLE 2

The Effect of Chloride on Kinetic Constants for Monoamine Oxidase in Beef Brain Mitochondria[a]

	Chloride Concentration		
	0 mM	100 mM	200 mM
Kynuramine (3)			
K_m	$4.38 \pm 0.16 \times 10^{-5}$	$4.44 \pm 0.22 \times 10^{-5}$	$4.31 \pm 0.17 \times 10^{-5}$
V_{max}	3.52 ± 0.32	2.74 ± 0.26	2.22 ± 0.14
Phenylethylamine (2)			
K_m	$2.34 \pm 0.28 \times 10^{-5}$	$2.78 \pm 0.22 \times 10^{-5}$	$3.37 \pm 0.16 \times 10^{-5}$
V_{max}	2.72 ± 0.12	2.43 ± 0.05	2.24 ± 0.01
Serotonin			
K_m	1.61×10^{-4}	1.62×10^{-4}	1.62×10^{-4}
V_{max}	4.43	3.52	2.93
Tyramine (4)			
K_m	$7.30 \pm 0.33 \times 10^{-5}$	$9.27 \pm 0.21 \times 10^{-5}$	$10.86 \pm 0.92 \times 10^{-5}$
V_{max}	7.18 ± 0.65	6.90 ± 0.70	6.27 ± 0.42

[a]All assays were carried out in 50 mM phosphate buffer, pH 7.4 at 37°C. K_m is expressed in molar concentration (M); V_{max} is in terms of nanomoles of product per minute per milligram protein. Values given are means ± SEM. The number in parentheses besides substrate name is the number of replicate determinations for each constant. No estimates of variation are given for serotonin, since there was only a single kinetic study performed on the effect of chloride with this substrate.

activity when heated in pH 7.4 phosphate buffer at 50°C. After 20 min of heating, approximately 90% of the activity toward 5-HT remained, whereas the activity remaining toward PEA was only about 20%.

Furthermore, to ascertain whether the pH of the mitochondrial suspension could be influencing the thermostability, since the two substrates showed different pH optima, the activities with PEA and 5-HT remaining after heating the mitochondria in pH 7.4 and 8.2 buffers were determined (Fig. 2). Although, at pH 8.2, the activity is less stable than at pH 7.4, at both pHs, it was found that there was practically no difference in the rate of MAO inactivation measured with the two substrates.

Thus, the pH was not a factor in our not being able to discern variations in thermostability with substrate. Concretely, these experiments showed that the thermostability criterion could not necessarily be considered a factor in differentiating the behavior of MAO activity with PEA as contrasted with 5-HT in the case of beef brain mitochondria as had been reported for the rat brain enzyme. While this may be a matter of species variation, studies with purified rabbit brain mitochondria yielded qualitatively similar results to that of bovine mitochondria (Achee and Gabay, unpublished observations).

C. Inhibition Studies

1. Selective MAO Inhibitors

As mentioned earlier, inhibitor sensitivity has played an important part in the characterization of MAO with respect to its possible dual nature and in the distinction of a specific PEA-deaminating activity. In addition to clorgyline and deprenyl, several other inhibitors have been shown to be selective in their action on MAO and numerous reports on the variable degrees of inhibition dependent on substrate have appeared (5,9,10,27,41). In our studies, we have contrasted the effects of clorgyline and deprenyl

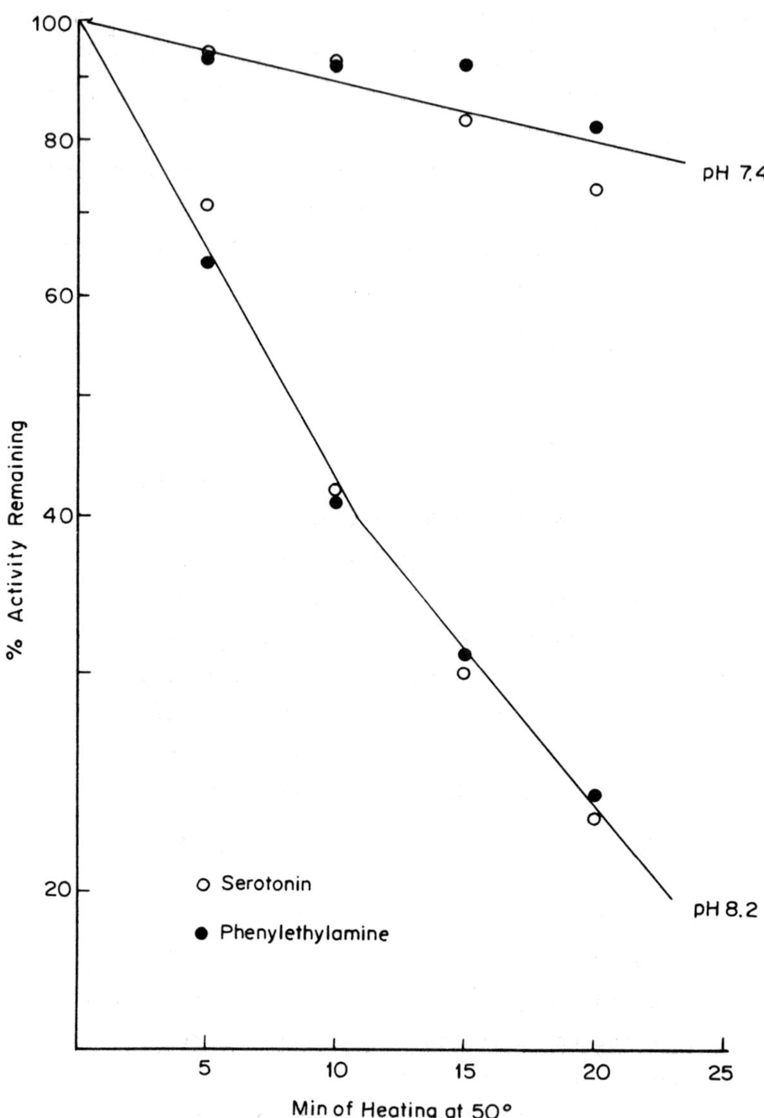

FIG. 2. Effect of pH on the thermostability of MAO at 50°C. Mitochondrial fractions were heated at 50°C in either 50 mM phosphate buffer, pH 7.4, or 10 mM pyrophosphate buffer, pH 8.2. The remaining activity was measured using 5-HT or PEA as substrates. The percent remaining activity was determined by comparison to control samples which had not been heated. The activities were measured in the same buffer systems.

TABLE 3

Rate of Heat Inactivation at 50°C of MAO in Bovine Brain Mitochondria[a]

	Percent Activity Remaining after Heating at 50°C for					
	10 min	20 min	30 min	40 min	50 min	60 min
Dopamine	86 ± 2.5	74 ± 6.2	62 ± 4.9	50 ± 3.2	47 ± 3.2	38 ± 3.5
Kynuramine	97 ± 1.0	84 ± 4.5	71 ± 4.0	61 ± 3.0	54 ± 7.5	46 ± 5.0
m-Iodobenzylamine	82	78	60	52	50	45
Phenylethylamine	96 ± 4.5	84 ± 0.5	76 ± 1.5	69 ± 2.0	63 ± 1.0	53 ± 2.0
Serotonin	82 ± 7.0	70 ± 2.4	57 ± 4.9	50 ± 6.4	48 ± 6.8	38 ± 7.5
Tryptamine	90 ± 2.5	77 ± 4.2	64 ± 4.0	52 ± 3.9	50 ± 4.0	40 ± 1.2
Tyramine	91 ± 0.9	75 ± 3.4	68 ± 3.2	54 ± 5.8	49 ± 5.2	42 ± 3.1

[a]The mitochondrial suspension (~1mg/ml) in 0.27 M sucrose was heated at 50°C for the indicated time intervals and assayed by standard procedures in 0.05 M phosphate buffer, pH 7.4 at 37°C. Activities were compared to an unheated control sample and values are the means ± SEM.

on PEA and 5-HT deamination. As shown in Fig. 3, results with clorgyline were as expected with 5-HT deamination being inhibited more strongly than PEA. The difference in effect between the two substrates may be quantitated by the I_{50} values (inhibitor concentration required to cause 50% inhibition), which were calculated to be 2.7 x 10^{-8} and 7.9 x 10^{-6} M for 5-HT and PEA, respectively.

With deprenyl, however, there was almost no difference in the effect on PEA and 5-HT deamination. As seen in Fig. 4, the dose response curves for the two substrates were almost superimposable, with calculated I_{50} values of 9.6 x 10^{-6} M for 5-HT and 7.1 x 10^{-6} M for PEA. These inhibition studies were performed using saturating substrate concentrations of 5 x 10^{-4} M for both substrates, and there was no preincubation of the enzyme with the inhibitor before initiating the reaction. In this manner, effectively, both substrate and inhibitor compete for enzyme. On the other hand, in studies in which the enzyme was preincubated with the inhibitor before addition of substrate, PEA deamination was inhibited by deprenyl to a slightly greater extent than was 5-HT. The I_{50} values with 15-min preincubation were 5 x 10^{-7} M with PEA as substrate and 4 x 10^{-6} M with 5-HT. From these experiments, it cannot be concluded that the enzyme-deaminating PEA does constitute a distinct molecular form from that metabolizing 5-HT. Some distinctions in the deamination of PEA may be made through the use of selective inhibitors. Nonetheless, studies on the kinetics of inactivation similar to those recently reported by Mantle et al. (20) are deemed necessary in this regard.

2. Tricyclic Antidepressant Drugs

In addition to the MAO I, we have been investigating the inhibition of MAO by the tricyclic compounds, imipramine, chlorimipramine, desipramine, and doxepin. The therapeutic action of the tricyclic antidepressants had been thought to be due only to amine uptake blockade, but it has become clear from several studies that they do inhibit MAO (11,12,15,29). This inhibitory action might also be a factor in their clinical efficacy. In particular, PEA deamination has been reported to be a preferential target of the tricyclics.

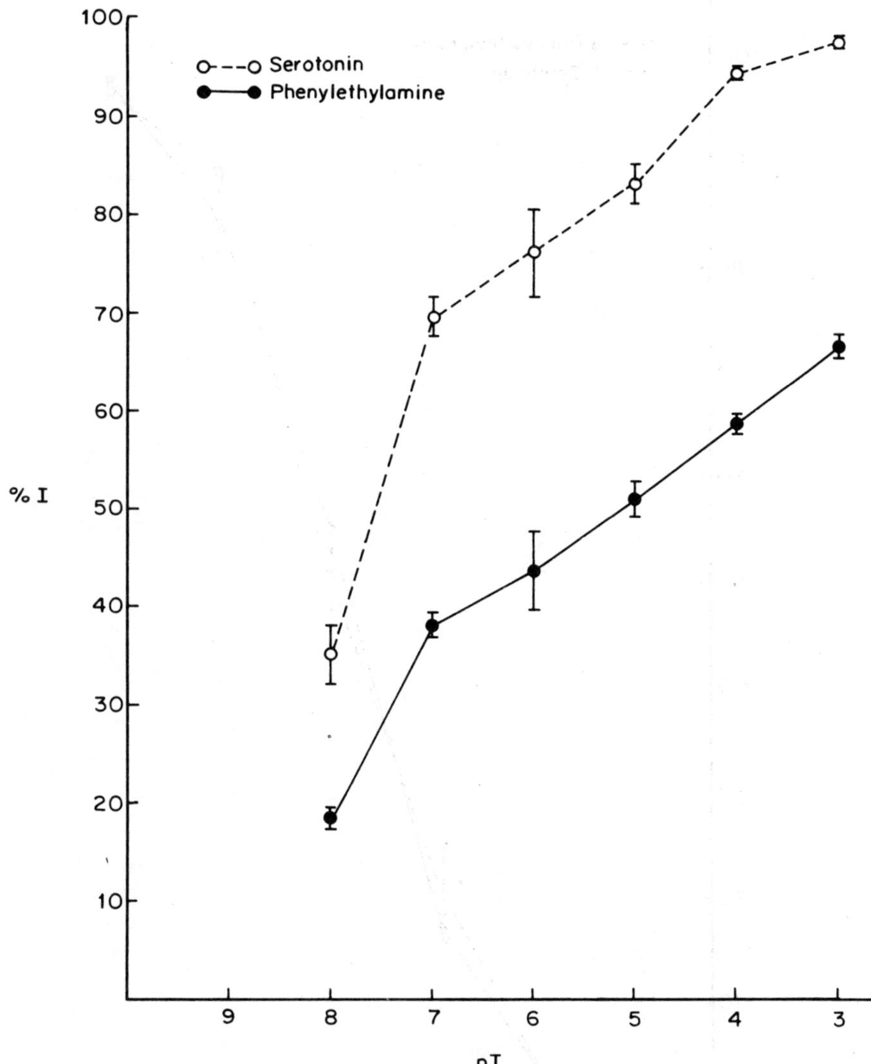

FIG. 3. The effect of clorgyline on the deamination of PEA and 5-HT by beef brain mitochondria with no pre-incubation of enzyme and inhibitor. Assay mixtures contained 5 x 10^{-4} M [^{14}C]substrate, 50 to 100 μg mitochondrial protein, 50 mM phosphate buffer, pH 7.4 and 10^{-8} to 10^{-3} M inhibitor in 0.2 ml. Assays were started by addition of enzyme, and incubations were carried out at 37°C for 20 min. The percent inhibition was determined by comparison to assays containing no inhibitor.

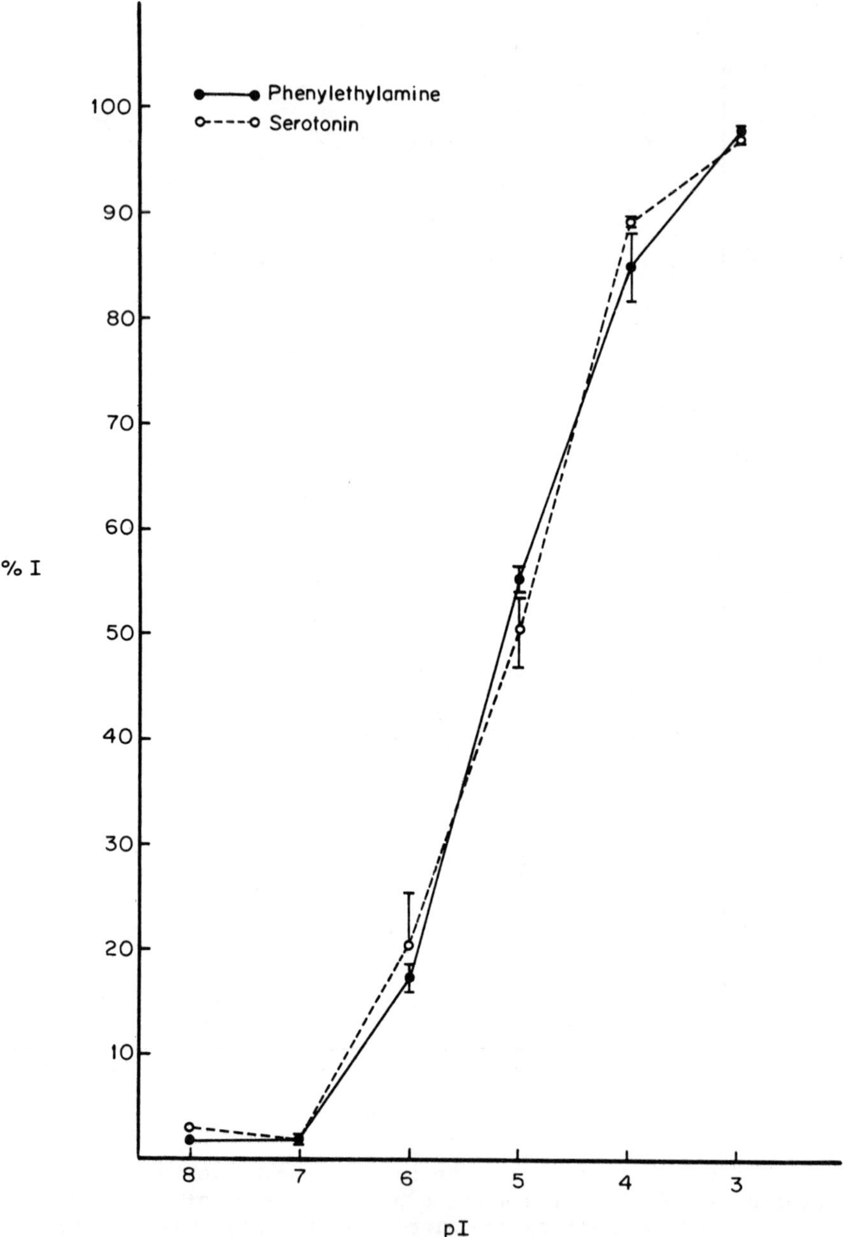

FIG. 4. Inhibition of PEA and 5-HT deamination by deprenyl with no pre-incubation of enzyme and inhibitor. Same conditions as for Fig. 3 except deprenyl was used as the inhibitor.

If this is so, then this would enhance the proposed PEA hypothesis of affective disorders. However, from an analysis of our findings on beef brain mitochondrial preparations, as summarized in Table 4, no marked differences could be observed in the degree of inhibition by the tricyclics with the various substrates used. The activities were measured using two different concentrations of the substrate, a saturating concentration and the K_m concentration (cf., 1,13). Using the pI_{50} values as an index of inhibitory efficiency, it is apparent that chlorimipramine is more effective in inhibiting MAO than the other tricyclics and that doxepin is the least effective. For any given inhibitor, using the substrate K_m concentration resulted in greater inhibition indicated by the higher pI_{50} values. These results do not necessarily rule out that these agents do not preferentially inhibit PEA deamination in vivo. It is noteworthy that substrate concentrations used in our experiments were higher than those used in other studies where such preferential inhibition of PEA deamination was reported (6,30,31).

D. Effect of pH on Ionization of MAO Substrates

That substrate structure such as phenyl-, indolyl-, and kynuramine could be differentiated by their pH optima has already been discussed in Section II.A. With respect to pH effects, one of the more striking studies in this regard was the effect of pH on the kinetic parameters. Phenylethylamine deamination, in particular, stood out as being markedly affected by increasing pH in showing pronounced substrate inhibition. This is illustrated in Fig. 5, which shows reciprocal plots for beef brain MAO with PEA and 5-HT as substrates. At pH 7.4, there was no noticeable substrate inhibition with either substrate. Serotonin, at pH 8.2, showed a slight amount of high substrate inhibition, which became more pronounced at pH 9.1. With PEA, however, there was a significant amount of high substrate inhibition at pH 8.2, which at pH 9.1 was markedly amplified. Thus, at the higher pH, it was not possible to obtain meaningful data in the same concentration range shown here.

TABLE 4

Inhibition of MAO in Beef Brain Mitochondria by Tricyclic Antidepressants[a]

Substrate	Substrate Concentrations (M)	pI_{50} Values			
		Chlorimipramine	Desipramine	Imipramine	Doxepin
Serotonin	a) 5×10^{-4}	3.50	3.18	3.06	2.93
	b) 1×10^{-4}	3.84	3.65	3.49	3.46
Dopamine	a) 5×10^{-4}	3.32	3.33	3.07	2.97
	b) 1.67×10^{-4}	3.70	3.62	3.34	3.40
Tyramine	a) 1×10^{-3}	3.14	2.47	2.69	2.54
	b) 6.25×10^{-5}	3.70	3.42	3.29	3.26
Tryptamine	a) 2×10^{-4}	2.98	2.70	2.69	2.60
	b) 1×10^{-5}	3.79	3.44	3.19	3.17
Phenylethylamine	a) 5×10^{-4}	3.10	2.88	2.80	2.66
	b) 2.5×10^{-5}	3.37	3.09	3.18	3.11

[a]Assays were carried out in 50 mM phosphate buffer, pH 7.4 at 37°C using (a) saturating substrate concentrations and (b) K_m substrate concentrations. Reaction was started by addition of enzyme. Percent inhibition was determined by comparison of inhibited samples with control samples assayed identical conditions in absence of inhibitor. pI_{50} values (negative logarithm of inhibitor concentration required to inhibit 50% of enzyme activity) were estimated from dose response plots of percent inhibition versus pI.

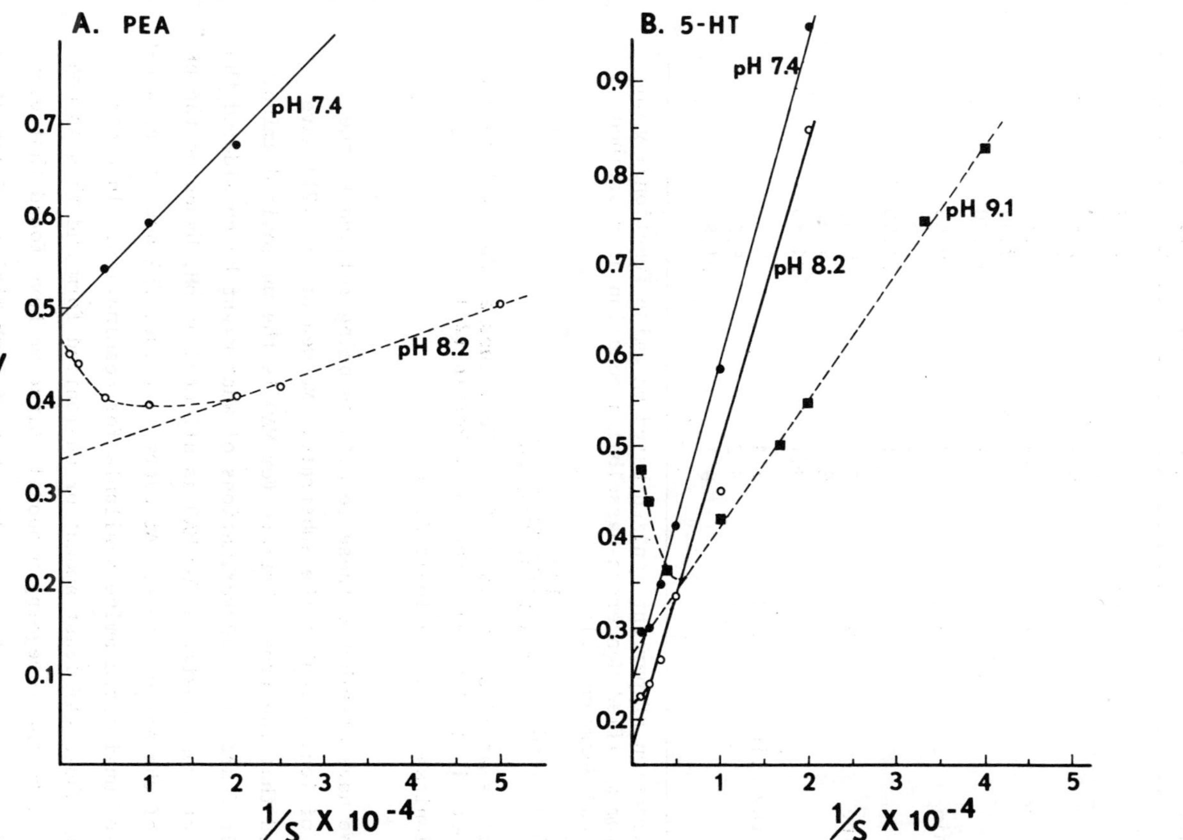

FIG. 5. Effect of pH on high substrate inhibition of MAO. A. Deamination of phenylethylamine in 50 mM phosphate buffer, pH 7.4 (●—●) and 10 mM pyrophosphate buffer, pH 8.2 (o--o). B. Deamination of 5-HT in 50 mM phosphate buffer, pH 7.4 (●—●) and 10 mM pyrophosphate buffer, pH 8.2 (o—o) and pH 9.1 (■---■).

TABLE 5

The Effect of pH on the Ionization of MAO Substrates[a]

Substrate	pKa[b]	pH	Percent Un-ionized (37°C)[c]
Tyramine	10.13	7.4	0.19
		8.2	1.16
		9.1	8.54
Tryptamine	9.83	7.4	0.37
		8.2	2.29
		9.1	15.70
Serotonin	9.60	7.4	0.63
		8.2	3.83
		9.1	24.03
Knuramine	9.58	7.4	0.66
		8.2	4.02
		9.1	24.88
Phenylethylamine	9.43	7.4	0.92
		8.2	5.56
		9.1	31.87

[a]Approximation of percent ionized based on amine functional group only from modified Henderson-Hasselbalch equation as described in Albert and Sergeant (2):

$$\% \text{ Ionized} = \frac{100}{1 + \text{Antilog (pH-pKa)}}$$

[b]pKa values at 37°C, estimated from pKa values at 25°C taken from McEwen et al. (22) using equation of Perrin (26).

[c]% Un-ionized = 100 - % Ionized.

We have attributed these results as being related to the extent of ionization of the substrate. McEwen et al. (21) have proposed that the true substrate for MAO is the un-ionized amine. In Table 5 are given approximations of what might be considered the amount of true substrate for MAO as affected by pH, based on the pK values of the amino groups. At every pH value, PEA has the greatest amount of un-ionized amine available for deamination. In contrast, tyramine shows the least amount of un-ionized form and this is consistent with the foregoing proposal, since we have found that even at pH 9.1, tyramine shows little substrate inhibition. Relevant to these results, it would seem justifiable to suggest a reappraisal of MAO substrate specificity and characteristics on the basis of the true K_m. The un-ionized concentration of substrate available for deamination might thus present an entirely different criterion of the enzyme characteristics.

III. CONCLUDING REMARKS

It is significant that PEA has become such an important compound in the study of MAO. The functionality of PEA as a neuromodulator or neurotransmitter has attracted much attention, leading to its increased use as a substrate of MAO, which, in turn, has aided in characterizing the subtle nature of MAO. Deamination by MAO represents, perhaps, the major pathway of PEA metabolism, and we have observed that some properties of PEA serve to distinguish it from the other biogenic amines. Unlike 5-HT, DA, norepinephrine, and tyramine, PEA is capable of crossing the blood-brain barrier (24), and the factors that contribute to this phenomenon may be responsible for the differentiation with respect to its metabolism. The fact that PEA has a low ionization potential favors its penetration into the brain, and we have reported that substrate ionization influences the activity of MAO. We regard this as a probable indication of the influence of the membrane lipid environment of the enzyme. Other studies have also suggested that the presence of phospholipid and other membrane constituents may be an important factor in the observed properties of MAO (16,25,35). Thus, it was shown that treatment of a rat liver MAO preparation with a chaotropic agent abolished the apparent multiplicity characteristics of the enzyme, both in terms of electrophoretic mobility and substrate variations (16). More recently, Williams and Lawson (38) have shown a close correlation between lipophilicity and inhibitory potency of propargylamines, such as clorgyline and pargyline. It was suggested that penetration through the lipid associated with MAO may be a factor in the selectivity of a particular inhibitor.

While the studies presented here have not answered the question posed earlier with respect to what fosters the property variations observed in MAO, whether different catalytic sites on a single form or multiple forms or modulation by lipid environment, these observations at the subcellular level using well-characterized and intact mitochondria could be important in gaining a truer understanding of what might be occurring in vivo. Further studies need to be carried out on substrate, inhibitor, and tissue variations of MAO, and undoubtedly the knowledge gained with respect to the distinct characteristics of MAO activity relative to PEA could contribute much to the elucidation of the role of PEA in the nervous system.

REFERENCES

1. F. M. Achee, G. Togulga, and S. Gabay (1974). Studies of monoamine oxidases: Properties of the enzyme in bovine and rabbit brain mitochondria. *J. Neurochem.*, 22:651-661.

2. A. Albert and E. P. Sergeant (1971). *The Determination of Ionisation Constants*. Chapman and Hall, London.

3. K. H. Beyer and H. S. Morrison (1945). Relation of structure to deamination of sympathomimetic amines. *Indust. Eng. Chem.*, 37:143-148.

4. A. A. Boulton (1974). Amines and theories in psychiatry. *Lancet*, ii:52.

5. A. J. Christmas, C. J. Coulson, D. R. Maxwell, and D. Riddel (1972). A comparison of the pharmacological and biochemical properties of substrate-selective monoamine oxidase inhibitors. *Br. J. Pharmacol.*, 45:490-503.

6. D. J. Edwards and M. L. Burns (1974). Effects of tricyclic antidepressants upon human platelet monoamine oxidase. *Life Sci.*, 15:2045-2058.

7. E. Fischer (1965). Monoamine oxidase inhibitors. *Lancet*, ii:245-246.

8. E. Fischer, H. Spatz, B. Heller, and H. Reggiani (1972). Phenylethylamine content of human urine and rat brain, its alterations in pathological conditions and after drug administration. *Experientia*, 28:307-308.

9. R. W. Fuller (1972). Selective inhibition of monoamine oxidase. *Adv. Biochem. Psychopharmacol.*, 5:339-354.

10. R. W. Fuller and B. W. Roush (1972). Substrate-selective and tissue-selective inhibition of monoamine oxidase. *Arch. Int. Pharmacodynam. Ther.*, 198:270-276.

11. S. Gabay and A. J. Valcourt (1968). Biochemical determinants in the evaluation of MAO inhibitors. *Rec. Adv. Biol. Psychiat.* 10:29-41.

12. S. Gabay, P. King, and F. M. Achee (1975). Characteristics of MAO activity in purified brain mitochondria with reference to the question of multiple forms. *Proc. 5th Int. Cong. Neurochem.*, Barcelona, p. 82.

13. S. Gabay, F. M. Achee, and G. Mentes (1976). Some parameters affecting the activity in purified bovine brain mitochondria. *J. Neurochem.*, 27:415-424.

14. M. Guggenheim and W. Loffler (1916). Das Schicksal Proteinogener Amine in Tienkorper. *Biochem. Z.*, 72:325-350.

15. A. E. Halaris, R. A. Lovell, and D. X. Freedman (1973). Effect of chlorimipramine on the metabolism of 5-hydroxytryptamine in the rat brain. *Biochem. Pharmacol.*, 22:2200-2202.

16. M. D. Houslay and K. F. Tipton (1973). The nature of the electo-phoretically separable multiple forms of rat liver monoamine oxidase. *Biochem. J.*, 135:173-186.

17. J. P. Johnston (1968). Some observations upon a few inhibitors of MAO in brain tissue. *Biochem. Pharmacol.*, 17:1285-1297.

18. J. Knoll and K. Magyar (1972). Some puzzling pharmacological effects of monoamine oxidase inhibitors. *Adv. Biochem. Psycho-pharmacol.*, 5:393-408.

19. P. Mantegazza and M. Riva (1963). Amphetamine-like activity of β-phenethylamine after a monoamine oxidase inhibitor in vivo. *J. Pharm. Pharmacol.*, 15:472-478.

20. T. J. Mantle, K. Wilson, and R. F. Long (1975). Studies on the selective inhibition of membrane-bound rat liver monoamine oxidase. *Biochem. Pharmacol.*, 24:2031-2038.

21. C. M. McEwen, Jr., G. Sasaki, and W. R. Lenz (1968). Human liver mitochondrial MAO. I. Kinetic studies of model inter-actions. *J. Biol. Chem.*, 243:5217-5225.

22. C. M. McEwen, Jr., G. Sasaki, and D. C. Jones (1969). Human liver mitochondrial MAO. II. Determinants of substrates and inhibitor specificities. *Biochemistry*, 8:3952-3962.

23. A. D. Mosnaim, E. E. Imwang, J. H. Sugerman, W. J. DeMartini, and H. C. Sabelli (1973). Ultraviolet spectrophotometric determination of 2-phenylethylamine in biological samples and its possible correlation with depression. *Biol. Psychiat.*, 6:235-257.

24. W. H. Oldendorf (1971). Brain uptake of radiolabeled amino acids, amines, and hexoses after arterial injection. *Am. J. Physiol.*, 221:1629-1639.

25. L. Oreland and B. Ekstedt (1972). Soluble and membrane-bound pig liver mitochondrial monoamine oxidase: Thermostability, tryptic digestability and kinetic properties. *Biochem. Pharmacol.*, 21:2479-2488.

26. D. D. Perrin (1964). The effect of temperature on pK values of organic bases. *Austral. J. Chem.*, 17:484-488.

27. N. Popov, H. Matthies, W. Lietz, Ohr. Thiemann, and E. Jassmann (1970). The effect of different substrates on the inhibition of rat brain and liver monoamine oxidase by arylalkylhydrazines. *Biochem. Pharmacol.*, 19:2413-2418.

28. D. Richter (1937). Adrenaline and amine oxidase. *Biochem. J.*, 31:2022-2028.

29. J. A. Roth and C. N. Gillis (1974a). Deamination of β-phenylethylamine by MAO. Inhibition by imipramine. *Biochem. Pharmacol.*, 23:2537-2545.

30. J. A. Roth and C. N. Gillis (1974b). Inhibition of lung, liver and brain MAO by imipramine and desipramine. *Biochem. Pharmacol.*, 23:1138-1140.

31. J. A. Roth and C. N. Gillis (1975). Some structural requirements for inhibition of Type A and B forms of rabbit MAO by tricyclic psychoactive drugs. *Molec. Pharmacol.*, 11:28-35.

32. H. C. Sabelli, A. D. Mosnaim, and A. J. Vazquez (1974). Phenylethylamine: Possible role in depression and antidepressive drug action. In: *Neurohumoral Coding of Brain Function*, edited by R. R. Drucker-Colin and R. D. Myers. Plenum, New York.

33. H. C. Sabelli and A. D. Mosnaim (1974). Phenylethylamine hypothesis of affective behavior. *Am. J. Psychiat.*, 131:695-699.

34. J. J. Schildkraut (1965). The catecholamine hypothesis of affective disorders: A review of supporting evidence. *Am. J. Psychiat.*, 122:509-522.

35. K. F. Tipton, M. D. Houslay, and N. J. Garrett (1973). Allotopic properties of human brain MAO. *Nature (New Biol.)*, 246:213-214.

36. N. Weiner (1960). Substrate specificity of brain amine oxidase of several mammals. *Arch. Biochem. Biophys.*, 91:182-188.

37. C. Whalley, A. D. Mosnaim, and H. C. Sabelli (1973). 2-Phenylethylamine in rabbit brain and liver. Its synthesis, metabolism and role in the action of imipramine, pargyline and marihuana. *The Pharmacologist*, 15:258.

38. C. H. Williams and J. Lawson (1975). MAO-III. Further studies of inhibition by propargylamines. *Biochem. Pharmacol.*, 24: 1889-1891.

39. H.-Y. T. Yang and N. H. Neff (1973a). β-Phenylethylamine: A specific substrate for Type B monoamine oxidase of brain. *J. Pharmacol. Exp. Ther.*, 187:365-371.

40. H.-Y. T. Yang and N. H. Neff (1973b). Monoamine oxidase. I. A natural substrate for Type B enzyme. *Fed. Proc.*, 32:797.

41. H. Yasuhara, S. Sho, and K. Kamijo (1972). Difference in actions of harmine on the oxidations of serotonin and tyramine by beef brain mitochondria. *Jap. J. Pharmacol.*, 22:439-441.

42. E. A. Zeller, L. A. Blanksma, W. P. Burkard, W. L. Pacha, and J. C. Lazanas (1959). *Ann. N.Y. Acad. Sci.*, 80:583-589.

Chapter 5

STUDIES ON THE UPTAKE OF PHENYLETHYLAMINE INTO THE SECRETORY VESICLES OF BRAIN SYNAPTOSOMES

Ruven Greenberg and Christopher E. Whalley

Department of Physiology
College of Medicine
University of Illinois at the Medical Center
Chicago, Illinois

I. INTRODUCTION

β-Phenylethylamine (PEA) has been hypothesized to be a neuromodulator, which is in part responsible for triggering or sustaining the state of wakefulness of an organism (39).

Phenylethylamine is unusual as compared with neurotransmitter amines: a) in that it readily crosses the blood-brain barrier (31, 34,45) and, therefore, could influence CNS activity by perfusion as well as by release subsequent to its synthesis within the CNS (40,48), and b) in that the endogenous brain concentration is several orders reduced (e.g., nanograms per gram) (10,21,37,51) as compared with other neurotransmitter amines, which are expressed in micrograms per gram concentrations.

The endogenous concentration of PEA varies within the large brain regions of the rat: a) Willner et al. (51) reported the highest concentrations in the pineal and pituitary glands (0.92 nM/g, 0.73 nM/g, respectively) and the next highest in caudate nucleus and hypothalamus (0.15 nM/g, 0.11 nM/g, respectively), followed by cerebellar and cerebral cortices and whole brain (0.068 nM/g, 0.059 nM/g, 0.014 nM/g, respectively), b) while Durden et al. (10) reported the highest concentration in the hypothalamus, the next highest in the caudate nucleus, in comparison with whole brain (0.209 , 0.066, and 0.008 nM/g, respectively).

The concentration of exogenous PEA is unequally distributed in the brain regions with the administration of milligram doses. Nakajima et al. (31) reported that the PEA of the brain of monoamine oxidase inhibitor (MAOI)-treated rabbits was concentrated above plasma level at 5 min subsequent to i.v. injection of PEA (10 mg/kg) with slightly higher concentrations in the caudate nucleus and cerebral cortex. Jackson and Smythe (22) reported that the i.p. injection of a massive dose of PEA (100 mg/kg) increased the concentration preferentially in corpus striatum and hypothalamus of rats; however, it was distributed uniformly in the brain after a lesser dose (1 mg PEA/kg, i.p.).

We now present data a) on the distribution of [1-^{14}C]PEA in rat brain at 15 sec after the intracarotid artery perfusion of 0.2 μM of [1-^{14}C]PEA , wherein the exogenous label was distributed uniformly in the large brain regions; and b) on the in vivo and in vitro uptake of radiolabeled PEA by the synaptosomes of the brain wherein the exogenous label was taken up in similar proportion by the synaptosomes prepared from various brain regions.

These data a) and b) afford no mechanism to explain the varying concentrations of endogenous PEA in brain regions based on differences in arterial perfusion and/or uptake with a near-physiologic dose of [1-^{14}C]PEA.

The subcellular distribution of PEA in whole brain of the rat has been reported to be localized predominantly in the supernatant and myelin subfractions subsequent to intraventricular injection (6) and subsequent to experimentally induced phenylketonuria (11). Contrarily, according to other reports, a significant portion of endogenous PEA (~13%) (5) and exogenous PEA (after in vitro uptake) (3) is in the synaptosomes.

In the latter report, Baldessarini and Vogt (3) demonstrated that the in vitro uptake of labeled PEA by the homogenates of rat brain: a) is not increased with time of incubation; b) is not temperature dependant; c) is not altered by pretreating the animals with pargyline or ouabain or imipramine; d) is localized significantly in the synaptosome fraction and may be released from the synaptosomes by disrupting them; e) is effective in inhibiting the uptake of norepinephrine (NE) (2).

Our in vitro data, in confirmation of Baldessarini and Vogt is consistent with uptake into synaptosomes by passive diffusion; however, in extension of their data, we now report a) that subsequent to uptake, the PEA label of the synaptosomes is concentrated 2 to 3x above incubate level with the majority of the label in the secretory vesicles (46); b) that there is a selective displacement of previously incorporated NE [e.g., and not serotonin (5-HT)] from the secretory vesicles subsequent to their uptake of PEA by passive diffusion; and c) that there is a 2 to 3x increased uptake of [^{3}H]PEA by synaptosomes and secretory vesicles prepared from the brain of rats pretreated (48, 24 hr earlier) with large doses of reserpine, at which time the amine-bearing secretory vesicles (e.g., NE, dopamine (DA), 5-HT) are decreased in number and depleted in amine content (8,25,26, 43). Conceivably, the increased uptake of PEA label by the reserpinized synaptosomes points to a novel and unanticipated major population of synaptosomes as a site of action of PEA and reserpine, wherein PEA may function as a displacer of transmitters or as a co-transmitter subsequent to sequestration within the secretory vesicles.

II. MATERIALS AND METHODS

Sixty male Holtzman rats, weighing 200 to 300 g, were used in the experiments. The radioisotopes used in the experiments were: [4-^{3}H(N)] β-phenylethylamine (sp. act. 4.58 Ci/mM, New England Nuclear, Boston, Mass., donated by Dr. A. A. Boulton, University Hospital, Saskatoon, Saskatchewan, Canada); [1-^{14}C] β-phenylethylamine hydrochloride (sp. act. 9.86 mCi/mM, New England Nuclear, Boston, Mass.); [^{3}H]H_2O (1 mCi/ml). The following compounds were obtained from Amersham/Searle Corp. (Arlington Heights, Ill.): D,L-[7-^{3}H]noradrenaline hydrochloride (sp. act. 8.0 Ci/mM); 5-hydroxy[6-^{3}H] tryptamine creatine sulfate (sp. act. 0.5 Ci/mM); L-[U-^{14}C]aspartic acid (sp. act. 280 mCi/mM); and L-[U-^{14}C]glutamic acid (sp. act. 290 mCi/mM).

A. Intracarotid Artery Injection Experiments

A bolus of [1-^{14}C]PEA was rapidly injected into the common carotid artery of rats in combination with [^{3}H]H_2O in order to determine the brain uptake index (BUI) as defined by Oldendorf (33) from whole brain as well as brain regions: a) the amount of [^{3}H]H_2O in the brain after 15 sec at which time the rat was decapitated defined the fraction of the injected bolus that entered the brain; b) the BUI is the brain uptake of carbon-14 tracer relative to the uptake of [^{3}H]H_2O tracer expressed as a percent; thus,

$$\frac{{}^{14}\text{C Tissue}/{}^{3}\text{H tissue}}{{}^{14}\text{C Injectant}/{}^{3}\text{H injectant}} \times 100 = \text{BUI}$$

For the BUI experiments, the rats were anesthetized either with sodium pentobarbital (50 mg/kg, i.p.) or with ketamine hydrochloride (132-155 mg/kg, i.m.). After exposure of the left common carotid artery, 0.2 ml of the injectant consisting of 0.26 to 0.94 μCi [1-^{14}C]PEA and 7.62 to 29.34 μCi [$^{3}H_2$]O in Krebs-Ringer solution (composition in meq/liter: Na^{+}, 147; K^{+}, 4; Ca^{++}, 2; Cl^{-}, 155) buffered at pH 7.56 with 4 mM N-2-hydroxyethylpiperazine (HEPES, Calbiochem, La Jolla, Calif.) was administered via a 27-gauge needle. Care was taken not to impede the blood flow in the artery. The rats were usually decapitated at 15 sec after injection; however, in the experiments where the rate of disappearance of [1-^{14}C]PEA was studied, the animals were decapitated at 30, 60, 120, or 600 sec after injection. After decapitation, the brains were removed as quickly as possible and frozen at -20°C until dissected. Whole brain was dissected according to the method of Oldendorf (33), while the large brain regions were dissected according to the method of Glowinski and Iverson (17). Each sample consisting of approximately 50 mg of brain tissue was digested in 0.5 ml NCS solubilizer (Amersham/Searle Corp., Arlington Heights, Ill.). After the addition of 12 ml of scintillation fluid [5 g 2,5-diphenyloxazole (PPO), 0.05 g 1,4-bis-2-(4-methyl-5-phenyloxazolyl)-benzene (POPOP) in 1 liter toluene], the tissue digests were counted on a Packard model 3375 liquid scintillation spectrometer set for counting dual label.

The Oldendorf technique was modified to study the distribution of [1-^{14}C]PEA in synaptosomes, in synaptic vesicles, and in membrane fragments after intracarotid injection. In these experiments, the bolus contained only [1-^{14}C]PEA in the buffered Krebs-Ringer solution. Furthermore, the rats were pretreated with pargyline hydrochloride (Abbott, Chicago, Ill., 75 mg/kg, i.p., 90 min prior to decapitation). The subfractions were prepared according to procedures described by Whittaker (49). The P_2 fraction (containing the synaptosomes) was lysed to prepare the secretory vesicle subfraction by adding 3 ml of 4°C H_2O and then shaking to release the synaptic vesicles, while other aliquots of the P_2 fraction were frozen and thawed (10x) to prepare the membrane fragment subfraction. The subfractions were harvested by centrifugation at 27,000 g for 10 min. The samples were digested as before and counted. Representative samples of the synaptic vesicle subfraction and the membrane fragment subfraction were examined and their organelle content confirmed by electron-microscopy subsequent to negative staining with 2% phosphotungstic acid after layering the 10% formalin preserved samples on Cu^{++}-coated grids according to the procedure of Whittaker et al. (50).

B. In Vitro Synaptosome Uptake Studies

Synaptosomes were also prepared as the S_1 fraction from whole rat brain or from large brain regions (17) by the technique described by Whittaker (49). For the experiments in 143 mM Na^+, 0.2 ml of the S_1 fraction was incubated (4) for 4 min at either 37 or 0°C with varying concentrations of [^{3}H] in 1.8 ml Krebs-Ringer bicarbonate (pH 7.4) containing 1.1 mM ascorbic acid, 0.16 mM disodium ethylenediamine tetraacetic acid (EDTA, Fisher, Fair Lawn, N.J.), and 11.1 mM D-glucose as described by Kuhar et al. (27). For the experiments with 0 mM Na^+ at 37°C, the following changes were made: a) tris(hydroxymethyl)aminomethane (Sigma, St. Louis, Mo.) was substituted for the bicarbonate buffer; b) 0.32 M sucrose replaced the 0.9% saline in order to maintain the isotonicity of the incubation medium; c) 0.16 mM ethylenebis(oxyethylenenitrilo)tetraacetic acid (EGTA, Eastman, Rochester, N.Y.) which has K^+ as the cation replaced EDTA. In all experiments, 7×10^{-5} M pargyline hydrochloride was used to prevent the oxidation of PEA by monoamine oxidase. After incubation with [^{3}H]PEA, the synaptosomes were either a) collected and washed (using 10 ml 0.9% saline) on membrane filters with 0.45 μm pores (Millipore Corp., Bedford, Mass.) followed by H_2O vacuum filtration as described by Horn et al. (19) or b) isolated by a standard centrifugation procedure (4). The results obtained by the two techniques were comparable. The [^{3}H]PEA containing synaptosomes were eluted from the millipore filters (19) by 12 ml of a scintillation cocktail containing Triton X-100 (5 g PPO, 0.9 g POPOP, 250 ml Triton X-100, 750 ml toluene). Synaptosomes isolated by centrifugation were digested as described and counted.

In other studies, we estimated the [^{3}H]PEA label associated with synaptosomes, synaptic vesicles, and membrane fragments prepared from the brain S_1 fractions which were incubated with varying concentrations of [^{3}H]PEA. Protein determinations of the synaptosomal aliquots (S_1 fraction, 1.5-2.0 mg protein/incubation) were done according to the method of Lowry et al. (29), and the results were expressed as uptake of [^{3}H]PEA in pM/mg protein.

The brain S_1 fractions were prepared from control rats and from rats pretreated with reserpine. The control rats received i.p. injections of 0.9% saline; the chronic reserpinized rats received i.p. injections of reserpine as serpasil (CIBA-Geigy Corp., Summit, N.J., 5 mg/kg) at 44 and 20 hr prior to sacrifice; and the acutely reserpinized rats received an i.p. injection of reserpine (5 mg/kg, 3 hr prior to sacrifice).

C. In Vitro Displacement Studies

In the in vitro studies to determine whether PEA displaces the amines NE and 5-HT or the excitatory amino acids, glutamic acid, and aspartic acid, the S_1 fraction was prepared as described previously and incubated for 4 min in Krebs-Ringer bicarbonate containing pargyline hydrochloride (7 x 10^{-5} M) at 37 or 4°C in the presence of either D,L-[^{3}H]NE, [^{3}H]5-HT, [^{14}C]aspartic acid, [^{14}C]glutamic acid at K_m concentrations. After cooling the incubate to 4°C, the P_2 fraction from both the 37 and 4°C incubation temperatures was centrifuged, washed with 4 ml of cold 0.9% saline, and then resuspended in fresh Krebs-Ringer bicarbonate in a final volume of 2 ml to which either 3.1 x 10^{-7} M PEA, 3.2 x 10^{-6} M PEA, or no PEA was added. The samples were then incubated for an additional 4 min at 37°C followed by cooling at 4°C. The P_2 fractions were harvested by centrifugation, washed, digested, and counted.

III. RESULTS

A. Intracarotid Artery Injection Experiments

The BUI values of [1-^{14}C]PEA relative to [^{3}H]H_2O by the large brain regions of the rat are listed in Table 1. The values were similar in cerebral cortex, forebrain, and hypothalamus; the uptake in the cerebellum was decreased.

TABLE 1

The BUI of [1-^{14}C]PEA in Whole Brain and in Brain Regions[a]

	N	BUI
[$^3H_2$]O reference		100
Whole brain	4	55.5 ± 8.1
Cerebral cortex	3	55.4 ± 13.6
Hypothalamus	3	56.9 ± 9.6
Forebrain[b]	3	64.7 ± 15.6
Cerebellum	3	30.9 ± 4.5

[a]The brain uptake index was performed according to the method of Oldendorf (33). The concentration of [1-^{14}C]PEA in the 0.2 ml injection bolus was 0.236 mM. N = number of rats used. The BUI values are expressed as $\overline{X}$ ± SD.

[b]The forebrain included the striatus, hippocampus, and midbrain, and did not include the cerebral cortex.

The percent of the total of the ^{14}C-label of the brain which was in the brain regions at 15 sec after the intracarotid artery injection was 62% in the cortex, 7% in the hypothalamus, 23% in the forebrain, 8% in the cerebellum.

The amounts administered per intracarotid artery injection ranged from 0.13 to 0.48 μM [1-^{14}C]PEA and 111 to 167 μM [^{3}H]H_2O. After 15 sec, 2.05 ± 1.3% of the ^{14}C-label and 3.7 ± 2.3% of the ^{3}H-label per gram of brain were recovered, corresponding to a BUI of 55.5 ± 8.1. After 10 min, 0.35 ± 0.3% of the ^{14}C-label and 0.95 ± 0.1% of the ^{3}H-label per gram were recovered corresponding to a BUI of 35.1 ± 23.8.

In the same animals the pituitary and the pineal gland uptake indexes (each sample consisted of three pooled glands) were 158 and 104 corresponding to 7.6% of the ^{14}C-label and 4.8% of the ^{3}H-label per gram of pituitary and 4.4% of the ^{14}C-label and 4.2% of the ^{3}H-label per gram of pineal.

A minority of the [1-^{14}C]PEA was present in the synaptosomes (S_1 fraction) prepared from the standard brain regions (see Table 2);

TABLE 2

The Percent of [1-^{14}C]-labeled PEA in the Synaptosomal Pellet (P_2) Subfractions of the Total Label Recovered in the Brain Regions of [1-^{14}C]PEA [a]

	N	[1-^{14}C]-labeled PEA in P_2 Subfraction
Whole brain	4	23.5 ± 2.6
Hypothalamus	4	28.2 ± 5.5
Cerebral cortex	4	27.0 ± 4.4
Forebrain	4	21.9 ± 3.2
Cerebellum	4	13.1 ± 2.0

[a]The rats were decapitated 15 sec after the injection of [1-^{14}C]PEA into the left common carotid artery. The rats were pretreated with pargyline HCl 75 mg/kg i.p. 90 min prior to sacrifice. N = number of rats used. The [1-^{14}C]-labeled PEA in the P_2 subfraction is expressed as $\overline{X}$ = SD.

ranging from 13.1 ± 2% in the cerebellar synaptosomes to 28.2 ± 5.5% in the hypothalamus synaptosomes of the total present in each brain region. The distribution of the label within the synaptosome subfractions subsequent to the intracarotid artery perfusion of [1-^{14}C]PEA is presented in Fig. 1. The major portion, 77% of the label of the synaptosomes, was in the secretory vesicles; 53% of the label of the secretory vesicles was released into the supernatant subsequent to freezing and thawing. In in vitro experiments (see the discussion that follows), the distribution of the labeled PEA in the synaptosome subfractions was similar.

B. In Vitro Synaptosome Uptake Studies

Figure 2 presents the in vitro uptake of [^{3}H]PEA at varying concentration by synaptosomes (S_1 fraction), prepared from the whole rat brain, incubated at 37°C in 143 and 0 mM NaCl and at 0°C in 143 mM NaCl. The uptake of the label for a given concentration of PEA was similar in all cases. The ^{3}H-labeled PEA of the synaptosomes was concentrated 2.8 ± 1.2x the incubate throughout the range of

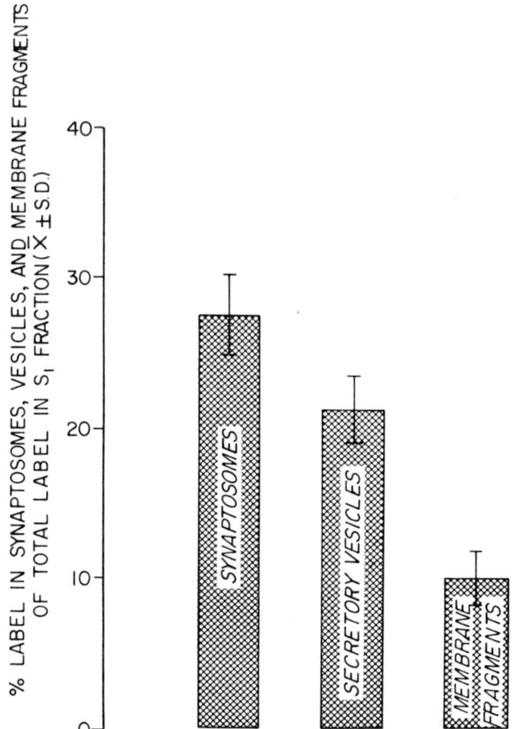

FIG. 1. The percentage label associated with synaptosomes (P_2 fraction), secretory vesicles, and membrane fragments of the total label associated with the S_1 fraction of rat brain following intracarotid artery injection of [1-^{14}C]PEA. The animals were pretreated with pargyline hydrochloride (75 mg/kg, i.p.) 90 min prior to the intracarotid injection of [1-^{14}C]PEA. The bars represent the $\overline{X} \pm SD$ of four rats.

concentrations of PEA. The volume of the synaptosomes was calculated based on the protein determination taken as 12% of the wet weight of the synaptosome subfraction. Figure 3 is a plot of V versus V/S of [^{3}H]PEA penetration kinetics of the same data represented in Fig. 2, where V equals velocity of uptake and S equals the injected concentration of [^{3}H]PEA. The straight line which was obtained is compatable

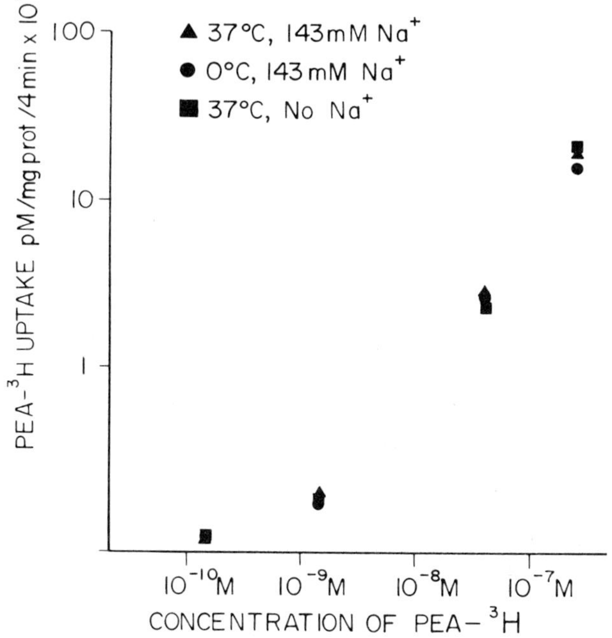

FIG. 2. The in vitro uptake of $[^3H]$PEA by synaptosomes (S_1 fraction) prepared from rat brain. The concentration of pargyline hydrochloride in the incubation media was 7 x 10^{-5} M. Each point represents the $\overline{X} \pm$ SD of 8 to 15 determinations.

with diffusional transport. The diffusion constant (32) for the in vitro uptake into the synaptosomes of the whole brain was 0.004 ml/mg protein/min.

Figure 4, in comparison with Fig. 2, presents the in vitro uptake of $[^3H]$PEA by synaptosomes prepared from the major brain regions rather than the whole brain. The data are compatible with uptake by passive diffusion; there was no significant difference in uptake of label by the brain regions.

The distribution of the label within the synaptosome subfractions subsequent to in vitro incubation with $[^3H]$PEA at varying

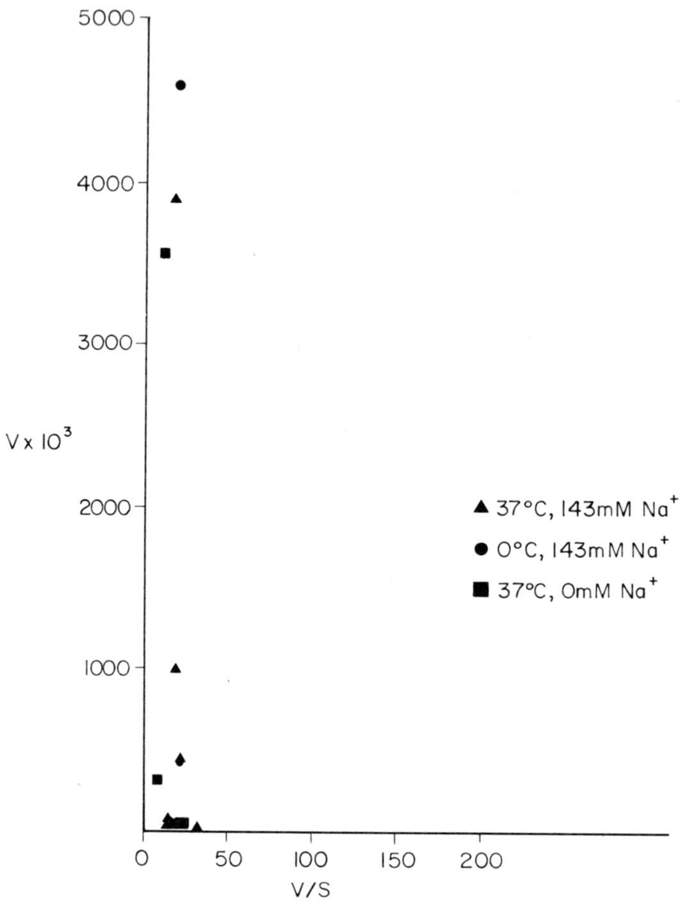

FIG. 3. Plot of V versus V/S of the in vitro uptake of [^{3}H]PEA by synaptosomes (S_1 fraction) prepared from rat brain. Data from Fig. 2 was used for the calculations. V (velocity) is expressed as p moles [^{3}H]PEA/mg protein/4 min, and S (substrate concentration of [^{3}H]PEA in the incubation media) is expressed in p mole/ml. The concentration of pargyline hydrochloride in the incubation media was 7 x 10^{-5} M. Each point represents $\overline{X} \pm$ SD of 8 to 15 determinations.

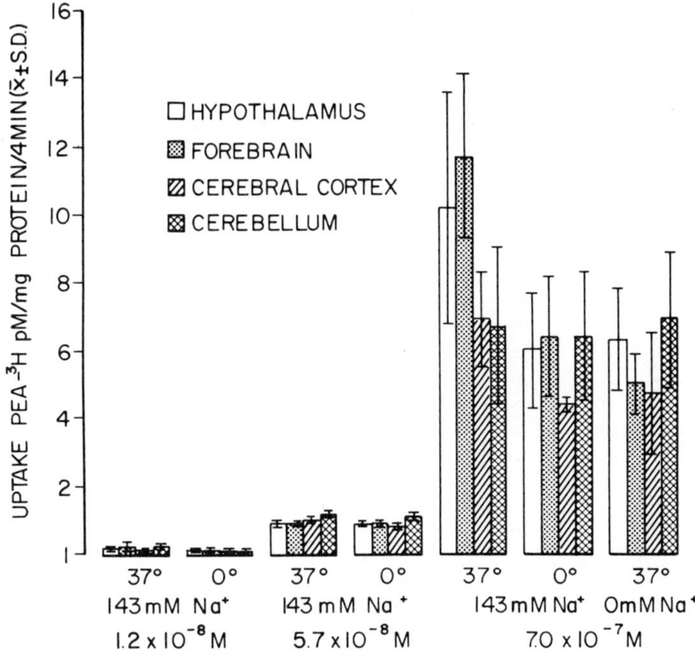

FIG. 4. The in vitro uptake of [^{3}H]PEA by synaptosomes (S_1 fraction) prepared from various regions of rat brain. The concentration of pargyline hydrochloride in the incubation media was 7 x 10^{-5} M. Each bar represents the $\overline{X}$ ± SD of at least four animals with at least four determinations per animal.

concentrations is presented in Table 3 (see control values). Approximately 60% of the label of the synaptosomes is in the secretory vesicles and 60% of the label of the secretory vesicles is released subsequent to freezing and thawing. The distribution of the label in the synaptic subfractions subsequent to in vitro incubation is similar to that observed subsequent to the intracarotid artery perfusions (Fig. 1).

Also in Table 3, we present the in vitro uptake of [^{3}H]PEA at varying concentrations by synaptosomes prepared from the whole brain of rats previously treated with large doses of reserpine in comparison

TABLE 3

The in Vitro Uptake of $[^3H]$PEA by Synaptosomes Prepared from Reserpine-Treated Rats: Including the Distribution of the Label within the Synaptosomal Subfractions[a]

Concentration of $[^3H]$PEA in Incubate		N	Uptake of $[^3H]$PEA (pM/mg protein/4 min)		
			Synaptosomes	Secretory Vesicles	Membrane Fragments
1.9×10^{-9} M	Control	5	0.038 ± 0.003	0.017 ± 0.003	0.012 ± 0.004
	Chronic reserpine	2	0.055 ± 0.004	0.059 ± 0.005	0.044 ± 0.003
	Acute reserpine	3	0.055 ± 0.005	0.023 ± 0.003	0.015 ± 0.005
1.8×10^{-8} M	Control	6	0.32 ± 0.06	0.22 ± 0.04	0.12 ± 0.03
	Chronic reserpine	3	0.93 ± 0.21	0.62 ± 0.03	0.37 ± 0.06
	Acute reserpine	3	0.37 ± 0.03	0.22 ± 0.01	0.06 ± 0.04
1.6×10^{-7} M	Control	3	2.67 ± 0.20	2.07 ± 0.24	1.00 ± 0.18
	Chronic reserpine	2	7.99 ± 0.58	5.72 ± 0.29	4.01 ± 0.32
	Acute reserpine	1	3.81 ± 0.40	1.96 ± 0.48	1.13 ± 0.21

[a]The synaptosomes (S_1 fraction) were incubated for 4 min at 37°C in Krebs-Ringer bicarbonate (pH 7.4), which contained pargyline HCl 7×10^{-5} M. The control rats were injected i.p. with 0.9% saline. The chronic reserpine treated rats were injected i.p. with 5 mg/kg reserpine 44 to 20 hr prior to sacrifice. The acute reserpine treated rats were injected i.p. with 5 mg/kg reserpine 3 hr prior to sacrifice. N = number of rats used; at least four aliquots per determination. The uptake values are expressed as $\overline{X}$ ± SD.

TABLE 4

The Percent Increased Release Above Control of Previously Incorporated Amines or Amino Acids by the Addition of PEA[a]

	N	3.1×10^{-7} M PEA		3.2×10^{-6} M PEA	
		37°C	4°C	37°C	4°C
DL[^{3}H]NE (8×10^{-9} M)	3	10.7 ± 4.2	8.9 ± 3.6	26.8 ± 9.0	21.5 ± 8.3
[^{3}H]5-HT (7×10^{-9} M)	4	9.4 ± 16.6	11.5 ± 18.1	13.3 ± 20.9	6.3 ± 16.4
[^{14}C]Glutamic acid (1×10^{-5} M)	3	15.0 ± 17.2	10.8 ± 12.6	10.6 ± 10.3	5.1 ± 5.0
[^{14}C]Aspartic acid (1×10^{-5} M)	2	4.6 ± 12.0	27.3 ± 8.6	6.4 ± 9.1	15.3 ± 14.4

[a]The amines and the amino acids were initially incorporated during incubation at 37° and at 4°C in the absence of PEA. The final incubation with two concentrations of PEA was performed at 37°C in order to determine whether or not displacement of the amines or amino acids occurred in comparison with the controls which did not have PEA added. Pargyline hydrochloride (7×10^{-5} M) was present in all incubations. N = number of experiments with at least three aliquots per determination. The percent release values are given as $\overline{X} \pm$ SD.

with the control synaptosomes. The distribution of the label within the synaptosome subfractions is also presented. The [^{3}H]PEA uptake by the synaptosomes from the acutely reserpinized rats and the distribution of the label in the subfractions of the synaptosomes is similar to the controls. The [^{3}H]PEA uptake by synaptosomes from the chronic reserpinized rats is increased on average 2.8x in the synaptosomes, 2.5x in the secretory vesicles, and 4x in the membrane fractions in comparison with controls. However, the distribution in the subcompartments is similar to controls.

Table 4 presents data indicative of the presence or absence of displacing effect by PEA upon the radio-labeled compounds previously incorporated by the synaptosomes. The incorporated [^{3}H]NE, [^{3}H]5-HT, [^{14}C]glutamic acid, and [^{14}C]aspartic acid was concentrated 3x, 3x, 5x, and 2x, respectively, by the 37°C control preparations during incubation for 4 min at 37°C in comparison with the control preparation incubated at 4°C. Phenylethylamine is considered to have had a displacing effect in proportion to the percent increased release above control of the ^{3}H- or ^{14}C-label into the supernatant during the 4-min re-incubation at 37°C. In three experiments at 3.2 x 10^{-6} M concentrations of PEA, there was a further increased release (above the release with the 3.1 x 10^{-7} M PEA) from the synaptosomes which had incorporated [^{3}H]NE at 37°C as well as from the synaptosomes which had incorporated [^{3}H]NE by passive diffusion at 4°C. The percent release from the synaptosomes which had previously incorporated [^{3}H]5-HT, [^{14}C]glutamic acid, and [^{14}C]aspartic acid was not similarly increased.

IV. DISCUSSION

The BUI of [1-^{14}C]PEA in comparison with [^{3}H]H_2O by the Oldendorf procedure (34) was 55.5 corresponding to the value of 67 obtained by Oldendorf. We have made several adaptations: a) to determine the BUI of the brain regions as well as the whole brain, b) to determine the percent of the total content of radioactive label of a brain region within the synaptosomes and other subfractions prepared from the tissue, and c) to determine the rate of disappearance with time of the radioactive label subsequent to uptake into the brain by the standard technique.

By means of a), we determined that the BUI values of the large brain regions reflects whole brain distribution of the perfusion via the carotid artery (Table 1). The BUI values were similar in cerebral cortex, forebrain, and hypothalamus and somewhat reduced in the cerebellum; we do not have an explanation for the reduced BUI by the cerebellum.

By means of b), we observed that 23.5% of the brain uptake of the radioactive label was in the synaptosomes (Table 2) (at 15 sec subsequent to the intracarotid artery injection). Similar amounts were in the synaptosomes prepared from the brain regions, except the cerebellum in which only 13.1% of its uptake was in the synaptosomes. The localization of 23.5% of the uptake within the synaptosomes corresponds with a 10 to 20 x increased concentration of label in these organelles which comprise approximately 1 to 2% of the brain volume as estimated by others (9).

As a further extension of b), we made subfractions of the synaptosomes subsequent to the uptake of [1-^{14}C]PEA (Fig. 1) and recovered 77% of the label in the secretory vesicles of the synaptosomes of which 53% could be released by freezing and thawing the secretory vesicles. The distribution of the labeled PEA within the subfractions of the synaptosomes corresponds well with the distribution of the classic transmitter amines such as NE (12,36) or 5-HT (35), and presumably, therefore, the exogenous PEA might interact with them as we will show below.

By means of c), we observed that the [1-^{14}C]-labeled PEA taken up by the whole brain and the large brain regions is lost rapidly: Approximately 2% of the injected bolus is present at 15 sec, and 0.35% at 10 min. The BUI values of [1-^{14}C]PEA in comparison with [^{3}H]H_2O at 10 min is decreased corresponding to an accelerated egress, either as PEA or metabolite of PEA. The half life of the [1-^{14}C]-labeled PEA of the brain was less than 2 min (47) corresponding with the short half-life observed by Wu and Boulton (53).

We have demonstrated that the in vitro uptake of radio-labeled PEA into synaptosomes of whole brain (Figs. 2 and 3) or synaptosomes of brain regions (Fig. 4) corresponds to penetration by a nonsaturable diffusion mechanism. There is no metabolic requirement; the uptake was similar at 37 and 0°C. There is no Na^+ dependance; the uptake at 37°C was similar in 143 mM NaCl or in 0.018 M Tris buffer in sucrose, which is isosomotic with 143 mM NaCl. The uptake by the cerebellar synaptosomes (Fig. 4) was not decreased, and the percent of label in the cerebellar synaptosomes after the intracarotid artery perfusion (Table 2) is not explained on this basis; nor is it due to the partial perfusion of the cerebellum by the vertebral artery (30).

Endogenous PEA is distributed unequally in the large brain regions (10,21,37,51). According to Willner et al. (51), it is increased several times in hypothalamus and caudate nucleus as compared with whole brain. Since the uptake of exogenous PEA by the large brain regions is similar a) in the intracarotid artery injection experiments and b) in the in vitro incubation experiments, it is unlikely that the increased concentration of endogenous PEA in the hypothalamus and caudate nucleus is a function of increased perfusion.

Most of the exogenous label of the synaptosomes is in the synaptic vesicles: a) 77% of the label of the synaptosomes is in the synaptic vesicles after the intracarotid artery injection (Fig. 1), and b) an average of 60% of the label of the synaptosomes is in the synaptic vesicles after the in vitro incubations (Table 3). We presume that the label is in some way sequestered within the secretory vesicles, otherwise it would have diffused into the supernatant during preparation of the synaptosome fraction or when the synaptosomes were lysed in cold distilled water in order to release the secretory vesicles from the nerve endings. After repeated (10x) freezing and thawing, approximately 47% of the labeled PEA of the synaptic vesicles remains in association with the membrane fragments and 53% of label is released into the supernatant. The general properties and chemical composition of the neuronal secretory vesicles and adrenal chromaffin granules are somewhat similar (20); we cannot at this time state whether the labeled PEA is bound proportionately to any of the known components of the synaptic vesicles. However, the sequestration of the PEA within the secretory vesicles is consistent with its possible role as a co-transmitter (23) available to be released upon nerve stimulation.

The preponderant localization of the labeled PEA to the secretory vesicles in conjunction with the displacement of previously incorporated [^{3}H]NE and lack of displacement of [^{3}H]5-HT (Table 4) affords insight into the selective mechanism by which PEA diminished the green fluorescence of noradrenergic fibers of the CNS, while the yellow fluorescence of serotonergic fibers was unaffected subsequent to the injection of massive doses of PEA to intact rats (15). There are mixed reports relative to the displacement of 5-HT by PEA from the brain of rats based on the chemical levels. Jackson and Smythe (22) reported a 14% depletion of 5-HT as compared with a 75% depletion of NE by PEA (100 mg/kg/1 hr, i.p.); however, Loo (28) reported a 50% depletion of brain 5-HT in weanling rats by a lesser dose of PEA (20 mg/kg, in divided doses 20 min apart, s.c.).

Previously PEA was shown to be effective as a releaser of NE from sympathetically innervated perfused organs (18,24), from isolated bovine chromaffin granules (42), and from isolated adrenergic nerve granules (13). On balance, as discussed by von Euler and Lishajko (13), PEA operates dually upon granules by direct release, by stoichiometric substitution for NE, and by inhibition of re-uptake of the NE released during the incubation; both of these processes are dependent upon diffusion and/or uptake of PEA across the plasma membrane prior to any action on the secretory vesicles.

The efficacy of PEA as a displacer of neurotransmitter amines from brain synaptosomes has not been clearly demonstrated. In a previous in vitro study (1), the concurrent addition of PEA and radio-labeled NE, DA, or 5-HT to synaptosome preparations diminished the uptake of radio-labeled NE and DA (and 5-HT to a lesser degree); by the procedure used, it was not possible to distinguish whether the PEA acted to displace and/or diminish the uptake of NE. By our procedure, the uptake and displacement of NE were separable; according to our data (Table 4), PEA selectively displaced NE from synaptosomes. It did not displace 5-HT, glutamic acid, or aspartic acid under the same conditions. We were not able to determine whether there was stoichiometric substitution of PEA for NE, since we did not obtain data relative to the amount of endogenous NE displaced.

The 2 to 3x increased uptake of PEA by the synaptosomes of chronic reserpinized rats (Table 3) suggests that PEA might be selectively incorporated into other (unknown) synaptosomes. Since as much as two-thirds of the increased uptake of the reserpinized synaptosomes is in the secretory vesicles (Table 3) and since under the conditions of the experiment, the NE, 5-HT, and DA secretory vesicles would be decreased in frequency as well as in amine content. We conceive a) that the remaining depleted amine-bearing secretory vesicles contain a more than 2 to 3x increased number of combining sites* for PEA or b) that reserpine has interacted with other unidentified secretory vesicles to affect the 2 to 3x increased number of combining sites. The latter alternative points to a novel and unanticipated population of synaptosomes as a site of action of PEA and reserpine.

Sabelli and associates (38) and Fischer and associates (16) have reported a decrease in PEA in the brain at 24 hr subsequent to large doses of reserpine. We did not obtain data on the PEA content of the brain subsequent to reserpine; however, the increased uptake of PEA by the synaptosomes of the chronic reserpinized rats suggests that there exists an increased and perhaps a different complement of combining sites.

Based on fluorescent microscopy data (15), exogenous NE is readily taken up by the noradrenergic neurons of rats at 24 hr subsequent to depleting doses or reserpine. However, the uptake is

---

*The term "combining sites" is used descriptively and not according to the strict meaning of the term.

different in that it is localized predominately in the neuronal cytosol with little or no fluorescence in the secretory vesicles. Based on our data (Table 3), the uptake of PEA by the reserpinized synaptosomes is enhanced 2 to 3x both in the cytosol and in the secretory vesicles subfraction, and presumably, therefore, PEA might interact at both sites.

We reported in the intracarotid artery injection results that the uptake of [1-^{14}C]-labeled PEA by the pituitary and pineal glands was increased 2 and 3x, respectively, in comparison with the whole brain. This observation is not unexpected since these glands are outside the blood-brain barrier (14,41,52), and in correlation with the very high endogenous levels of PEA within these glands (51). We suggest that a) within the pineal, the PEA might act either as a displacer of neurotransmitter amines or as a co-transmitter and b) within the pituitary, the PEA might act to displace the hypophyseal peptides from their secretory vesicles.

V. SUMMARY

The intracarotid artery quick injection of [1-^{14}C]PEA in association with [^{3}H]H_2O according to the method of Oldendorf (33) was adapted to determine the uptake of labeled PEA by the brain regions and the synaptosomes and secretory vesicles prepared from them. The PEA uptake was increased several-fold in the pineal gland and in the pituitary gland. The labeled PEA was uniformly distributed in the large brain regions, except the cerebellum in which the uptake was reduced. Of the content of the PEA label of the large brain region 21.9 to 28.2% was localized to the synaptosomes corresponding to a 10 to 20x increased concentration in these organelles, which make up only 1 to 2% of the brain volume. Two-thirds of the content of the synaptosomes were localized in the secretory vesicles. The in vitro uptake of [1-^{14}C]PEA by the synaptosomes prepared from the whole brain and from the large brain regions was by passive diffusion. There was no evidence of high affinity transport; there was no evidence of Na^+ dependence; there was no evidence of enhanced uptake by any one of the large brain regions relative to the others. Notwithstanding, the labeled PEA was concentrated 2 to 3x above gradient in the synaptosomes. Interestingly, the labeled PEA was also concentrated 2 to 3 times within the secretory vesicles of chronic reserpine-treated rats, which presumably have a reduced complement of secretory vesicles.

In other experiments, PEA selectively displaced previously incorporated [^{3}H]NE, while failing to displace previously incorporated [^{3}H]5-HT, [^{14}C]glutamic acid, and [^{14}C]aspartic acid from the secretory vesicles of brain synaptosomes. The sequestration of labeled PEA within secretory vesicles of brain synaptosomes corresponds with its reported role as a displacer of neurotransmitters and permits speculation as to whether it may also be released as a co-transmitter from stimulated neurons.

Note Added in Proof: Previously incorporated [^{14}C]DA by synaptosomes was displaced by PEA as was [^{3}H]NE (see Table 4); while [^{3}H]γ-aminobutyric acid (GABA) previously incorporated was not displaced by PEA.

ACKNOWLEDGMENTS

This work was supported in part by PHS GRSG #FR5369.

REFERENCES

1. R. Ashkenazi and B. Haber (1974). Inhibition of synaptosomal biogenic amine uptake by sympathominetic amines (abstract). In: *Society for Neurosciences: Fourth Annual Meeting*, p. 123.

2. R. J. Baldessarini and M. Vogt (1971). Uptake and release of norepinephrine by rat brain tissue fractions prepared by ultrafiltration. *J. Neurochem.*, 18:951-962.

3. R. J. Baldessarini and M. Vogt (1971). The uptake and subcellular distribution of aromatic amines in the brain of the rat. *J. Neurochem.*, 18:2519-2533.

4. J. P. Bennett, W. J. Logan, and S. H. Snyder (1972). Amino acid neurotransmitter candidates: Sodium-dependent high-affinity uptake by unique synaptosomal fractions. *Science*, 178:997-999.

5. A. A. Boulton and G. B. Baker (1975). The subcellular distribution of β-phenylethylamine, p-tyramine and tryptamine in rat brain. *J. Neurochem.*, 25:477-481.

6. A. A. Boulton, P. H. Wu, and S. Philips (1972). Binding of some primary aromatic amines to certain rat brain particulate fractions. *Can. J. Biochem.*, 50:1210-1218.

7. A. S. V. Burgen and L. L. Iverson (1965). The inhibition of nonadrenaline uptake by sympatomimetic amines in the rat isolated heart. *Br. J. Pharmacol.*, 25:34-49.

8. A. Carlsson, N. A. Hillarp, and B. Waldeck (1968). Analysis of the Mg^{++}-ATP dependent storage mechanism in the amine granules of the adrenal medulla. *Acta Physiol. Scand.*, 59(suppl. 215): 1-38.

9. F. Clementi, V. P. Whittaker, and M. N. Sheridan (1965). The yield of synaptosomes from the cerebral cortex of guinea pigs estimated by a polystyrene bead "tagging" procedure. *Z. Zellforsch.*, 72:126-138.

10. D. A. Durden, S. R. Philips, and A. A. Boulton (1973). Identification and distribution of β-phenylethylamine in the rat. *Can. J. Biochem.*, 51:995-1002.

11. D. J. Edwards and K. Blau (1973). Phenylethylamines in brain and liver of rats with experimentally induced phenylketonuria-like characteristics. *Biochem. J.*, 132:95-100.

12. U. S. von Euler and F. Lishajko (1965). Effects of drugs on the storage granules of adrenergic nerves. In: *Proceedings of the Second International Pharmacological Meeting, Vol. 3*, edited by G. B. Koelle, W. W. Douglas, and A. Carlsson. Macmillan, New York.

13. U. S. von Euler and F. Lishajko (1968). Effect of directly and indirectly acting sympathomimetic amines on adrenergic transmitter granules. *Acta Physiol. Scand.*, 73:78-92.

14. K. Fuxe (1964). Cellular localization of monoamines in the median eminence and the infundibular stem of some mammals. *Z. Zellforsch.*, 61:710-724.

15. K. Fuxe, H. Grobecker, and J. Jonsson (1967). The effect of β-phenylethylamine on central and peripheral monoamine-containing neurons. *Eur. J. Pharmacol.*, 2:202-207.

16. E. Fischer, H. Spatz, B. Heller, and H. Reggiani (1972). Phenylethylamine content of human urine and rat brain: its alteration in pathological conditions and after drug administration. *Experientia*, 28(3):307-308.

17. J. Glowinski and L. L. Iverson (1966). Regional studies of catecholamines in the rat brain I. the disposition of [^{3}H]-norepinephrine and [^{3}H]-dopa in various regions of the brain. *J. Neurochem.*, 13:655-669.

18. H. W. Haag, A. Philippu, and H. J. Schumann (1961). Freisetzung von Brenzcatechinaminen aus der isoliert durchstromten Nebenniere durch Tyramin und β-Phenylathylamin. *Experientia*, 17:187-188.

19. A. S. Horn, J. T. Coyle, and S. H. Snyder (1971). Catecholamine uptake by synaptosomes from rat brain. Structure-activity relationships of drugs with differential effects or dopamine and norepinephrine neurons. *Mol. Pharmacol.*, 7:66-80.

20. L. L. Iverson and B. A. Callingham (1971). Adrenergic transmission. In: *Fundamentals of Biochemical Pharmacology*, edited by Z. M. Bacq. Pergamon Press, Oxford.

21. E. E. Inwang, A. D. Mosnaim, and H. C. Sabelli (1973). Isolation and characterization of phenylethylamine and phenylethanolamine from human brain. *J. Neurochem.*, 20:1469-1473.

22. D. M. Jackson and D. M. Smythe (1973). The distribution of β-phenylethylamine in discrete regions of the rat brain and its effect on brain noradrenaline, dopamine and 5-hydroxytryptamine levels. *Neuropharmacology*, 12:663-668.

23. G. Jaim-Etcheverry and L. M. Zieher (1974). Localizing serotonin in central and peripheral nerves. In: *The Neurosciences: Third Study Program*, edited by F. O. Schmitt and F. G. Worder. The MIT Press, Cambridge, Massachusetts.

24. J. Jonsson, H. Grobecker, and P. Holtz (1966). Effect of β-phenylethylamine on content and subcellular distribution of norepinephrine in rat heart and rat brain. *Life Sci.*, 5:2235-2246.

25. N. Kirshner (1962). Uptake of catecholamines by a particulate fraction of the adrenal medulla. *J. Biol. Chem.*, 237:2311-2317.

26. R. L. Klein and A. K. Thureson-Klein (1974). Pharmacomorphological aspects of large dense-core vesicles. *Fed. Proc.*, 33(10):2195-2206.

27. M. J. Kuhar, R. M. Roth, and G. K. Aghajanian (1972). Synaptosomes from the forebrain of rats with midbrain raphe lesions: Selective reduction of serotonin uptake. *J. Neurochem.*, 13: 655-669.

28. Y. H. Loo (1974). Serotonin deficiency in experimental hyperphenylalaninemia. *J. Neurochem.*, 23:139-147.

29. O. H. Lowry, N. J. Rosebrough, A. L. Farr, and R. S. Randall (1951). Protein measurement with the Folin phenol reagent. *J. Biol. Chem.*, 193:265-275.

30. D. B. Moffat (1961). The development of the posterior cerebral artery. *J. Anat.*, 95:485-494.

31. T. Nakajima, Y. Kakimoto, and I. Sano (1964). Formation of β-phenylethylamine in mammalian tissue and its effect on motor activity in the mouse. *J. Pharmacol. Exp. Ther.*, 143:319-324.

32. K. D. Neame and T. G. Richards (Eds.) (1972). *Elementary Kinetics of Membrane Carrier Transport*. Wiley, New York.

33. W. H. Oldendorf (1970). Measurement of brain uptake of radiolabeled substances using a tritiated water internal standard. *Brain Res.*, 24(6):372-376.

34. W. H. Oldendorf (1971). Brain uptake of radiolabeled amino acids, amines and hexoses after arterial injection. *Am. J. Physiol.*, 221:1629-1639.

35. A. Pellegrino de Iraldi, L. M. Zieher, and G. Jaim-Etcheverry (1968). Neuronal compartmentation of 5-hydroxytryptamine stores. *Adv. Pharmacol.*, 6A:257-270.

36. L. T. Potter and J. Axelrod (1963). Properties of norepinephrine storage particles of the rat heart. *J. Pharmacol. Exp. Ther.*, 142:299-305.

37. J. M. Saavedra (1974). Enzymatic isotopic assay for and presence of β-phenylethylamine in brain. *J. Neurochem.*, 22:211-216.

38. H. C. Sabelli, W. J. Giardina, A. D. Mosnaim, and N. H. Sabelli (1973). A comparison of the functional roles of norepinephrine, dopamine and phenylethylamine in the central nervous system. *Acta Physiol. Pol.*, 24(1):33-40.

39. H. C. Sabelli and A. D. Mosnaim (1974). Phenylethylamine hypothesis of affective behavior. *Am. J. Psychiat.*, 131(6): 695-699.

40. H. C. Sabelli, W. A. Pedemonte, C. Whalley, A. D. Mosnaim, and A. J. Vazquez (1974). Further evidence for a role of 2-phenylethylamine in the mode of action of Δ^9-tetrahydrocannabinol. *Life Sci.*, 14:149-156.

41. T. Samorajski and B. H. Marks (1962). Localization of tritiated norepinephrine in mouse brain. *J. Histochem. Cytochem.*, 10: 392-399.

42. H. J. Schumann and A. Philippu (1962). Release of catechol amines from isolated medullary granules by sympathomimetic amines. *Nature*, 193:890-891.

43. A. D. Smith (1973). Mechanisms involved in the release of noradrenaline from sympathetic nerves. *Br. Med. J.*, 29(2): 123-129.

44. S. H. Snyder, A. B. Young, J. P. Bennett, and A. H. Mulder (1973). Synaptic biochemistry of amino acids. *Fed. Proc.*, 32(10):2039-2047.

45. C. Whalley and R. Greenberg (1974). In vivo and in vitro uptake of β-phenylethylamine-1-^{14}C (PEA) by rat brain synaptosomes (abstract). *Physiologist*, 17:357.

46. C. Whalley and R. Greenberg (1975). Major uptake of phenylethylamine (PEA) into the secretory vesicles of synaptosomes (abstract). In: *Society for Neurosciences: Fifth Annual Meeting*, p. 392.

47. C. Whalley and R. Greenberg (1977) (in preparation).

48. C. Whalley, A. D. Mosnaim, and H. C. Sabelli (1973). 2-phenylethylamine in rabbit brain and liver: Its synthesis, metabolism and role in the action of imipramine, pargyline and marihuana (abstract). *Pharmacologist*, 15(2):258.

49. V. P. Whittaker (1963). The separation of subcellular structures from brain tissue. *Biochemical Society Symposium No. 23*, 109-126.

50. V. P. Whittaker, J. A. Michaelson, and R. J. A. Kirkland (1964). The separation of synaptic vesicles from nerve-ending particles ('synaptosomes'). *Biochem. J.*, 90:293-303.

51. J. Willner, H. F. LeFevre, and E. Costa (1974). Assay by multiple ion detection of phenylethylamine and phenylethanolamine in rat brain. *J. Neurochem.*, 23:857-859.

52. C. W. Wilson, A. W. Murray, and E. Titus (1962). The effects of reserpine on uptake of epinephrine in brain and certain areas outside the bloodbrain barrier. *J. Pharmacol. Exp. Ther.*, 135: 11-16.

53. P. H. Wu and A. A. Boulton (1975). Metabolism, distribution and disappearance of injected β-phenylethylamine in the rat. *Can. J. Biochem.*, 53:42-50.

Chapter 6

β-PHENYLETHYLAMINE: A METABOLICALLY AND PHARMACOLOGICALLY ACTIVE AMINE

Stephen R. Philips

Psychiatric Research Division
University Hospital
Saskatoon, Saskatchewan
Canada

I. INTRODUCTION

Phenylethylamine (PEA) has long been known to possess pharmacological properties. As long ago as 1910, Barger and Dale (2) concluded that its physiological action was of the sympathomimetic type. Much of the interest in phenylethylamine has arisen as a result of its discovery in blood (1) and urine (27); PEA was the first compound with known direct physiological activity shown to exhibit a greater turnover in phenylketonuric patients than in normals.

More recently, abnormal excretion of PEA has been linked to depression. Although the amounts which have been claimed to be excreted have varied widely, ranging from 350 to 450 μg/day (22,23, 35) to less than 50 μg/day (10,18), it was observed in all cases that much less phenylethylamine was excreted by patients suffering with endogenous depression than normal subjects. Two of the most recent studies, however, have been unable to confirm this observation (46,47). Furthermore, the much more specific analytical techniques employed in these latter surveys indicated mean daily excretion levels of only about 10 μg/day or less.

Urinary PEA has been claimed to be excreted in greater than normal amounts by manic and by some schizophrenic patients, and to be increased from below normal toward normal levels in patients with endogenous depression who were being treated with antidepressive drugs (21,23). In connection with the observation that urinary PEA levels are elevated in schizophrenic patients, it is interesting to note that a reduced monoamine oxidase (MAO) activity has been reported in blood platelets from such patients (36).

Phenylethylamine is known to possess weak sympathomimetic properties (2,30). Its pharmacological effects have been shown to be similar to those of amphetamine, especially after pretreatment of the animals with a monoamine oxidase inhibitor (MAOI) (19,33). The effects have been claimed by some to be indirect and to be due to the release of catecholamines (11,45). Indeed, in the rat, injected PEA has since been shown to reduce the noradrenaline content of both heart and brain (28). Others, however, have attributed to PEA a direct action, independent of the release of sympathomimetic amines (19,20). In an iontophoretic study, in which the effects of norepinephrine (NE) and PEA on cortical neurons were compared, Giardina and co-workers (25) observed that the two amines had different effects on the spontaneous spike rate in some cells in which they were tested consecutively. The results showed that the effects of PEA were not due solely to either the direct or the indirect activation of NE receptors. It was suggested that PEA might act both presynaptically to release catecholamines and post-synaptically as a partial agonist, and might modulate central synaptic transmission.

Several theories regarding the physiological role of PEA have been advanced in recent years. Fischer, Saavedra, and Heller (20) suggested that the amine might represent a neurohumoral ergotropic agent, the ergotropic behavior of animals being that which exerts itself in an animal exposed to an emergency situation, and being principally of a sympathetic character. Sabelli et al. (43) suggested an excitatory role for PEA in the modulation of behavior, and later developed the PEA hypothesis of affective behavior (44), which stated that PEA was a neuromodulator responsible for triggering or sustaining a state of alertness and excitement. Thus, a decrease in the brain level of PEA might play a major role in some forms of endogenous depression, while an increase, or the activation of specific PEA receptors in brain neurons might be responsible for the actions of antidepressant and stimulant drugs.

Dewhurst (13) proposed that cerebral amines be divided into two categories, according to their cerebral actions. Type A, or excitant amines, were lipid-soluble and included such compounds as PEA and tryptamine. A specific type A receptor mediated actions of the type A amines. Responses evoked by these compounds were those of the alert state. Type C, or depressant amines, were relatively insoluble in lipid and included such compounds as the catecholamines. Responses, mediated through a specific receptor, were those of drowsiness or sleep, and locomotor activity was reduced. Dewhurst proposed that some forms of depressive illness were caused by a deficiency of type A amines or an insensitivity of the type A receptor, while some forms of mania were due to an excess of type A amines or hypersensitivity of the receptor.

Finally, a recent hypothesis by Boulton (4) suggested that PEA may function as a neurotransmitter, either in the conventional sense, or in a subthreshold sense. In the latter case, it may function as an activator of regions in the synaptic area, the activation being insufficient to cause propagation of the nervous impulse, but sufficient to maintain the synapse in a state of readiness for firing by a different neurotransmitter. The precise physiological function of this amine awaits further study.

II. IDENTIFICATION AND QUANTITATIVE ANALYSIS OF PHENYLETHYLAMINE

A. Introduction

Asatoor and Dalgliesh (1), who first reported the presence of PEA in blood, measured the whole amine content of blood and urine by spectrophotometric assay of the 2,4-dinitrophenyl derivatives. Since then, others have used procedures which have involved extraction followed by chromatographic separation and spectrophotometric determination (18), spectrophotofluorometric determination of the fluorescent derivative following reaction with p-dimethylaminocinnamaldehyde (23), or extraction, separation by thin layer chromatography, and identification by infrared and mass spectrometry (35). Reported levels of PEA have varied widely.

In mammalian tissues, PEA has been claimed to be present both in very low, and in relatively high concentrations. Nakajima, Kakimoto, and Sano (37) estimated the PEA content of rabbit tissues to be about 1 ng/g or less. On the other hand, Fischer, Spatz, Heller, and Reggiani (21) reported that PEA occurred in normal rat brain at a concentration of 0.492 μg/g, while Mosnaim and Inwang (34) reported that human brain contained 0.18 μg/g and rabbit brain 0.40 μg/g. The latter figure could not be confirmed by Boulton and co-workers (8), who reported only 0.44 ng/g in rabbit brain.

The analytical procedure used to determine PEA in rabbit brain, and essentially that which is described in this chapter, involves the mass spectrometric determination of the amine as its dansyl derivative through the use of a deuterated internal standard. Full details of this procedure were first published by Durden, Philips, and Boulton in 1973 (15), although the quantitative mass spectrometric integrated ion current technique was originally described in 1967 by Jenkins and Majer (26), and first used by Boulton and Majer (9) to determine p-tyramine in rat brain. The concentration of PEA in rat whole brain determined by Durden et al. (15) (1.8 ng/g) has since been confirmed by two independent studies. Saavedra (42), using an enzymatic isotopic assay, reported that rat brain contained PEA at a concentration of 1.5 ng/g. The assay involved β-hydroxylation of PEA by dopamine (DA) β-hydroxylase, followed by transfer of [^{3}H]methyl from [^{3}H]methyl-S-adenosyl-1-methionine to the terminal group of phenylethanolamine by phenylethanolamine-N-methyl transferase. The product was extracted into an organic solvent, and the radioactivity was assayed. A value of 0.014 nmol/g (1.7 ng/g) was obtained by Willner, LeFevre, and Costa (51) who prepared the pentafluoropropionyl derivative of tissue PEA and quantitated the amine by gas chromatography-mass spectrometry with multiple ion detection.

As well as describing the mass spectrometric integrated ion current technique of amine analysis, this chapter discusses the effect of some drugs on PEA levels in the rat, and reviews the metabolic and clinical studies which have been undertaken in our laboratories.

B. Isolation of the Amine Fraction

Male Wistar rats (175-250 g) were stunned and decapitated, and the brain, heart, kidneys, liver, lungs, and spleen quickly removed and chilled in ice-cold physiological saline. Blood was collected in a beaker containing three drops of 1% heparin solution. Tissues were pooled for analysis as previously reported (15), then homogenized in approximately 15 ml of 0.4 N perchloric acid. Deuterated PEA hydrochloride (1,1-dideutero-2-phenylethylamine hydrochloride) was added as internal standard, and an amine fraction was isolated (15).

The distribution of PEA in subcellular fractions of the rat brain has been reported by Boulton and Baker (6). After removal of the cerebella, brains from 4 to 20 animals were pooled and homogenized in 0.32 M sucrose. Subcellular fractionation was performed essentially according to the procedure of Whittaker and Barker (50). Particulate samples were diluted to approximately 15 ml with 0.32 M sucrose, and the deuterated internal standard was added. Protein was precipitated by a dropwise addition of concentrated perchloric acid, with stirring, to a final concentration of 0.4 N. The amine fraction was obtained as described.

C. Preparation and Chromatographic Separation of Dansyl Phenylethylamine

Dansyl amines were prepared by dissolving the dried amine extract in 1.0 ml of sodium carbonate (10% w/v) and adding 1.5 to 2.0 ml of 5-dimethylamino-1-naphthalene sulfonyl (dansyl) chloride reagent (1 mg/ml in acetone) to the solution. After overnight reaction at room temperature, sodium carbonate was precipitated by adding 10 ml of acetone. The reaction mixture was centrifuged and the supernatant evaporated to dryness under reduced pressure at 45°C. Dansyl amines were eluted from the flask with ethyl acetate, reduced to small volume under a stream of nitrogen, and transferred to a 20 x 20 cm glass plate coated with silica gel (Brinkmann Instruments, Ltd., Rexdale, Ont.). Four tissue samples, as well as a dansyl PEA standard, were applied to a single plate. Dansyl PEA was purified by successive unidimensional separations in the solvent systems chloroform : butyl acetate, 4 : 1 (v/v), benzene : triethylamine, 8 : 1 (v/v), and carbon tetrachloride : triethylamine, 5 : 1 (v/v), then prepared for mass spectrometric analysis in a manner identical to that described for p-tyramine (41).

D. Mass Spectrometric Procedures

Mass spectra and integrated ion current (IIC) profiles were recorded on an AEI MS-902S mass spectrometer operated at 70 eV with a source temperature of 250 ± 10°C. The instrument was equipped with a direct insertion inlet and a Massmaster mass indicator. Spectra and IIC profiles were obtained using a resolution of 2000 and 10,000, respectively. At a resolution of 10,000, interfering background contamination arising from the chromatogram, the solvents and the mass spectrometer itself was reduced or rejected, but high sensitivity was retained.

The linearity and sensitivity of the IIC procedure was determined by measuring different concentrations of synthetic dansyl PEA dissolved in redistilled ethanol (15). In the range 10^{-11} to 10^{-7}g, an excellent linear relationship was obtained between the amount of dansyl PEA introduced into the mass spectrometer and the area enclosed by the integrated ion current curves (Fig. 1). When standard solutions of PEA were subjected to derivatization and chromatographic separation in a manner identical to that of the tissue homogenates, a linear relationship was obtained in the range 10^{-10} to 10^{-8}g. Picogram quantities of PEA as its dansyl derivative could be detected, but the lack of reproducibility between samples was such that precise quantitative evaluation was not feasible in the absence of an internal standard. The lower detection limit was set by the capabilities of the mass spectrometer amplifier and by the level of impurities in the solvents, while the upper limit was dictated by the onset of suppression of the heptacosafluoro-tri-n-butylamine reference peak by the sample ($\sim 10^{-5}$g).

To measure tissue levels of PEA, the ion currents of the dansyl PEA and dansyl deuterated PEA present in the sample were recorded and the IIC curves of the two compounds compared. Details of the mass spectrometric IIC procedure for analysis of tissue PEA through the use of an internal standard have been published previously (15).

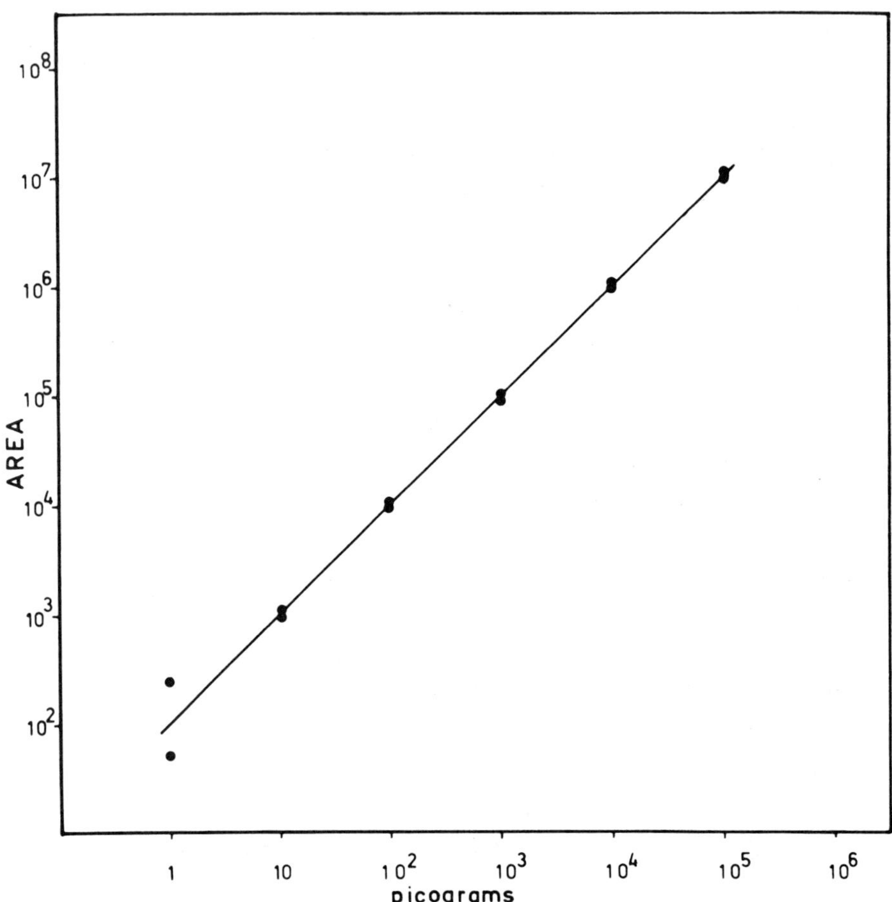

FIG. 1. Typical calibration curve of dansyl PEA. The area enclosed by the IIC curve is in arbitrary units. Reproduced by permission of the National Research Council of Canada from the *Canadian Journal of Biochemistry,* Volume 51, 1973. pp. 995-1002.

E. Evaluation of the Integrated Ion Current Technique

Use of a deuterated internal standard to establish tissue concentrations of PEA ensures that all variations in the analytical procedure are taken into account in every analysis, since the standard is identical in all respects, except mass, to the substance being analyzed. Correction for recovery is inherent in every analysis; thus, there is no need to refer to a standard curve or to analyze duplicate samples, one of which contains a known supplement of amine.

Proof of identity of the compound eluted from the dansyl PEA zone on thin layer chromatograms was obtained by comparing the spectrum of the dansyl compound isolated from a tissue extract supplemented with deuterated PEA with the spectra of dansyl PEA and dansyl deuterated PEA (Fig. 2). In order to obtain a satisfactory mass spectrum of the dansyl derivative isolated from tissue, a kidney taken from a rat treated with the MAOI catron was chosen. The derivatization and extensive chromatographic purification, followed by analysis at precise masses with the use of an internal standard, as well as the mass spectral evidence indicated in Fig. 2 would appear to leave little doubt that PEA was analyzed.

During the development of an analytical procedure for PEA, it became apparent to us that significant difficulties were involved in obtaining reliable measurements of very low levels of some of the amines present in complex biological systems such as tissues and body fluids. In our early attempts to measure this amine, an extract from a rat brain homogenate was chromatographed on paper, dansylated, separated in two solvents on thin layers of silica gel, and measured spectrophotometrically. Phenylethylamine appeared to be present in amounts ranging from 100 to 200 ng/g. A refinement of the procedure, in which an amine fraction isolated on Dowex AG 50W-X2 was dansylated and chromatographed in three separate solvent systems on thin layers of silica gel, reduced these values to approximately 20 ng/g. The currently employed procedure,

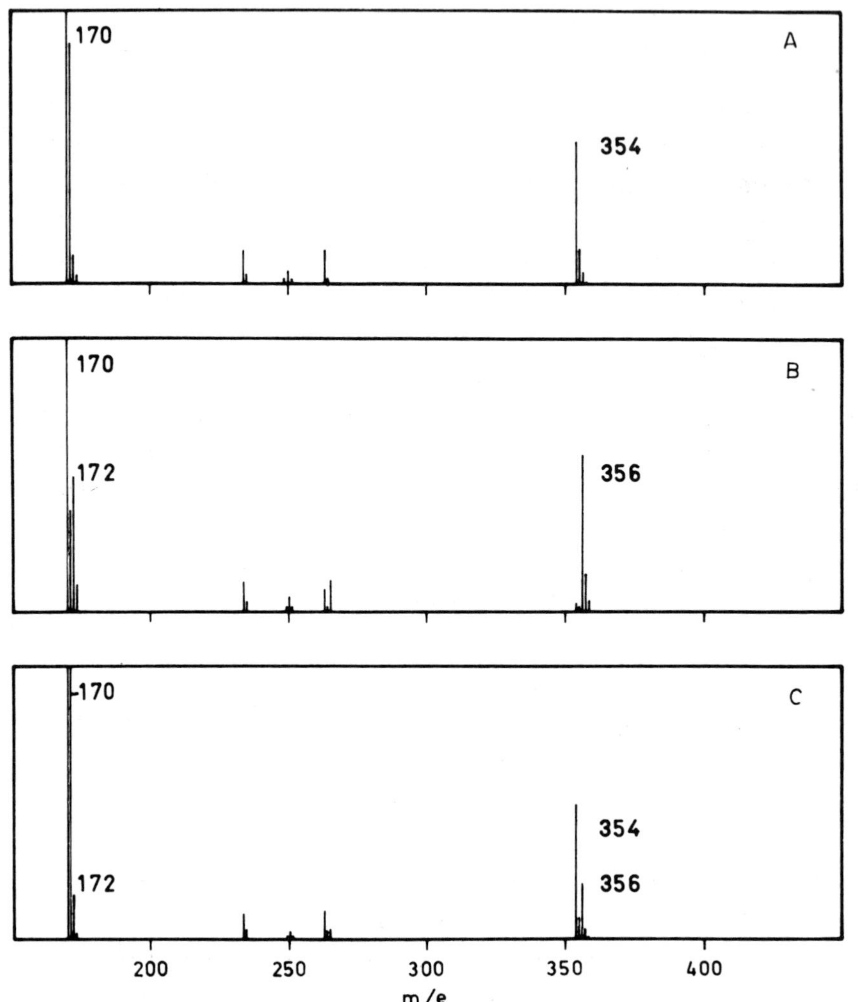

FIG. 2. Mass spectra of dansyl PEA: (A) Dansyl PEA, (B) dansyl-1,1-dideutero-2-PEA, (C) dansyl derivative isolated from rat kidney in presence of catron, and supplemented with 1,1-dideutero-2-PEA. Reproduced by permission of the National Research Council of Canada from the *Canadian Journal of Biochemistry,* Volume 51, 1973. pp. 995-1002.

involving extensive chromatographic purification of the dansyl derivative followed by high resolution mass spectrometric analysis in the presence of a deuterated internal standard, yielded a value of 1.8 ng/g for PEA in rat brain, a figure which is in excellent agreement with values of 1.5 ng/g obtained by Saavedra (42) using an enzymatic isotopic assay, and 0.014 nmol/g (1.7 ng/g) obtained by Willner and co-workers (51) using an assay involving gas chromatography-mass spectrometry with multiple ion detection.

Before an aberration in the concentration of a constituent of tissue or body fluids can be linked conclusively to a clinical state, it is essential that that compound, and only that compound, be measured in any analytical procedure. It has become obvious that attempts to determine low levels of amines in biological systems by procedures involving crude extractions, followed by limited paper or thin layer chromatographic separation, are likely to yield anomalously high values. Even mass spectrometric estimations, if performed on crude extracts without prior chemical modification and purification, will be unlikely to yield reliable data. The molecule ion and other ions of underivatized amines are found at relatively low m/e values. The mass spectrum is complex at these m/e values due to the presence of molecule ions from other low molecular weight compounds and fragment ions from higher molecular weight compounds present in the extract. Interference from these ions would be difficult to avoid without using extremely high resolution and consequent loss of sensitivity. The presently described procedure, in which the dansyl derivative of PEA is analyzed, ensures that the molecule ion, with a m/e value of 354, occurs in a region of the spectrum which is relatively free of other ions. The probability that the IIC profile for dansyl PEA arises from an unresolved interfering ion is, therefore, very low, and thus the specificity of the procedure is extremely high.

III. DISTRIBUTION OF PHENYLETHYLAMINE IN CEREBRAL REGIONS, CEREBRAL SUBCELLULAR FRACTIONS, AND PERIPHERAL TISSUES

Phenylethylamine levels observed in various tissues of the rat ranged from 1.8 ng/g in the brain to 20.5 ng/g in the kidney (Table 1). In the brain PEA was distributed more or less homogeneously (Table 2), except in the caudate nucleus, a region known to contain substantial decarboxylase activity in a number of species (3,29,31). In this area, PEA occurred in significantly greater amounts than in the other regions examined.

On subcellular fractionation, a substantial portion of the amine was found to be associated with the particulate fractions, both in untreated and pargyline-treated brain (6). Its distribution among the primary particulate fractions, P_1 (nuclear debris), P_2 (crude mitochondria), and P_3 (microsomes), and between the P_2 subfractions, P_2A (myelin), P_2B (synaptosomes), and P_2C (mitochondria), is shown in Fig. 3. Of the primary particulate fractions, P_2 contained the greatest proportion of PEA, while the amine was associated with the subfractions in the order $P_2B > P_2A > P_2C$. In brain samples taken from animals pretreated with pargyline, the proportion of PEA in P_1 was decreased, while that in P_3 was increased somewhat with respect to the untreated case. In the P_2 subfractions, less PEA was associated with the P_2C fraction isolated from brains of pargyline-treated animals, while more was found in the P_2A fraction. The reason for the nonproportional increases observed in some fractions in the presence of pargyline is not apparent. It may reflect (1) the localization in various subfractions of MAO having different substrate specificities (24,39); (2) the effects of different binding properties of different fractions (5); or (3) in the case of the supernatant S_3, the fact that PEA may flood the brain from other parts of the body. The latter may well occur, since we have shown (Philips and Boulton, unpublished observations) that PEA levels in blood and peripheral tissues are markedly elevated by treatment with pargyline, and since PEA is known to cross the blood-brain barrier with ease (40).

TABLE 1

Distribution of Phenylethylamine in the Rat[a]

Tissue	Phenylethylamine (ng/g)
Blood	1.9 ± 0.5 (4)
Brain	1.8 ± 0.4 (7)
Heart	5.7 ± 3.1 (9)
Kidney	20.5 ± 2.2 (8)
Liver	2.0 ± 0.7 (9)
Lung	4.0 ± 1.4 (8)
Spleen	4.7 ± 2.7 (9)

[a]Results expressed as mean ± SD; number of experiments in brackets.

TABLE 2

Distribution of Phenylethylamine in Rat Brain[a]

Region	Phenylethylamine (ng/g)
Caudate nucleus	4.21 ± 0.77 (8)
Cerebellum	2.12 ± 0.33 (6)*
Hypothalamus	2.10 ± 0.44 (6)*
Rest	1.35 ± 0.22 (6)**
Stem	1.64 ± 0.18 (5)**

*$p < 0.05$.

**$p < 0.01$.

[a]Results expressed as mean ± SE; number of experiments in brackets. Brain region values significantly different from caudate nucleus are indicated.

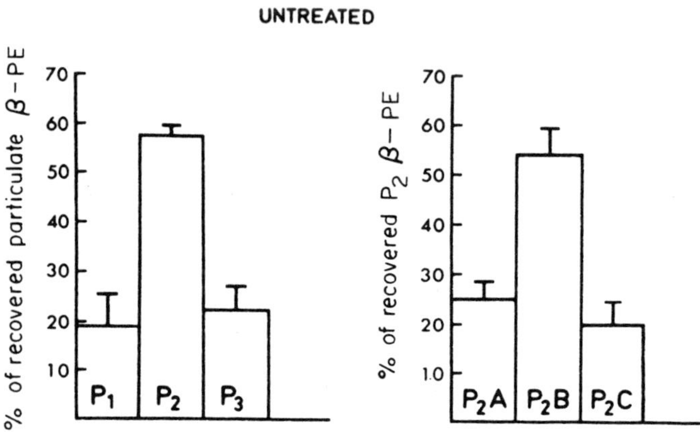

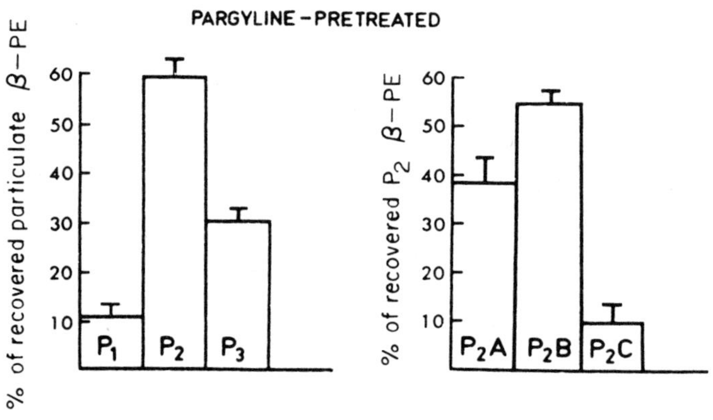

FIG. 3. Distribution of phenylethylamine between various particulate subcellular fractions isolated in the presence and absence of pargyline. Values represent mean ± SD (n = 3 or 4 experiments). Reproduced by permission of Pergamon Press Ltd. from the *Journal of Neurochemistry,* Volume 25, 1975. pp. 477-481.

IV. PHENYLETHYLAMINE AND THE EFFECTS OF MONOAMINE OXIDASE INHIBITORS

In comparison to the classical neurotransmitters, NE, and 5-HT, PEA is present in tissues of the rat at such low concentrations that a first response may be to discount it as a significant neurochemical agent. There is much evidence, however, to indicate that PEA exhibits an exceptionally rapid metabolism. Some of this evidence arose as a result of studies involving MAOI.

Three MAOI were used in our experiments. Male Wistar rats, 175 to 250 g, were injected i.p. with saline solutions of pargyline hydrochloride (75 mg/kg body weight), iproniazid phosphate (100 mg/kg), or catron (15 mg/kg) 4 hr before sacrifice. Tissues from a single animal, or a pool of brain regions from two animals, were analyzed by the mass spectrometric ICC technique. Phenylethylamine concentrations, listed in Tables 3 and 4 for several tissues and brain regions of the rat, were markedly elevated by all three MAOI, but particularly by pargyline and iproniazid (Table 3). β-Phenylethylhydrazine, a compound having a structure very similar to that of catron, is known to inhibit decarboxylation (14). It may be that catron, as well as

TABLE 3

The Effect of Monoamine Oxidase Inhibitors on the Phenylethylamine Concentration in Some Tissues of the Rat[a]

	Phenylethylamine (ng/g)		
Tissue	Pargyline	Iproniazid	Catron
Blood	33 ± 6 (6)	36 ± 5 (3)	—
Brain	185 ± 28 (6)	114 ± 22 (4)	—
Heart	91 ± 10 (6)	89 ± 11 (4)	67 ± 10 (6)
Kidney	773 ± 66 (6)	967 ± 149 (4)	429 ± 62 (5)
Liver	487 ± 88 (6)	316 ± 29 (3)	242 ± 57 (6)
Lung	414 ± 80 (5)	522 ± 86 (4)	197 ± 22 (5)
Spleen	270 ± 33 (5)	224 ± 31 (4)	127 ± 17 (5)

[a]Results expressed as mean ± SE; number of experiments in brackets.

TABLE 4

The Effect of Monoamine Oxidase Inhibitors on the Phenylethylamine Concentration in Brain Regions of the Rat[a]

Region	Phenylethylamine (ng/g) Pargyline	Phenylethylamine (ng/g) Iproniazid
Caudate nucleus	213 ± 33 (4)	227 ± 9 (4)
Cerebellum	136 ± 16 (4)	176 ± 22 (4)
Hypothalamus	139 ± 23 (4)	150 ± 6 (4)***
Rest	173 ± 18 (4)	218 ± 32 (4)
Stem	152 ± 14 (4)	199 ± 22 (4)

***$p < 0.001$

[a]Results expressed as mean ± SE; number of experiments in brackets. Brain region values significantly different from caudate nucleus value are indicated.

being a MAOI, also acts as a decarboxylase inhibitor, thereby reducing the rate of synthesis of PEA from phenylalanine, and resulting in relatively small increases in tissue PEA compared to those observed after the administration of pargyline and iproniazid, which do not inhibit decarboxylase.

Increases in the PEA concentrations induced by MAO blockade varied considerably from one tissue to another. The endogenous concentration of a compound must reflect both the rate of synthesis and the rate of degradation. Smythe and Temple (48) observed that MAO activity in tissue, measured using PEA as substrate, was greatest in the liver, and successively less in the kidney, spleen, and heart. This, in conjunction with the rate of synthesis, as determined by the decarboxylase activity (12), may partially explain the differential peripheral increases observed.

It is likely that other factors are also involved. In kidney, a 40-fold increase in PEA was observed in the presence of pargyline, while in spleen, which has a lower MAO activity, the increase was about 60-fold. The relatively small increase observed in kidney may have been due to the passage of some of the amine into the urine

(16). Tissue PEA levels also appear to be determined to some extent by the ability of the tissue to concentrate the amine from blood, as all tissues taken from MAO-inhibited animals exhibited a higher PEA concentration than did blood. Although this in itself does not necessarily indicate that the amine is concentrated by the tissues, other experiments by Wu and Boulton (53) and by Philips (unpublished observations) showed that such a concentration does occur. Wu and Boulton (53), using male Wistar rats pretreated i.p. with pargyline 45 min before an i.v. injection of [^{3}H]PEA, found that 30 min after its injection, [^{3}H]PEA was most concentrated in the kidneys and lungs, and least concentrated in the heart. No radioactivity could be detected in the blood.

Phenylethylamine is distributed relatively homogeneously throughout the normal rat brain, except in the caudate nucleus, which is somewhat enriched in this amine. In the presence of MAO I, however, significant differences between the PEA levels in the caudate nucleus and most other regions disappear (Table 4). Similar findings have been reported by Edwards and Blau (17). Thus, PEA, which passes freely across the blood-brain barrier (40), appears to be concentrated by the brain more or less equally in all regions.

The dramatic increases in tissue PEA levels which have been measured in the presence of MAOI clearly illustrate that the amine possesses a very vigorous turnover and oxidation rate. Four hours after administration of pargyline and iproniazid, PEA levels in whole brain were elevated approximately 100- and 60-fold, respectively. By contrast, Spector and co-workers (49) found that in rabbit brain stem, NE and 5-HT concentrations were increased only twofold by iproniazid (100 mg/kg), and that the maximum levels were not attained until approximately 24 hr after injection of the drug. Thus, although PEA is normally present in only minute amounts in tissues, its known pharmacological properties, as well as its rapid turnover rate, and consequently its potential for marked concentration change in the event of a change in the rate of degradation, make it an interesting and potentially significant neurochemical agent.

V. METABOLISM OF PHENYLETHYLAMINE

Phenylethylamine appears to arise by decarboxylation of phenylalanine (32); tissue levels of the amine are known to be increased in vivo by administration of phenylalanine, both in the presence and absence of MAOI (42). By examining the effect of the MAOI, pargyline hydrochloride, and the peripheral decarboxylase inhibitor, Ro 4-4602, on the urinary excretion of PEA, Dyck and Boulton (16) provided further evidence that PEA is synthesized by decarboxylation of phenylalanine. They found that while pargyline produced a significant increase in urinary PEA, Ro 4-4602 reduced the excretion to a level significantly below that of control animals.

Metabolic evidence of the exceptionally vigorous turnover rate of PEA in the rat has been obtained by Wu and Boulton (53) through the use of the radioactively labeled compound. At time intervals of 15 to 180 min after an i.v. or intraventricular injection of labeled amine, the animals were sacrificed, and several tissues were removed for the analysis of either PEA or the acidic and neutral metabolites of PEA. In tissues isolated from pargyline-treated rats 30 min after an i.v. injection of labeled amine, most of the label remained as unchanged PEA; only traces of labeled phenyethanolamine could be detected. There was, however, some conversion to m- and p-tyramine and octopamine. In the absence of pargyline treatment, the major metabolite of injected PEA was phenylacetic acid. However, the presence of the labeled intermediate, phenylacetaldehyde, could not be demonstrated.

Radioactive PEA was lost rapidly from most tissues after i.v. administration, both in the presence and absence of pargyline. Clearance was approximately exponential in the presence of pargyline, and the half-life of PEA was calculated in the usual way. In the absence of the drug, clearance from all tissues except brain and lung was nonexponential, and the half-life was estimated as described by Wu and Boulton (52). The half-life values, which ranged from 1 min in the spleen to 74 min in the lung in untreated animals, and from 30 min in the lung to 45 min in the heart and liver in pargyline-treated animals, are much shorter than the corresponding values for the classical neurotransmitters. Together with the rapid increase in tissue PEA levels observed after MAO I, they indicate that the amine has a very active metabolism.

The recently reported hydroxylation of PEA to the m- and p-isomers of tyramine is one of the metabolic pathways which is not blocked by MAOI, and may account for the rapid clearance of PEA from body tissues in the presence of pargyline. Tyramines labeled with tritium and deuterium have been identified in rat urine after i.p. injection of the appropriately labeled PEA to animals treated with pargyline. Tritium-labeled tyramines were measured by liquid scintillation counting, while deuterated compounds were identified and quantitated by mass spectrometry. Hydroxylation of tetra-deuterated PEA in the p-position accounted for 0.74% of the PEA dose in the presence of pargyline, while m-hydroxylation occurred to the extent of 0.12% of the dose. Only a trace (0.008%) of deuterated PEA was converted to tetradeutero o-tyramine. Hydroxylation of PEA may prove to be one of the main synthetic routes of p-tyramine and has been suggested as a mechanism for detoxification of PEA (7).

VI. CLINICAL STUDIES

Because of the known pharmacological properties of PEA, and because of its claimed association with depression, mania, schizophrenia, and phenylketonuria, as evidenced by its reputed abnormal excretion in the urine in these disease states, an extensive project was undertaken in an attempt to establish possible relationships between mental illness and the excretion of PEA (47). Every second patient admitted to the psychiatric ward of a general hospital was selected on a random basis. A 24-hr urine sample was collected from each patient beginning the day after admission, and during that day the patient was interviewed by a psychiatrist and asked to complete a rating scale (BPRS: Brief Psychiatric Rating Scale) which was intended to assess the presence and severity of selected symptoms. A total of 129 samples were collected from the psychiatric population; in addition, 15 samples were obtained from normal volunteers.

On completion of the study, patient histories were obtained from medical records, and discharge diagnoses and drug therapies were recorded.

Immediately after collection of the urine samples, the volumes were measured, the pH recorded, and the samples frozen and stored at -16°C until analysis. At that time, each sample was thawed, and any solid matter was suspended before removal of duplicate aliquots (10 ml). The internal deuterated standard (usually 1 μg) was added, and the pH adjusted to 6.8 to 7.2. Particulate matter was removed by centrifugation at 1000g for 2 to 3 min. The supernatant was decanted and percolated through a column of ion exchange resin as previously described. Dansyl derivatives were prepared, and the dansyl PEA isolated and quantitated by the integrated ion current technique.

In normal subjects, PEA excretion was 4.9 ± 1.0 μg/24 hr (mean ± SE, range 1.4 to 16) or 4.6 ± 1.2 μg/g creatinine (mean ± SE, range 1.3 to 20). Among the psychiatric population, the amine was excreted in amounts of 35 ± 10 μg/24 hr (range 0.21-1100) or 45 ± 13 μg/g creatinine (range 0.29 to 850). When excretion was expressed as micrograms per gram creatinine, males were found to excrete less PEA than females, but no correlation with age could be established. If excretion values were expressed as μg/24 hr, however, excretion rates decreased with age; males still excreted less PEA than females. The presence of high urinary PEA was related significantly to the absence of hostility, but unrelated to the presence of other BPRS symptoms, diagnostic groups or psychoses. A one-way analysis of variance showed no significant differences in PEA excretion between groups of psychotic depressives, neurotic depressives, and other diagnoses. Significant differences were found between depressed females and depressed males, other females, and other males. Males always excreted less PEA than females, but only depressed females excreted less than other females. In general, however, this study, in agreement with that of Schweitzer et al. (46), did not substantiate earlier reports of reduced PEA excretion in depression.

Finally, a comparison of the normal and psychiatric populations showed that PEA excretion in the low amine range (≤ 20 μg), when expressed as micrograms per gram of creatinine, was significantly higher among psychiatric patients (P = 0.002, Student's t test), a result which may be related to the antidepressant drugs taken by many of the patients.

Slingsby (47) concluded that the large number of factors which affect urinary excretion of amines in the psychiatric population as a whole, as well as the fact that certain diagnoses were more often assigned to patients of a particular age or sex, might prevent use of analysis of variance in studies of this kind. The wide range in urinary excretion values, coupled with the nonnormal frequency distribution of these values, eliminated the possibility of using statistical techniques which assume a normal distribution. She suggested that while the analysis of small numbers of patients in a particular diagnostic group might have led to the discovery of abnormalities with respect to the normal population, such abnormalities would not necessarily have been related to diagnosis. In view of her findings, it seems advisable to reassess previous work which purported to indicate that abnormalities in urinary PEA excretion were related to certain diagnoses.

For the future, it seems sensible to undertake longitudinal studies in particular patients, and metabolic studies in humans in general as a more reasonable approach to the problem of investigating possible relationships between amine excretion and mental illness.

VII. CONCLUSIONS

Reports that urinary PEA excretion is decreased in depressed patients were not supported by a clinical study performed in our laboratories. This does not necessarily indicate, however, that the function of PEA in the CNS remains unaltered in depression or other mental or drug-induced aberrant states. The total urinary output of any substance might easily mask any changes in its cerebral metabolism, since it seems likely that cerebral metabolism will contribute only a fraction of the total amount of the substance excreted. A decrease or increase in brain PEA levels might well influence behavior dramatically but be undetectable in the urine. This amine after all is highly permeable, exerts marked behavioral effects, and possesses significant pharmacological and physiological properties.

Although PEA has been shown to be present in very low concentrations in tissues of the rat and rabbit, it has been demonstrated in our laboratories and by others that the metabolism is very rapid, as evidenced by both the very short half-life in various tissues of the rat, and the rapid increase in concentration observed when MAO, one of the main degradative enzymes of the amine, is blocked. Indeed, it has been previously suggested (38) that the turnover rates of pharmacologically active compounds may be a more significant index of their effectiveness than the endogenous concentrations.

ACKNOWLEDGMENTS

I sincerely thank Dr. A. A. Boulton for his many helpful suggestions and critical reading of the manuscript; Dr. D. A. Durden for his supervision of, and numerous time-saving refinements in the mass spectrometric analyses; Dr. B. A. Davis for his synthesis of the deuterated internal standards; Mr. H. Miyashita, Mrs. N. M. Choo, Mr. N. F. Binder, and Miss E. E. Johnson for their skilled technical assistance; and Mrs. J. Sweeting for her typing of the manuscript. I also express my appreciation to the Psychiatric Services Branch, Province of Saskatchewan, and the Medical Research Council of Canada for continuing financial support.

REFERENCES

1. A. M. Asatoor and C. E. Dalgliesh (1959). Amines in blood and urine. *Biochem. J.*, 73:26P.
2. G. Barger and H. H. Dale (1910). Chemical structure and sympathomimetic action of amines. *J. Physiol. (Paris)*, 41:19-59.
3. D. F. Bogdanski, H. Weissbach, and S. Udenfriend (1957). The distribution of serotonin, 5-hydroxytryptophan decarboxylase and monoamine oxidase in brain. *J. Neurochem.*, 1:272-278.

4. A. A. Boulton (1976). Cerebral aryl alkyl aminergic mechanisms. In: *Trace Amines and the Brain,* edited by E. Usdin and M. Sandler. Marcel Dekker, New York.

5. A. A. Boulton and G. B. Baker (1974). The subcellular distribution of some monoamines following their intraventricular injection into the rat. *Can. J. Biochem.,* 52:288-293.

6. A. A. Boulton and G. B. Baker (1975). The subcellular distribution of β-phenylethylamine, p-tyramine and tryptamine in rat brain. *J. Neurochem.,* 25:477-481.

7. A. A. Boulton, L. E. Dyck, and D. A. Durden (1974). Hydroxylation of β-phenylethylamine in the rat. *Life Sci.,* 15:1673-1683.

8. A. A. Boulton, A. V. Juorio, S. R. Philips, and P. H. Wu (1975). Some arylalkylamines in rabbit brain. *Brain Res.,* 96:212-216.

9. A. A. Boulton and J. R. Majer (1970). Mass spectrometry of crude biological extracts. Absolute quantitative detection of metabolites at the submicrogram level. *J. Chromatogr.,* 48: 322-327.

10. A. A. Boulton and L. Milward (1971). Separation, detection and quantitative analysis of urinary β-phenylethylamine. *J. Chromatogr.,* 57:287-296.

11. J. H. Burn and M. J. Rand (1958). The action of sympathomimetic amines in animals treated with reserpine. *J. Physiol.,* 144: 314-336.

12. V. E. Davis and J. Awapara (1960). A method for the determination of some amino acid decarboxylases. *J. Biol. Chem.,* 235: 124-127.

13. W. G. Dewhurst (1968). New theory of cerebral amine function and its clinical application. *Nature,* 218:1130-1133.

14. B. Dubnick, G. A. Leeson, and G. E. Phillips (1959). In vivo inhibition of serotonin synthesis in mouse brain by β-phenylethylhydrazine, an inhibitor of monoamine oxidase. *Fed. Proc.,* 18:218.

15. D. A. Durden, S. R. Philips, and A. A. Boulton (1973). Identification and distribution of β-phenylethylamine in the rat. *Can. J. Biochem.,* 51:995-1002.

16. L. E. Dyck and A. A. Boulton (1975). The effect of some drugs on the urinary excretion of some aryl alkyl amines in the rat. *Res. Commun. Chem. Pathol. Pharm.,* 11:73-77.

17. D. J. Edwards and K. Blau (1973). Phenethylamines in brain and liver of rats with experimentally induced phenylketonuria-like characteristics. *Biochem. J.,* 132:95-100.

18. E. Fischer, B. Heller, and A. H. Miro (1968). β-Phenylethylamine in human urine. *Arzneim.-Forsch.,* 18:1486.

19. E. Fischer, R. I. Ludmer, and H. C. Sabelli (1967). The antagonism of phenylethylamine to catecholamines on mouse motor activity. *Acta Physiol. Lat. Amer.*, 17:15-21.

20. E. Fischer, J. M. Saavedra, and B. Heller (1968). Effects of catecholamines, adrenergic substances and their blocking agents on the searching behavior of mice. *Arzneim.-Forsch.*, 18:780-786.

21. E. Fischer, H. Spatz, B. Heller, and H. Reggiani (1972). Phenethylamine content of human urine and rat brain, its alterations in pathological conditions and after drug administration. *Experientia*, 28:307-308.

22. E. Fischer, H. Spatz, R. S. F. Labriola, E. M. R. Casanova, and N. Spatz (1973). Quantitative gas chromatographic determination and infrared spectrographic identification of urinary phenethylamine. *Biol. Psychiat.*, 7:161-165.

23. E. Fischer, H. Spatz, J. M. Saavedra, H. Reggiani, A. H. Miro, and B. Heller (1972). Urinary elimination of phenethylamine. *Biol. Psychiat.*, 5:139-147.

24. R. W. Fuller and B. W. Roush (1972). Substrate-selective and tissue-selective inhibition of monoamine oxidase. *Arch. Int. Pharmacodyn. Ther.*, 198:270-276.

25. W. J. Giardina, W. A. Pedemonte, and H. C. Sabelli (1973). Iontophoretic study of the effects of norepinephrine and 2-phenylethylamine on single cortical neurons. *Life Sci.*, 12(Part 1):153-161.

26. A. E. Jenkins and J. R. Majer (1967). Mass spectrometry of metal chelates. I. Detection on the picogram scale. *Talanta*, 14:777-783.

27. J. B. Jepson, W. Lovenberg, P. Zaltzman, J. A. Oates, A. Sjoerdsma, and S. Udenfriend (1960). Amine metabolism studied in normal and phenylketonuric humans by monoamine oxidase inhibition. *Biochem. J.*, 74:5P.

28. J. Jonsson, H. Grobecker, and P. Holtz (1966). Effect of β-phenylethylamine on content and subcellular distribution of norepinephrine in rat heart and brain. *Life Sci.*, 5:2235-2246.

29. R. Kuntzman, P. A. Shore, D. Bogdanski, and B. B. Brodie (1961). Microanalytical procedures for fluorometric assay of brain dopa-5HTP decarboxylase, norepinephrine and serotonin, and a detailed mapping of decarboxylase activity in brain. *J. Neurochem.*, 6:226-232.

30. A. M. Lands and J. I. Grant (1952). The vasopressor action and toxicity of cyclohexylethylamine derivatives. *J. Pharmacol. Exp. Ther.*, 106:341-345.

31. K. G. Lloyd and O. Hornykiewicz (1972). Occurrence and distribution of aromatic L-amino acid (L-Dopa) decarboxylase in the human brain. *J. Neurochem.*, 19:1549-1559.

32. W. Lovenberg, H. Weissbach, and S. Udenfriend (1962). Aromatic L-amino acid decarboxylase. *J. Biol. Chem.*, 237:89-93.

33. P. Mantegazza and M. J. Riva (1963). Amphetamine-like activity of β-phenethylamine after monoamine oxidase inhibition in vivo. *J. Pharm. Pharmacol.*, 15:472-478.

34. A. D. Mosnaim and E. E. Inwang (1973). A spectrophotometric method for the quantification of 2-phenylethylamine in biological specimens. *Anal. Biochem.*, 54:561-577.

35. A. D. Mosnaim, E. E. Inwang, J. H. Sugerman, and H. C. Sabelli (1973). Identification of 2-phenylethylamine in human urine by infrared and mass spectroscopy and its quantitation in normal subjects and cardiovascular patients. *Clin. Chim. Acta*, 46: 407-413.

36. D. L. Murphy and R. J. Wyatt (1972). Reduced monoamine oxidase activity in blood platelets from schizophrenic patients. *Nature*, 238:225-226.

37. T. Nakajima, Y. Kakimoto, and I. Sano (1964). Formation of β-phenylethylamine in mammalian tissue and its effect on motor activity in the mouse. *J. Pharmacol. Exp. Ther.*, 143:319-325.

38. N. H. Neff, S. H. Ngai, C. T. Wang, and E. Costa (1969). Calculation of the rate of catecholamine synthesis from the rate of conversion of tyrosine-^{14}C to catecholamines. *Mol. Pharmacol.*, 5:90-99.

39. N. H. Neff and H.-Y. T. Yang (1974). Another look at the monoamine oxidases and the monoamine oxidase inhibitor drugs. *Life Sci.*, 14:2061-2074.

40. W. H. Oldendorf (1971). Brain uptake of radiolabeled amino acids, amines, and hexoses after arterial injection. *Am. J. Physiol.*, 221:1629-1639.

41. S. R. Philips, D. A. Durden, and A. A. Boulton (1974). Identification and distribution of p-tyramine in the rat. *Can. J. Biochem.*, 52:366-373.

42. J. M. Saavedra (1974). Enzymatic isotopic assay for and presence of β-phenylethylamine in brain. *J. Neurochem.*, 22:211-216.

43. H. C. Sabelli, W. J. Giardina, A. D. Mosnaim, and N. H. Sabelli (1973). A comparison of the functional roles of norepinephrine, dopamine, and phenylethylamine in the central nervous system. *Acta. Physiol. Pol.*, 24:33-40.

44. H. C. Sabelli and A. D. Mosnaim (1974). Phenylethylamine hypothesis of affective behaviour. *Am. J. Psychiat.*, 131: 695-699.

45. J. L. Schmidt and W. W. Fleming (1963). The structure of sympathomimetics as related to reserpine-induced sensitivity changes in the rabbit ileum. *J. Pharmacol. Exp. Ther.*, 139: 230-237.

46. J. W. Schweitzer, A. J. Friedhoff, and R. Schwartz (1975). Phenethylamine in normal urine: Failure to verify high values. *Biol. Psychiat.*, 10:277-285.

47. J. M. Slingsby (1975). Factors affecting urinary excretion of arylalkylamines in a randomly selected psychiatric population. Thesis, University of Saskatchewan.

48. D. B. Smythe and D. M. Temple (1971). The enzymatic control of β-phenylethylamine concentration in mammalian tissues. *Proc. Austral. Biochem. Soc.*, 4:74.

49. S. Spector, D. Prockop, P. A. Shore, and B. B. Brodie (1958). Effect of iproniazid on brain levels of norepinephrine and serotonin. *Science*, 127:704.

50. V. P. Whittaker and L. A. Barker (1972). The subcellular fractionation of brain tissue with special reference to the preparation of synaptosomes and their component organelles. In: *Methods of Neurochemistry, Vol. 2*, edited by R. Fried. Marcel Dekker, New York, pp. 1-52.

51. J. Willner, H. F. LeFevre, and E. Costa (1974). Assay by multiple ion detection of phenylethylamine and phenylethanolamine in rat brain. *J. Neurochem.*, 23:857-859.

52. P. H. Wu and A. A. Boulton (1974). Distribution, metabolism, and disappearance of intraventricularly injected p-tyramine in the rat. *Can. J. Biochem.*, 52:374-381.

53. P. H. Wu and A. A. Boulton (1975). Metabolism, distribution, and disappearance of injected β-phenylethylamine in the rat. *Can. J. Biochem.*, 53:42-50.

Chapter 7

β-PHENYLETHYLAMINE: IS THIS BIOGENIC AMINE RELATED TO NEUROPSYCHIATRIC DISEASES?

Juan M. Saavedra

Laboratory of Clinical Science
National Institute of Mental Health
Bethesda, Maryland

I. INTRODUCTION

β-Phenylethylamine (PEA) is a biogenic amine formed in mammalian tissues by decarboxylation of the dietary aminoacid L-phenylalanine by L-aromatic amino acid decarboxylase (28).

When injected parenterally in high doses, PEA produces a number of interesting pharmacological effects, among those a depletion of endogenous catecholamines in the rat brain and peripheral tissues (21,26) and when administered to animals pretreated with monoamine oxidase inhibitors (MAOI), clearcut amphetaminelike effects (29).

On the basis of the excitatory pharmacological effects of PEA, and the reports of decreased elimination of this compound in the urine of depressed patients, this amine has been postulated as an endogenous "ergotropic" factor (10,17). High PEA levels in mammalian brain and urine have been reported by several groups (18,31,32). Involvement of "β-phenylethylaminergic" mechanisms, including the participation of specific PEA "receptors" (22) and "release mechanisms," in the mechanism of action of amphetamine, tricyclic antidepressants, marijuana and other drugs, have been also postulated (31,32).

Alterations in PEA metabolism have been also postulated in a number of psychiatric and neurologic disorders, such as phenylketonuria (34), mental depression (15,16), and schizophrenia (20).

The development of more rigorous methodology, however, has challenged the hypothesis of PEA as a central naturally occurring stimulant amine with a physiological role in the CNS. New specific microtechniques have recently shown PEA to be present in mammalian brain at levels two orders of magnitude lower than previously reported (11,36,42) and 1000 times lower than those required to produce pharmacological effects when exogenous PEA is administered (7). These new techniques also showed the PEA is excreted in the human urine in amounts far lower than those previously reported (39). In addition, preliminary evidence could not confirm the reported changes in the elimination of this amine in depressive states (Beckmann, Saavedra, and Goodwin, unpublished observations). Many of the drug effects on PEA metabolism, as well as some pharmacological effects of this amine, could not be replicated in other laboratories, or can be explained by mechanisms other than direct actions of PEA on specific receptors.

At the present time, the existence of conflicting data from different laboratories does not allow us to postulate a role for PEA in brain function or neuropsychiatric diseases. The evidence so far, however, is strongly suggestive of this amine playing a minor role in the CNS of mammals.

We critically analyze some of the experimental evidence for and against the participation of PEA in brain function and neuropsychiatric disorders.

II. β-PHENYLETHYLAMINE AS A PHYSIOLOGICALLY ACTIVE COMPOUND IN MAMMALIAN CENTRAL NERVOUS SYSTEM

When considering the presence of PEA in tissues, two different questions should be answered, namely whether the compound is actually present and can be identified as authentic PEA, and whether the methods used are capable of measuring this compound with consistency, sensitivity, reproducibility, in an strict quantitative way, and, above all, specifically.

The presence of PEA in a number of mammalian tissues has been established beyond doubt. A number of sophisticated techniques have been used; the compound has been isolated, recrystalized, subjected to different chromatographic techniques, derivatized, and rechromatographed, and its mass-spectometric characteristics found identical with those of the authentic compound (11-13,23,30-32,39,42) both in brain tissue and human urine.

Very important discrepancies appear, however, when the methods to actually measure this compound are compared. The analysis of the literature shows that a number of different techniques have been used for the quantitation of PEA in mammalian tissues and body fluids. These methods included the use of paper chromatography and colorimetry (33), thin layer chromatography and fluorometry (6), biological techniques (24), a number of spectrophotometric techniques (30-32,41), dansyl derivation and mass spectrometry (8), gas-liquid chromatography with electron-capture detection (12,39), mass-spectrometry (11,42), and enzymatic-isotopic techniques (36) (Table 1).

In general, the use of earlier methods, based on chromatographic-colorimetric techniques, showed relatively small levels of PEA in human urine, and failed to detect endogenous PEA in mammalian brain. This was probably due to the relative insensitivity of these methods.

The development of highly specific and more sensitive techniques based on mass-spectrometry allowed the detection of endogenous PEA in mammalian brain. The levels reported, however, were two orders of magnitude lower than those published by the use of less specific methods, based on spectrophotometry or colorimetry (Table 1).

TABLE 1

Methods for Quantification of β-Phenylethylamine

Authors	Method	Sensitivity[a]	Values Reported	
			Brain (rat)	Human Urine[b] (normal subjects)
Jepson et al. (25)	Paper chromatography	?	—	< 20 μg/day
Perry (35)	Paper chromatography	?	—	0 μg/day
Oates et al. (34)	Paper chromatography	?	—	< 3 μg/day
Nakajima et al. (33)	Paper chromatography and colorimetry	?	Not detectable (rabbit)	—
Fischer et al. (15)	Paper chromatography and colorimetry	?	—	33 μg/day
Creveling et al. (8)	Dansyl derivation and mass spectrometry	30-40 ng	— (heart of mice)	—
Jackson (24)	Biological	?	— (human blood)	—
Boulton and Milward (6)	Thin layer chromatography and fluorometry	100 ng	—	47 μg/day (free) 34-365 μg/day (conjugated)
Edwards and Blau (12)	Gas-liquid chromatography with electron-capture detection	20 ng	Not detectable (rat)	—

Spatz and Spatz (41)	Spectrophotometric	?	—	239 μg/day
Fischer et al. (18)	Spectrophotometric, thin layer chromatography	?	492 ng/g (rat)	336 μg/day
Mosnaim and Inwang (30) and Inwang et al. (23)	Spectrophotometric	5 μg	—	292 μg/liter (free)
Mosnaim et al. (30-32)	Spectrophotometric	?	400 ng/g (rabbit) 110 ng/g (cat) 180 ng/g (fetal human brain)	453 μg/day (free) 886 μg/day (total)
Fischer et al. (19)	Gas chromatography	?	—	400 μg/day
Durden et al. (11)	Mass spectrometry	0.1-1 ng	1.8 ng/g (rat)	—
Saavedra (36)	Enzymatic-isotopic	0.5-1 ng	1.5 ng/g (rat)	—
Willner et al. (42)	Mass spectrometry	?	1-2 ng/g	—
Boulton and Baker (3)	Mass spectrometry	0.1-1 ng	1.8 ng/g	—
Schweitzer et al. (39)	Thin layer chromatography and gas-liquid chromatography with electron-capture detection	1 μg	—	10.3 μg/day

[a]When possible, values for sensitivity have been calculated from original tables and figures and represent only an approximation. For many methods, the sensitivity could not be calculated for lack of data.

[b]Some papers reported values for free and conjugated PEA, the latter probably as glucuronide.

Since the first group of methods utilized internal standards throughout the procedures, losses of PEA, if they occurred, were automatically corrected and could not be the source of the large differences observed when compared with the methods based on spectrophotometry. More likely, the latter represent nonspecific techniques, and some closely related compounds or impurities contaminated the results. Furthermore, large species differences in endogenous PEA are not likely to occur, since the rabbit presents the same order of magnitude of PEA levels as the rat, not only in the whole brain but also in localized brain areas (less than 5 ng/g) (5, Saavedra and Tallman, unpublished observations).

We have recently developed a sensitive and specific double enzymatic isotopic method to measure PEA. The method is based on the conversion of PEA to phenylethanolamine, and later to N-methyl phenylethanolamine, by sequential incubation in the presence of dopamine-β-hydroxylase (E.C. 1.14.2.1) and phenylethanolamine-N-methyltransferase (Fig. 1). The transfer of [^{3}H]methyl groups from [^{3}H]methyl-S-adenosyl-L-methionine to the terminal N group of phenylethanolamine results in the formation of a radioactive product which is isolated by selective extraction into organic solvents and quantitated. The use of internal standards of authentic PEA allows correction of recoveries and losses during extraction procedures (36). The specificity of this procedure is shown in Table 2.

This method has been proven useful in the detection of endogenous PEA in mammalian brain, in the study of the regional distribution of this amine in the CNS, and in the analysis of the effects of drugs in its formation and metabolism (36). The values

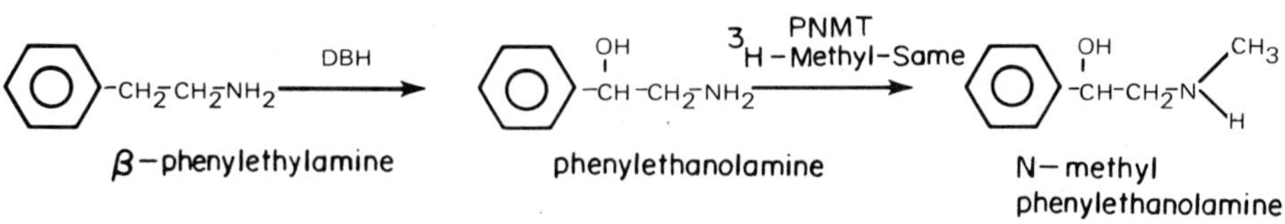

$$C_6H_5-CH_2-CH_2-NH_2 \xrightarrow{\text{DBH}} C_6H_5-\underset{\displaystyle OH}{CH}-CH_2-NH_2 \xrightarrow[\ ^3\text{H-Methyl-Same}]{\text{PNMT}} C_6H_5-\underset{\displaystyle OH}{CH}-CH_2-N(CH_3)H$$

β-phenylethylamine — phenylethanolamine — N-methyl phenylethanolamine

FIG. 1. β-Hydroxylation and N-methylation of PEA.

TABLE 2

Specificity of the β-Phenylethylamine Assay[a,b]

Amine	No DBH Added	Complete System (cpm)	Difference
β-Phenylethylamine	350	41,900	41,550
Phenylethanolamine	110,000	110,000	0
Octopamine	300	300	0
Normetanephrine	320	330	10
Tyramine	320	330	10
Dopamine	300	300	0
Noradrenaline	310	330	20
Serotonin	280	300	20
Tryptamine	260	240	0
Amphetamine	280	280	0

[a]β-Phenylethylamine (10 ng) and other amines (100 ng) were carried through the entire extraction and incubation procedure.

[b]From Saavedra (36).

obtained for PEA with the enzymatic-isotopic method are identical to those found after mass-spectrographic analysis.

In the rat brain, PEA is found to be unevenly distributed (Table 3) with the highest concentration in hypothalamus (8.5 ng/g) and the lowest in cerebral cortex (3.7 ng/g). Similar results have been recently reported (42). However, no specific PEA tract or cells have been yet identified, and when the subcellular distribution of PEA in brain tissue is studied (3), this amine is found to be equally distributed between particulated and supernatant fractions. Furthermore, it is of note that whole brain PEA represent about 1% of the concentrations of other biogenic amines such as the catecholamines and serotonin. These observations disagree with the formation and storage of this amine in a specific tracts, cells or cellular components in the CNS.

TABLE 3

Effects of Drugs on the Distribution of β-Phenylethylamine in Specific Regions of the Rat Brain[a]

Region	No Treatment	Pargyline	Pargyline plus L-Phenylalanine
Hypothalamus	8.5 ± 2	81 ± 6*	204 ± 12
Brainstem (pons and medulla oblongata)	6.9 ± 1.2	94 ± 7*	186 ± 10*
Striatum	6.0 ± 0.8	84 ± 6*	282 ± 16*
Hippocampus	6.0 ± 1	80 ± 9*	211 ± 19*
Cerebellum	5.1 ± 1	98 ± 10*	132 ± 13**
Midbrain	4.5 ± 0.9	23 ± 5*	164 ± 10*
Cerebral cortex	3.7 ± 0.5	43 ± 8*	184 ± 9*
Whole brain	5.0 ± 0.9	75 ± 13*	189 ± 14*

*Statistically significant, $P < 0.01$.

**Statistically significant, $P < 0.05$.

[a]Pargyline (75 mg/kg, i.p.) was injected 90 min, and L-phenylalanine methylester HCl (0.5 g/kg, i.p.), 60 min, before killing. Results are presented as means ± SEM for groups of five animals.

We have also detected endogenous PEA in peripheral tissues of the rat (Table 4). The lowest concentration was detected in the liver (2.8 ng/g) and the highest in the kidney (12.7 ng/g).

β-Phenylethylamine is mainly metabolized in tissues by monoamine oxidase (MAO). A type B MAO, which specifically deaminates PEA, has been detected in mammalian brain (44). Table 5 shows the time course of the accumulation of PEA after inhibition of MAO. The increase of PEA level is linear for at least 90 min, and indicates that this amine may have a very fast turnover rate. No difference was found in the brain regional distribution of PEA under these conditions (Table 4).

The rapid destruction of PEA by MAO can explain the very low concentrations of this amine normally found in tissues, and the fact that for some time, PEA was detected in animal tissues only after a pretreatment with MAOI (13,33).

TABLE 4

Endogenous β-Phenylethylamine Content in Rat Tissues

Tissue	PEA Content (ng/g)
Brain	5.5 ± 1.2
Liver	2.8 ± 0.8
Heart	3.3 ± 0.4
Spleen	4.0 ± 1.0
Lung	9.3 ± 1.5
Kidney	12.7 ± 0.9

Groups of six adult Sprague-Dawley rats were used. Each determination was performed on an individual tissue sample. Results are presented as means ± SEM.

TABLE 5

Time Course of the Accumulation of Brain β-Phenylethylamine after Inhibition of Monoamine Oxidase[a]

Time (min)	Brain PEA Content (ng/g)
0	5.7 ± 0.5
5	9.0 ± 0.8
10	13.0 ± 1.5
30	28.0 ± 3.5
60	48.7 ± 3.0
90	72.2 ± 6.0

[a]Pargyline, 75 mg/kg, was injected i.p. Results are presented as mean ± SEM, for groups of five animals.

When injected intracisternally (Table 6), PEA disappears from the brain at an extremely fast rate. Almost none of the PEA remains in brain tissue after 30 min, and two-thirds of the compound have already disappeared during the first 5 min after the injection. The pretreatment with MAOI (Table 6) only partially delays the disappearance of PEA from brain. These results confirm earlier reports (43).

An analysis of the material remaining in the brain tissue reveals that only 50% of the remaining radioactivity can be identified as authentic PEA (Table 7). The formation of two PEA derivaties, namely phenylethanolamine (37) and N-methyl PEA (38) can be demonstrated. About 40% of the radioactivity was in the form of unidentified metabolites. It is of note that recently Boulton et al. (5) demonstrated the formation of tyramine from PEA.

The effects of a number of compounds on brain PEA levels are summarized in Table 8. As expected, the administration of the amino acid precursor L-phenylalanine increased PEA levels. This increase was potentiated when the animals were treated with MAOI, and antagonized by the administration of decarboxylase inhibitors (Table 8).

The administration of the phenylalanine hydroxylase inhibitor, p-chlorophenylalanine (27) also increased PEA levels in brain (Table

TABLE 6

Disappearance from Rat Brain of 12 μg of Intracerebrally Injected [^{14}C]β-Phenylethylamine in Control Rats and in Rats Pretreated with a Monoamine Oxidase Inhibitor[a]

Time after Injections (min)	Nontreated Animals		Pretreated with Pargyline	
	Total cpm	Percent Total	Total cpm	Percent Total
0	345,804	100	325,905	100
5	123,405	35.7	256,060	78.6
10	75,006	21.7	144,800	44.4
30	2,016	0.6	20,328	6.2

[a]Pargyline, 75 mg/kg, i.p., administered 60 min before the β-[^{14}C]-phenylethylamine injection, was used as MAOI. Results are expressed as mean values of six independent determinations.

TABLE 7

Metabolism of Radiolabeled β-Phenylethylamine in the Rat Brain[a]

	cpm	Percent of Total Radioactivity
Total radioactivity	20,600	100
β-Phenylethylamine	10,950	53.2
Phenylethanolamine	890	4.3
N-methyl phenylethylamine	460	2.2
Unidentified metabolites	8,300	40.3

[a]β-[^{14}C]phenylethylamine (0.20 μCi), was injected intraventricularly into rats pretreated with the MAOI pargyline, 75 mg/kg, i.p., 60 min before the PEA injection. The rats were killed 30 min later, and the PEA and its metabolites were extracted and separated by thin layer chromatography. Results are expressed as mean values obtained from four different chromatograms, utilizing two different solvent systems.

TABLE 8

Effect of Drugs on the Endogenous β-Phenylethylamine Content of the Rat Brain[a]

Treatment	β-Phenylethylamine (ng/g)
None	1.5 ± 1
Phenylalanine	4.6 ± 2**
Pargyline	75.5 ± 8**
Pargyline + phenylalanine	630 ± 15*
Pargyline + phenylalanine + NSD-1055	57 ± 6*
p-Chlorophenylalanine	21.6 ± 4*
p-Chlorophenylalanine + phenylalanine	60.3 ± 12*†

[a]Phenylalanine methylester hydrochloride (1 g/kg) was injected 60 min; pargyline hydrochloride (75 mg/kg) 90 min, NSD-1055 (200 mg/kg) 2 hr; and p-chlorophenylalanine methylester (375 mg/kg) 48 and 24 hr before killing. Values represent mean values ± SEM for groups of six rats. Significance versus control values (Student's t test): *$p < 0.01$; **$p < 0.05$; †significant versus p-chlorophenylalanine treatment. From Saavedra (36).

8). The combined administration of p-chlorophenylalanine and phenylalanine resulted in a potentiation of this effect (Table 8).

These results, together with those of Boulton and co-workers (4), allow us to postulate the following pathways for PEA formation and metabolism (Fig. 2).

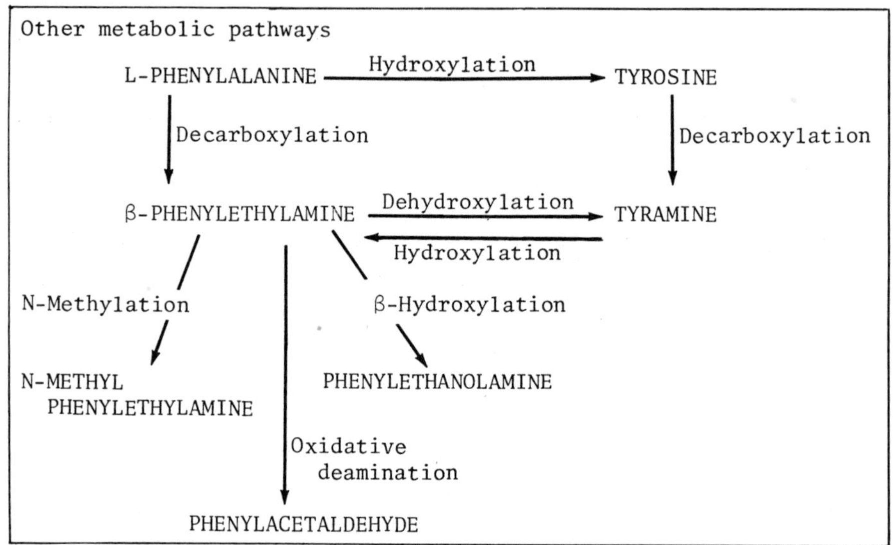

FIG. 2. Metabolism of PEA.

III. THE INVOLVEMENT OF β-PHENYLETHYLAMINE IN THE ETIOLOGY OF NEUROPATHOLOGICAL STATES IN MAN

Several groups have postulated that a defect in PEA metabolism is the basis for the production of depression in man (6,14,15,31,32).

Almost all the experimental data to support this theory refer to studies of the excretion of PEA in the urine. Unfortunately, these studies must be criticized on the basis of poor experimental design, absence of adequate controls, or failure to consider very important aspects of PEA physiology and metabolism in humans.

A great number of studies have been published concerning PEA levels in human urine (6,15,18,19,25,31,32,34,35,39). Earlier studies (25,34,35) reported that the amine could not be measured in the urine. This failure to detect PEA was very probably due to the low sensitivity of the methods utilized. The use of more sensitive techniques allowed later the detection and quantitation of urinary

PEA. The PEA levels reported in these studies, however, differed widely, with values as low as 10 μg/day (39) and as high as 453 μg/day (31,32). The reason for this 45-fold difference in amine levels is not immediately apparent. It is possible to speculate, however, that since a wide variety of techniques for the measurement of PEA was used, some of the methods may have not been totally specific for this amine, as it is the case for studies with brain material. Thus, low values, like those recently reported by Schweitzer et al. (39), are more likely to be close to the real normal PEA levels, and more so since these authors utilized internal standards to correct for losses of PEA that would occur during the detection procedure.

A number of very important points have to be emphasized when analyzing the urinary studies for PEA. Most of the studies presented data not corrected for creatinine values, or for density values. It is difficult to critically consider these data, since it is well known that great differences in the amount and density of the urinary output constantly occur. Data described as micrograms of PEA in the urine per day, or per hour, are meaningless unless adequately corrected.

None of the studies published so far consider the pH of the urine, or made any attempt to study the characteristics of the elimination of PEA in the urine when pH variations occur. This is surprising, in view of the well known characteristic of amphetamine, a compound closely related to PEA, which is eliminated in the urine by a pH-dependent mechanism (1,2,9). Finally, the effects of the dietary intake in the PEA urinary excretion were not adequately studied. It is known that a number of foodstuffs contain large quantities of PEA. It is then conceivable that dietary variations could affect PEA levels in the urine by a double mechanism: by additions of excess PEA contained in specific foodstuffs, or by drastic changes in urinary pH which would secondarily affect the reabsorption, metabolism, and excretion of this amine.

On the basis of these interesting but highly controversial findings, theories for the participation of PEA in depression (6,15, 18,20) and also schizophrenia (20) has been postulated. β-Phenylethylamine has been reported to be decreased in the urine of depressed patients (6,15,18,20) and increased in the urine of schizophrenic patients (20).

However, one recent study (39) failed to confirm increased levels of PEA in the urine of schizophrenics, and preliminary observations (Saavedra, Beckmann, and Goodwin, unpublished results) indicate that the PEA levels in the urine of depressive patients are not different from those of normal, healthy controls.

Furthermore, we should consider that urinary determinations of PEA may very probably not reflect changes in specific areas of the brain, like the ones which are likely to occur in human neuropathology, since PEA is very rapidly deaminated in blood platelets and liver. The brain PEA probably represents only a small fraction of the total body PEA. Furthermore, urinary PEA is more likely to reflect PEA metabolism in kidneys, since this organ contains the highest PEA levels and very high levels of L-amino acid decarboxylase.

Studies on PEA in humans, therefore, should be focused in the future in the study of the presence and metabolism of this compound in blood, or preferentially cerebrospinal fluid, and specific brain regions. The difficulties of these studies are obvious, but they represent the only route to clarify the role, if any, of this interesting amine.

We have recently identified PEA to be present in the plasma of normal subjects, in rather small amounts (Table 9), as well as the type B MAO in the platelets of the same control individuals (Table 10). Studies on the possible variations in PEA levels in plasma of depressive and schizophrenic patients are currently in progress.

It has been recently reported (16,40) that treatment of depressive patients with D-phenylalanine was effective in more than 50% of the cases of endogenous depression, while treatment with L-phenylalanine was not effective. Such a therapeutic effect, if confirmed, would be of great value in the treatment of severe endogenous depression. It would not support, however, a PEA theory of depression, since the amino acid decarboxylase is stereospecific for the L-form of the amino acids (28) and administration of D-phenylalanine does not result in increased amounts of PEA in the body (Fernandez-Pardal, personal communication).

TABLE 9

Endogenous β-Phenylethylamine Content in Human Plasma[a]

Subject	cpm	β-Phenylethylamine (ng/ml)
1	1608	0.6
2	2670	1
3	5350	2
4	4830	1.8
5	1370	0.5
6	2180	0.8
Mean ± SEM		1.12 ± 0.26

[a]Blanks were 250 cpm. Two nanograms of authentic PEA added to the plasma gave 2680 cpm. Each value is the mean of two separate determinations.

TABLE 10

Type B Monoamine Oxidase Activity in Human Platelets[a]

Subject	Type B Monoamine Oxidase Activity (nmol/mg protein/hr incubation)
1	2.81
2	1.90
3	2.05
4	2.20
5	1.95
6	2.30
Mean ± SEM	2.20 ± 0.14

[a]The substrate used to measure type B MAO was PEA.

IV. CONCLUSION

β-Phenylethylamine has been unambiguously identified as a normal constitutent of mammalian tissues, including brain. There is evidence that the metabolism of PEA in brain tissue is extremely fast. This amine may be related to the catecholamine and tyrosine metabolism through mechanisms of ring-dehydroxylation and hydroxylation.

However, these data do not support per se the existence of an important role for PEA in brain function. The relative influence of changes in PEA levels with respect to the metabolism of other brain amines is unknown. Furthermore, the brain levels of this amine are several orders of magnitude lower than those needed to produce pharmacological effects. Many studies on effects of drugs on PEA levels in brain should be re-evaluated, due to the existence of fundamental discrepancies in the methodology used by different laboratories to measure actually the compound in tissues. In some cases, the authors seem to have identified the presence of PEA in tissues beyond doubt, but utilize other techniques to actually measure the compound, without subjecting them to the same rigorous proof of specificity. Furthermore, mechanisms of storage and release of PEA by nerve terminals, similar to those of other putative neurotransmitters, have not been demonstrated, and the fundamental evidence for the presence of specific PEA receptors is still lacking.

With respect to the human studies, urinary determinations of PEA levels cannot at the present time be considered as proof of the involvement of this amine in human pathology.

The availability of specific methodology and a more realistic approach to the study of PEA metabolism could clarify the role, if any, of this compound in human neuropathology.

REFERENCES

1. A. H. Beckett and M. Rowland (1965). Urinary excretion kinetics of amphetamine in man. *J. Pharm. Pharmacol.*, 17:628-639.

2. A. H. Beckett, J. A. Salmon, and M. Mitchard (1969). The relation between blood levels and urinary excretion of amphetamine under controlled acidic acid and under fluctuating urinary pH values using (^{14}C) amphetamine. *J. Pharm. Pharmacol.*, 21:251-258.

3. A. A. Boulton and G. B. Baker (1975). The subcellular distribution of β-phenylethylamine, p-tyramine and tryptamine in rat brain. *J. Neurochem.*, 25:477-481.

4. A. A. Boulton, L. E. Dyck, and D. A. Durden (1975). Hydroxylation of β-phenylethylamine in the rat. *Life Sci.*, 15:1673-1683.

5. A. A. Boulton, A. V. Juorio, S. R. Philips, and P. H. Wu (1975). Some arylalkylamines in rabbit brain. *Brain Res.*, 96:212-216.

6. A. A. Boulton and L. Milward (1971). Separation, detection and quantitative analysis of urinary β-phenylethylamine. *J. Chromatogr.*, 57:287-296.

7. I. Cohen, J. F. Fischer, and W. H. Vogel (1974). Physiological disposition of β-phenylethylamine, 2,4,5-trimethoxyphenylethylamine, 2,3,4,5,6-pentamethoxyphenylethylamine and β-hydroxymescaline in rat brain, liver and plasma. *Psychopharmacologia*, 36:77-84.

8. C. R. Creveling, K. Kondo, and J. W. Daly (1968). Use of dansyl derivatives and mass spectrometry for identification of biogenic amines. *Clin. Chem.*, 14:302-309.

9. J. M. Davis, I. J. Kopin, L. Lemberger, and J. Axelrod (1971). Effects of urinary pH on amphetamine metabolism. *Ann. N.Y. Acad. Sci.*, 179:493-501.

10. W. B. Dewhurst (1968). New thoery of cerebral amine function and its clinical application. *Nature*, 218:1130-1133.

11. D. A. Durden, S. R. Philips, and A. A. Boulton (1973). Identification and distribution of β-phenylethylamine in the rat. *Can. J. Biochem.*, 51:995-1002.

12. D. J. Edwards and K. Blau (1972). Analysis of phenylethylamines in biological tissues by gas-liquid chromatography with electrocapture detection. *Anal. Biochem.*, 45:387-402.

13. D. J. Edwards and K. Blau (1973). Phenethylamines in brain and liver of rats with experimentally induced phenylketonuria-like characteristics. *Biochem. J.*, 132:95-100.

14. E. Fischer (1975). The phenethylamine hypothesis of thymic homeostasis. *Biol. Psychiat.*, 10:667-673.

15. E. Fischer, B. Heller, and A. Miro (1968). β-phenylethylamine in human urine. *Arzneim.-Forsch.*, 18:1486.

16. E. Fischer, B. Heller, M. Nachon, and H. Spatz (1975). Therapy of depression by phenylalanine. *Arzneim.-Forsch.*, 25:132.

17. E. Fischer, R. I. Ludmer, and H. C. Sabelli (1967). The antagonism of phenylethylamine to catecholamines on mouse motor activity. *Acta Physiol. Lat. Am.*, 17:15-21.

18. E. Fischer, H. Spatz, B. Heller, and H. Reggiani (1972). Phenthylamine content of human urine and rat brain, its alterations in pathological conditions and after drug administration. *Experientia*, 15:307-308.

19. E. Fischer, H. Spatz, R. S. F. Labriola, E. M. R. Casanova, and N. Spatz (1973). Quantitative gas-chromatographic determination and infrared spectrographic identification of urinary phenethylamine. *Biol. Psychiat.*, 7:161-165.

20. E. Fischer, H. Spatz, J. M. Saavedra, H. Reggiani, A. Miro, and B. Heller (1972). Urinary elimination of phenethylamine. *Biol. Psychiat.*, 130-147.

21. K. Fuxe, H. Grobecker, and J. Jonsson (1967). The effect of β-phenylethylamine on central and peripheral monoamine-containing neurons. *Eur. J. Pharmacol.*, 2:202-207.

22. W. J. Giardina, W. A Pedemonte, and H. C. Sabelli (1973). Iontophoretic study of the effects of norepinephrine and 2-phenylethylamine on single cortical neurons. *Life Sci.*, 12:153-161.

23. E. E. Inwang, A. D. Mosnaim, and H. C. Sabelli (1973). Isolation and characterization of phenylethylamine and phenylethanolamine from human brain. *J. Neurochem.*, 20:1469-1473.

24. D. M. Jackson (1975). β-Phenylethylamine and locomotor activity in mice. Interaction with catecholaminergic neurons and receptors. *Arzneim.-Forsch.*, 25:622-626.

25. B. Jepson, W. Lovenberg, P. Zaltman, S. Oates, A. Sjoerdsma, and S. Udenfriend (1960). Amine metabolism studied in normal and phenylketonuric humans by monoamineoxidase inhibition. *Biochem. J.*, 74:5P.

26. J. Jonsson, H. Grobecker, and P. Holtz (1966). Effect of β-phenylethylamine on content and subcellular distribution of norepinephrine in rat heart and brain. *Life Sci.*, 5:2235-2246.

27. B. Koe and A. Weissman (1966). p-Chlorophenylanine: A specific depletor of brain serotonin. *J. Pharm. Exp. Ther.*, 154:499-516.

28. W. Lovenberg, H. Weissbach, and S. Udenfriend (1962). Aromatic L-amino acid decarboxylase. *J. Biol. Chem.*, 237:89-93.

29. P. Mantegazza and M. Riva (1963). Amphetamine-like activity of β-phenylethylamine after a monoamineoxidase inhibitor in vivo. *J. Pharm. Pharmacol.*, 15:472-478.

30. A. D. Mosnaim and E. E. Inwang (1973). A spectrophotometric method for the quantification of 2-phenylethylamine in biological specimens. *Anal. Biochem.*, 57:561-577.

31. A. D. Mosnaim, E. E. Inwang, J. H. Sugerman, W. J. DeMartini, and H. C. Sabelli (1973). Ultraviolet spectrophotometric determination of 2-phenylethylamine in biological samples and its possible correlation with depression. *Biol. Psychiat.*, 6:235-257.

32. A. D. Mosnaim, E. E. Inwang, J. H. Sugerman, and H. C. Sabelli (1973). Identification of 2-phenylethylanine in human urine by infrared and mass spectroscopy and its quantification in normal subjects and cardiovascular patients. *Clin. Chim. Acta*, 46: 407-413.

33. T. Nakajima, Y. Kakimoto, and I. Sano (1964). Formation of β-phenylethylamine in mammalian tissue and its effect on motor activity in the mouse. *J. Pharm. Exp. Ther.*, 143:319-325.

34. J. A. Oates, P. Z. Nirenberg, B. Jepson, A. Sjoerdsma, and S. Udenfriend (1963). Conversion of phenylalanine to phenylethylamine in patients with phenylketonuria. *Proc. Soc. Exp. Biol.*, 112:1078-1081.

35. T. L. Perry (1962). Urinary excretion of amines in phenylketonuria and mongolism. *Science*, 136:879.

36. J. M. Saavedra (1974). Enzymatic isotopic assay for and presence of β-phenylethylamine in brain. *J. Neurochem.*, 22:211-216.

37. J. M. Saavedra and J. Axelrod (1973). Demonstration and distribution of phenylethanolamine in brain and other tissues. *Proc. Nat. Acad. Sci. USA*, 70:769-772.

38. J. M. Saavedra, J. T. Coyle, and J. Axelrod (1973). The distribution and properties of the nonspecific N-methyltransferase in brain. *J. Neurochem.*, 20:743-752.

39. J. W. Schweitzer, A. J. Friedhoff, and R. Schwartz (1975). Phenethylamine in normal urine: failure to verify high values. *Biol. Psychiat.*, 10:277-285.

40. H. Spatz, B. Heller, M. Nachon, and E. Fischer (1975). Effects of D-phenylalanine on clinical picture and phenethylaminuria in depression. *Biol. Psychiat.*, 10:235-239.

41. H. Spatz and N. Spatz (1972). Spectrophotofluorometric determination of beta-phenylethylamine in blood and urine. *Biochem. Med.*, 6:1-6.

42. J. Willner, H. Lefevre, and E. Costa (1974). Assay by multiple ion detection of β-phenylethylamine and phenylethanolamine in rat brain. *J. Neurochem.*, 23:857-859.

43. P. H. Wu, and A. A. Boulton (1975). Metabolism, distribution and disappearance of injected β-phenylethylamine in the rat. *Can. J. Biochem.*, 53:42-50.

44. H. Y. T. Yang, and N. H. Neff (1973). β-Phenylethylamine: A specific substrate for type B monoamineoxidase in brain. *J. Pharm. Exp. Ther.*, 187:365-371.

Chapter 8

DIFFERENCES IN THE ELECTRICALLY INDUCED RELEASE OF ^{3}H-LABELED β-PHENYLETHYLAMINE, TYRAMINE, OCTOPAMINE, AND NOREPINEPHRINE FROM RAT NEOCORTICAL SLICES

María Christina Saldate and Fernando Orrego

Department of Biochemistry
National Institute of Cardiology
Mexico City, Mexico

I. INTRODUCTION

β-Phenylethylamine (PEA) has been shown to produce in the mammalian brain important electrophysiological and behavioral effects (21,30, 31,32,38) that are independent of its weak effects as an indirect-acting sympathomimetic amine (12,16,34). This amine is normally present in rather low concentrations in the human brain and in that of all other mammalian species studied (10,15,17,21,29), where it is formed mainly by decarboxylation of phenylalanine and shows a very rapid turnover rate (11,21,29,38). A correlation has been also established between urinary excretion of PEA and the occurrence of depression or mania in human patients, where PEA production is decreased or increased, respectively (30,31). On the basis of these and other findings, a theory implicating PEA in the pathogenesis of manic-depressive psychoses has been put forward (30,31). However, the biochemical role of PEA in neural tissues has not been defined, and both a modulator (nontransmitter) (30) and a neurotransmitter role (5) have been proposed for it. The former is based on indirect evidences, and the latter, on subcellular distribution studies. No studies are, at present, available on electrically induced release of this amine that could help to clarify more precisely its functional role.

p-Tyramine is also normally present in brain (27), where its biosynthesis and metabolism have been studied both in vivo and in vitro (4,6,28,35,39). A small electrically induced release of this amine has been reported to occur from slices of different brain regions (1,3), but it is not known whether such release is calcium dependent and, therefore, whether tyramine may subserve a transmitter role. Also, although tyramine is readily converted to octopamine in brain both in vivo and in tissue slices (6,28,39), no release of this amine could be demonstrated when the slices were electrically stimulated (1,3). This situation is the opposite from what is found in peripheral sympathetic neurons (13).

The present experiments were performed with the intention of defining whether PEA and tyramine behave as transmitter substances under the experimental conditions used, and whether conversion of tyramine into octopamine, and subsequent release of the latter, occurred. In addition, the electrically induced release of [^{3}H]norepinephrine (NE) from reserpinized, monoamine oxidase (MAO)-inhibited brain cortex slices, a situation where impairment of NE retention by synaptic vesicles occurs (41), was also studied. For this we have mainly used controlled electrical stimulation of superfused rat neocortical thin slices. This procedure is able to distinguish clearly between the release of transmitter substances such as NE, 5-hydroxytryptamine (5-HT), acetylcholine (ACh) and dopamine (DA) (24,25,33, and our unpublished results) and that of nontransmitter substances (25,26). The former are released by low applied potentials, and their release shows an absolute calcium dependency, while the latter are only released by higher potentials and show no calcium dependency.

II. MATERIALS AND METHODS

The procedures utilized for obtaining, incubating, superfusing, and stimulating the slices have been described in detail (25). In brief, outermost slices, about 0.35 mm thick, were obtained with blade and blade-guide from neocortices of 120 g Wistar rats. They were then incubated in 1 ml of a balanced glucose-salt solution with the appropriate radioactive substances for 30 to 60 min in a Dubnoff incubator (100 min^{-1}), under 5% CO_2 in O_2 at 37°C. They were then placed in McIlwain silver-wire, quick-transfer electrodes and superfused with a nonradioactive medium. Efflux is expressed as a fractional rate constant, and each fraction represents a 2-min efflux period. Electrical stimuli were 50 Hz sine-wave currents of 10-sec duration. Applied potentials were monitored with a Tektronix 532 oscilloscope.

The concentrations of radioactive substances used during the incubation were: [^{3}H]PEA: 2.2 x 10^{-6} M; [^{3}H]tyramine: 2.04 x 10^{-6}M; and [^{3}H]NE: 5.4 x 10^{-7} M. When drugs were present during the incubation, their concentrations were: deprenyl: 20 μg/ml; pargyline: 20 μg/ml; and nialamide: 10^{-4} M. These drugs were added to the incubation medium 5 min prior to the radioactive test substances. Reserpine (8 mg/kg, i.p.) was injected 18 hr prior to sacrifice. Octopamine formation was measured by the specific and sensitive procedure of Friedman and Kaufman (14), that involves periodate oxidation followed by toluene extraction, exactly as described by Creveling and Daly (9). Statistical calculations of differences between pre- and poststimulation samples were done by Student's t test (paired variates).

[1,1-2,2-^{3}H]PEA (spec. act. 4.58 Ci/mmol was custom synthetized by New England Nuclear and purified by thin layer chromatography on sheets of aluminum oxide with tert-amyl alcohol, 20%

aqueous methylamine (4 : 1) as solvents. Its radiochemical purity was greater than 99%. [G-^{3}H]Tyramine (spec. act. 4.93 Ci/mmol, purity greater than 97%) and DL[7-^{3}H]NE (spec. act. 9.3 Ci/mmol) were from New England Nuclear. [^{3}H]PEA, [^{3}H]tyramine, deprenyl, and pargyline (Abbott Laboratories) were kind gifts of Dr. A. D. Mosnaim; nialamide (Pfizer) was a gift of Dr. A. Gómez; and reserpine was donated by CIBA-Geigy de México.

III. RESULTS

A. Efflux of β-Phenylethylamine

Initially, the efflux of PEA was studied from unstimulated slices in the presence of the monoamine oxidase inhibitors (MAOI) deprenyl or pargyline, the former being rather selective for the B form of the enzyme, that preferentially metabolizes PEA (22). Tissue desaturation curves indicated that the efflux of PEA could be resolved into at least three different first-order components (Fig. 1). The initial one, with a $t_{1/2}$ of 6 min is generally accepted as due to the washout of extracellular material (7,33), and is also present in comparable experiments done with other biogenic amines and with different amino acids (25,33, and our unpublished results). This component was followed by a second unusually rapid one with a $t_{1/2}$ of 12 min, that was responsible for the efflux of most of the PEA present in the slice. Finally, a third slower component was also apparent when more than 80% of PEA had been lost from the slice. Its half-life could not be measured accurately in experiments of this duration.

When the slices were electrically stimulated with applied potentials that are able to release considerable amounts of transmitter substances (24,25,33), no enhanced efflux of [^{3}H]PEA was observed (Fig. 2), except at 3V, where nontransmitter substances also begin to be released. Similarly, no change in the efflux rate of [^{3}H]PEA was seen when the slices were depolarized by raising the potassium concentration to 46 mM (Fig. 3).

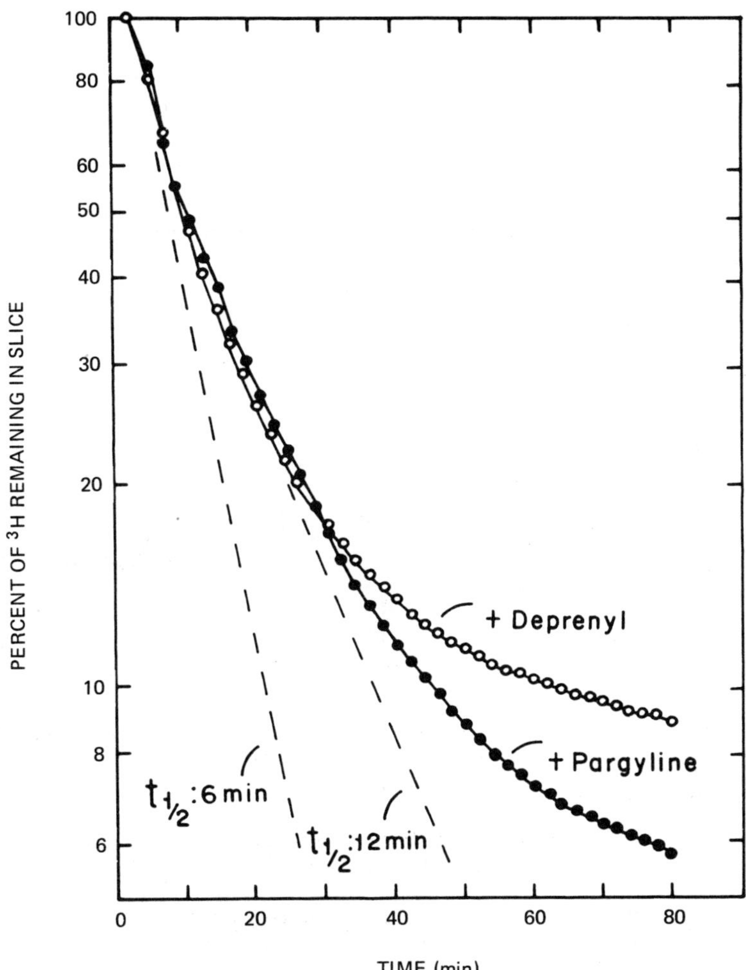

FIG. 1. Spontaneous Efflux of [^{3}H]PEA. The slices were initially incubated in 2.2 x 10^{-6} M [^{3}H]PEA for 60 min, in the presence of deprenyl (20 μg/ml) or pargyline (20 μg/ml), and then superfused with nonradioactive medium. The radioactivity remaining in the slice at time t, was estimated by adding the one found in the homogenized slice to that release into the medium between time t and the end of the experiment.

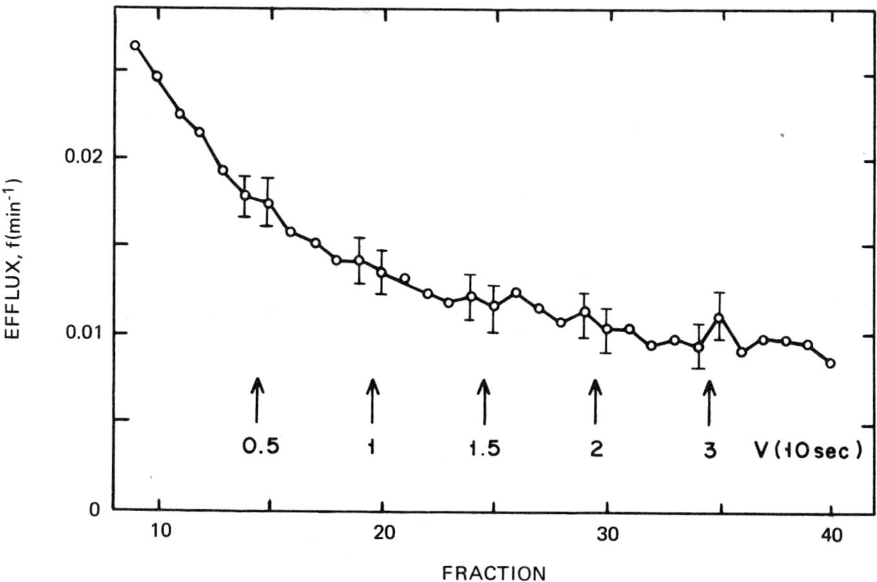

FIG. 2. Effect of electrical stimulation on release of [^{3}H]PEA. Efflux is expressed as a fractional constant,

$$f = \frac{\text{cpm released during collection period (c.p.)}}{\text{cpm in slice at beginning of c.p x duration of c.p.}}$$

Each fraction represents a 2-min collection period. At the arrows, stimuli of the indicated potentials (sine-wave, 50 Hz) were applied for 10 sec. Results are mean values ± SEM (when indicated) of six experiments. Differences between pre- and post-stimulation values are not statistically significant, except following 3 V, where $P \cong 0.05$.

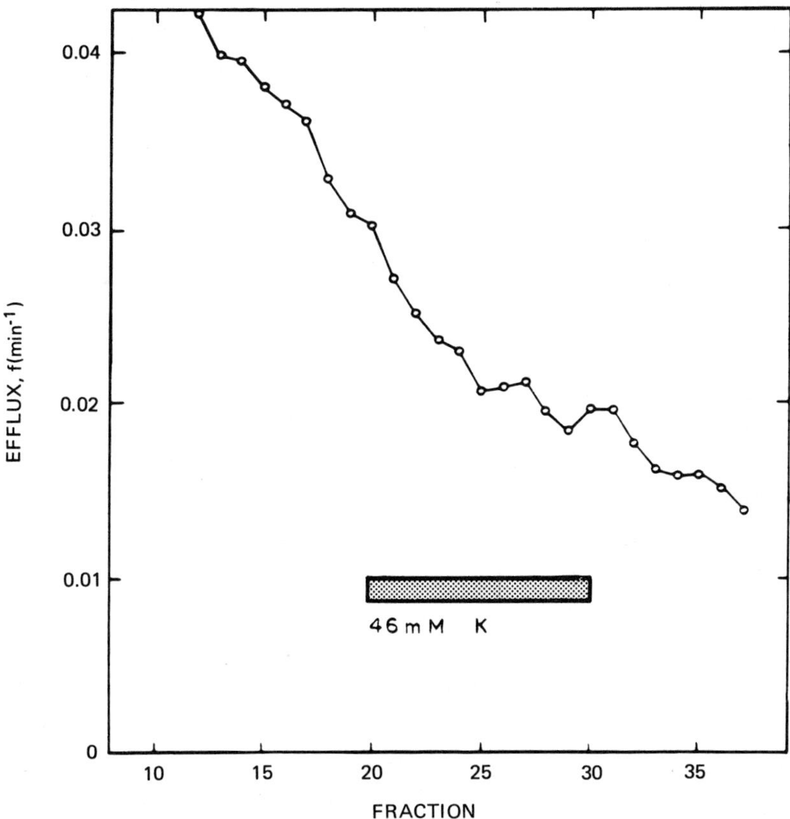

FIG. 3. Effect of elevated extracellular potassium on [^{3}H]PEA efflux. Conditions are the same as in Fig. 2, except that deprenyl (20 μg/ml) was present. From fraction 20 to 29 inclusive potassium was raised from 6.2 to 46 mM.

B. [^{3}H]Tyramine Efflux and Metabolism

The spontaneous efflux of tyramine, in the presence or absence of MAOI, could be resolved into two first-order components (Fig. 4). A fast initial one ($t_{1/2}$: 5.5 min) was of a similar nature as the one seen with PEA, as well as a second slower one with a half-life of 38 min in the presence of nialamide. Although this second component was considerably slower than the second PEA component, it is, nevertheless, much more rapid than the one seen previously with NE or 5-HT (25,33).

Electrical stimulation of the slices, performed under conditions identical to those used previously, now showed marked differences with respect to what was seen with [^{3}H]PEA (Fig. 5). Stimuli of 1 to 3 V induced considerable increases in the efflux of tritium labeled material, and such induced release was entirely calcium dependent. This is very similar to what has been previously seen with ^{3}H-labeled NE, 5-HT, ACh, and DA, except for one difference. In the case of the previously studied amines, the induced efflux increased with increasing applied potentials, up to about 4 V (24,25,33). In the present case, however, such increase is not seen, and the efflux peak after the 3 V stimulus is slightly smaller than, and not significantly different from those seen after 1.5 or 2 V. The absence of release seen with 0.5 V in the present case, but not with the amines previously studied, where a small peak is present, may be only a reflection of the high resting efflux of [^{3}H]tyramine that obscures small changes in efflux.

The contribution of [^{3}H]octopamine, formed from [^{3}H]tyramine, both to the resting and to the electrically induced efflux was also studied (Table 1). In the fraction immediately preceding stimulation, octopamine represented about 7% of the radioactivity. This increased to 18% after stimulation, and of the extra material released by the stimulus, 42% was accounted for by [^{3}H]octopamine and about 58% by [^{3}H]tyramine. The increase in [^{3}H]metabolites was negligible. Since no electrically induced efflux occurred in the absence of calcium ions, this indicated that octopamine was also released by a calcium-dependent process.

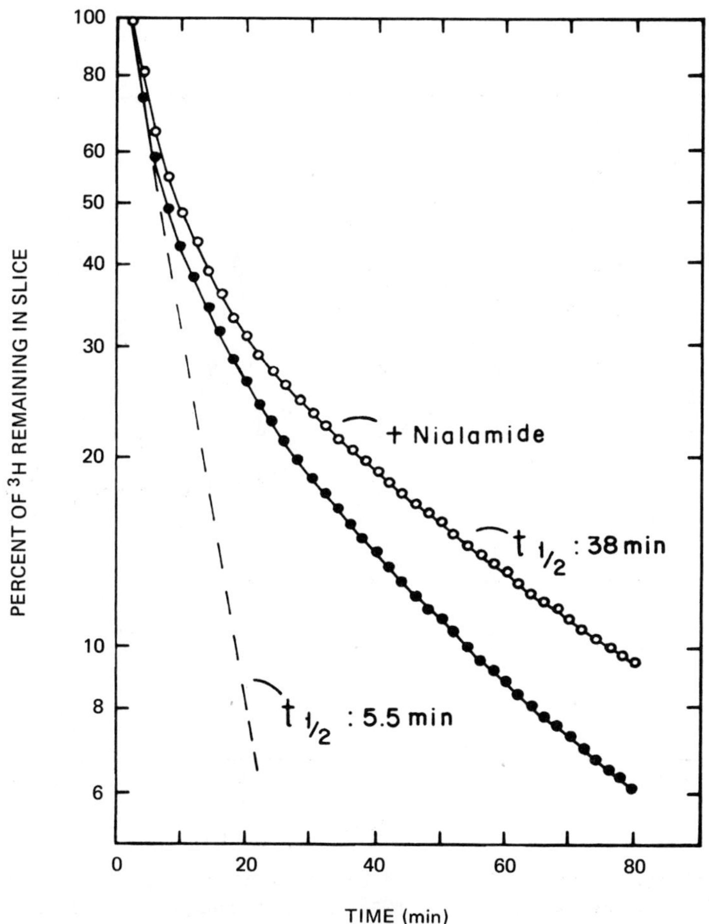

FIG. 4. Spontaneous efflux of [^{3}H]tyramine. Initial incubation was with 2.04 x 10^{-6} M [^{3}H]tyramine for 60 min. Other conditions were the same as in Fig. 1. Nialamide (10^{-4} M) was present during the initial incubation (one experiment), and curve without drugs is the mean of four experiments.

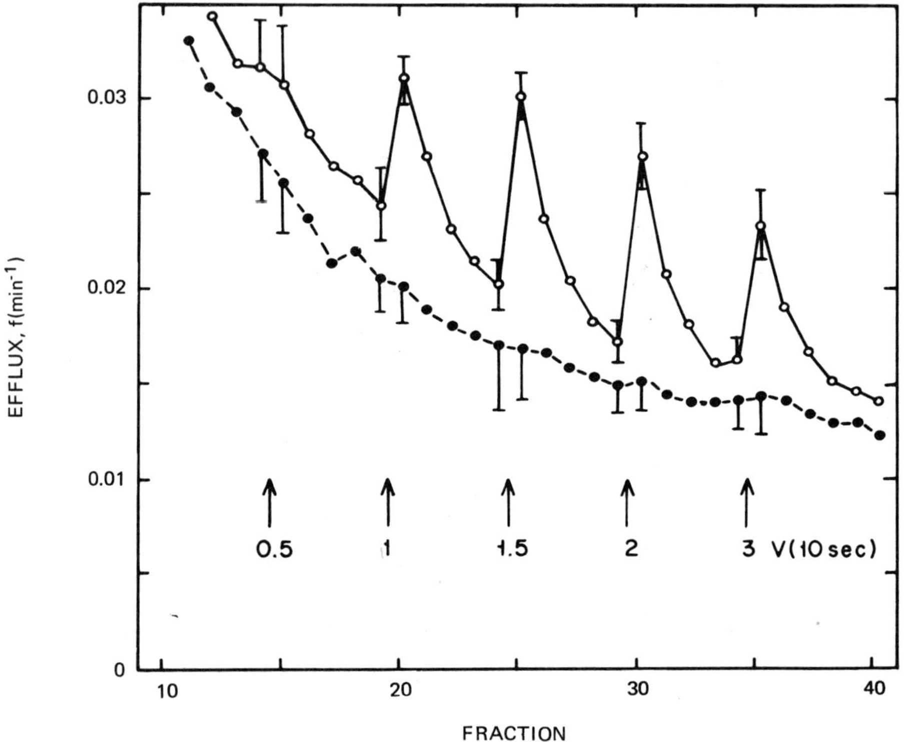

FIG. 5. Electrically induced release of [^{3}H]tyramine plus octopamine. Initial incubation was with 2.04 x 10^{-6} M [^{3}H]tyramine for 60 min in normal medium in all cases. Open circles: superfusion with normal medium; closed circles: superfusion with medium to which no calcium was added. At the arrows sine-wave (50 Hz) stimuli of the indicated potentials were applied for 10 sec. Differences between pre- and poststimulation values were statistically significant ($P < 0.01$ or less) only for the calcium-containing experiments with stimuli of 1 V or higher. Each curve is the mean (± SEM) of five experiments.

TABLE 1

Effect of Stimulation on Efflux of Tyramine, Octopamine, and Metabolites[a]

		Tyramine	Octopamine	Metabolites
Fraction 1	cpm	143,063	970	2762
	(% of total)	(97.5)	(0.66)	(1.9)
Fraction 24 (prestimulation)	cpm	1600	122	34
	(% of total)	(91.1)	(7.0)	(1.9)
Fraction 25 (poststimulation	cpm	2103	484	38
	(% of total)	(80.1)	(18.4)	(1.4)
Increase by stimulation	cpm	503	362	4
	%	31.4	297	11.7

[a]The slices were initially incubated with 2.04 x 10^{-6} M [^{3}H]tyramine for 60 min and then superfused with nonradioactive medium. Each fraction represents a 2-min efflux period. After fraction 24 had been collected, the slices were stimulated with 1.5 V (50 Hz) for 10 sec. Octopamine was estimated by periodate oxidation followed by toluene extraction of the p[^{3}H]hydroxybenzaldehyde formed (9,14). Blanks treated in the same manner, except that an equivalent volume of water replaced the periodate, were used to measure toluene-extractable deaminated metabolites. These blanks were substracted from the periodate-treated values. Material not extractable by toluene following periodate oxidation was assumed to be all tyramine. Counts per minute of the aqueous phase were corrected by a channels ratio procedure to the same counting efficiency as that in toluene. Results are the mean of duplicate measurements that did not differ by more than 5%.

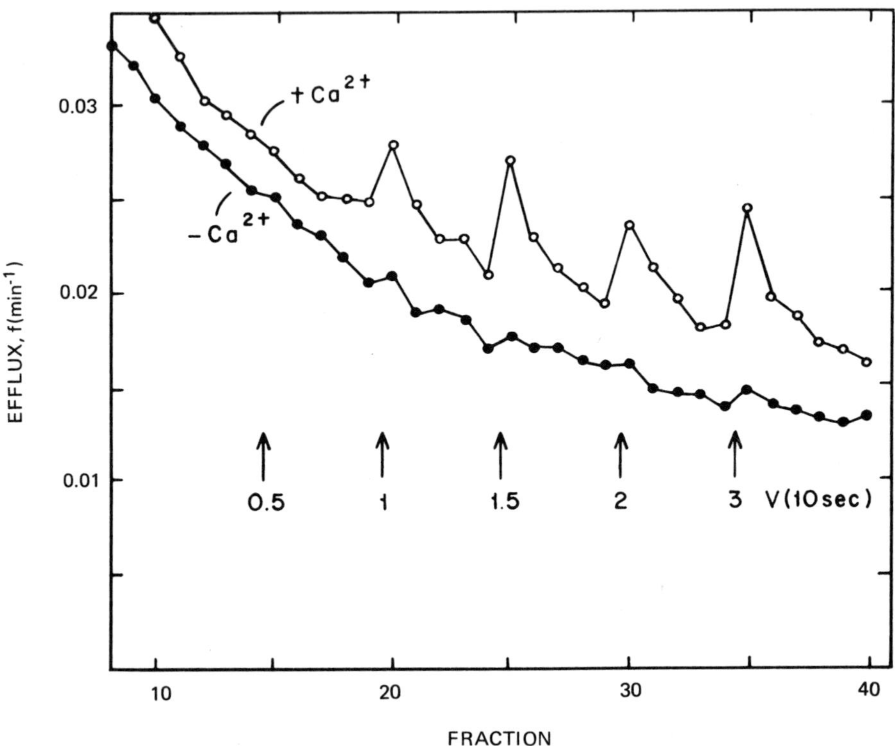

FIG. 6. Effect of electrical stimulation on release of $[^3H]NE$ from reserpinized, MAO-inhibited slices. Slices were obtained from rats reserpinized (8 mg/kg, i.p.) 18 hr prior to sacrifice. Initial incubations were for 30 min in the presence of 10^{-4} M nialamide and 5.4 x 10^{-7} M DL-$[^3H]NE$. Other conditions are as in Fig. 5. Each curve is the mean of five experiments, and presence or absence of calcium is indicated. Difference between pre- and poststimulus efflux is significant ($P < 0.02$ or less), only in the calcium-containing points following stimuli of 1 V or higher. Areas of the efflux peaks induced by 2- or 3-V stimuli are not statistically different from the one seen after 1.5 V.

C. Electrically Induced Release of $[^3H]$Norepinephrine from Reserpinized and MAO-Inhibited Slices

The resting efflux of $[^3H]$NE from slices obtained from reserpinized animals (8 mg/kg, i.p., 18 hr prior to sacrifice), and treated with 10^{-4} M nialamide in vitro, was considerably higher (Fig. 6) than the one seen when no drugs are used (25). Electrical stimulation of these drug-treated slices was also able to induce small but highly significant increases in efflux with 1- to 3-V stimuli. The area of the induced efflux peaks, however, was only 20% or less, than the one from untreated slices. It was also observed that the peaks after the 2- and 3-V stimuli were not significantly larger than the one seen after 1.5 V. A marked calcium dependency was also found in these experiments (Fig. 6).

IV. DISCUSSION

The extremely rapid efflux kinetics of $[^3H]$PEA, found in the present experiments, is entirely consistent with previous reports showing that radioactive PEA is rapidly lost from the brain after its intraventricular injection (4,35), and also with the fact that it is not actively transported by cells (2), and that it readily crosses different membranes, including the blood-brain barrier by simple diffusion (2,21,23).

The results obtained with electrical stimulation, as well as with elevated extracellular potassium, give very clear evidence against a transmitter role for PEA, since no induced release was evidenced under conditions where transmitter substances are readily released. This is also compatible with the lack of an adenosine triphosphate (ATP)-dependent accumulation of PEA by adrenomedullary storage particles (37) that represent a useful model of the adrenergic synaptic vesicle, since secretion requires a substance to be located in the interior of such vesicles (43). High concentrations of PEA are able to release NE from synaptic vesicles obtained from splenic nerves (34). This is probably mediated by diffusion of PEA into the interior of the vesicles, where it displaces NE from its storage sites. Experimental conditions in which very high concentrations of PEA are attained may, thus, predictably lead to penetration of PEA into synaptic vesicles, and to a transmitter-like release when depolarizing stimuli are used. But, because of the very low PEA concentrations normally present in brain (10,15,17,21,29), such a situation is most unlikely to occur in vivo. Subcellular distribution studies have shown that PEA is, to a considerable extent, a soluble substance (4,5,11), and the small amount present in synaptosomal fractions may, therefore, be interpreted as being due to the soluble cytoplasm known to be contained in those particles. On the other hand, a rapid and continuous efflux of PEA from the cells that synthetize it seems well adapted for a modulator substance that is involved in a tonic process such as mood.

The relatively rapid efflux of tyramine from the slices is similar to the less effective retention of this amine, as compared to NE, previously found in central and peripheral tissues (8,13,18, 19,20,40,44). Such lower retention seems to be correlated with a lesser ability of adrenergic synaptic vesicles or adrenomedullary storage particles to retain this amine (18,20,36,37). With the sensitive procedure used in this work, conversion of tyramine to octopamine could be readily demonstrated, thus confirming previous work performed in vivo (6) and in unstimulated brain cortex slices (28,39).

The transmitterlike release of both tyramine and octopamine, induced by electrical stimulation, suggests that both these amines were in the interior of synaptic vesicles at the time of stimulation. Octopamine is formed exclusively inside noradrenergic neurons (8,18, 39,40) and is more efficiently retained than tyramine, but less than NE, by the synaptic vesicles of those cells (13,18-20). The relatively greater stimulus-induced release of octopamine, compared to tyramine, may be a reflection of a greater concentration, in relative terms, of the former amine inside the secretory vesicles. Aside from tyramine being transported into neocortical adrenergic axon terminals, a desmethylimipramine-resistant uptake has been shown to exist in brain cortex slices (39) but not in the heart (40). This may be an indication of tyramine uptake into dopaminergic axon terminals. Part of the electrically induced release of tyramine may thus be derived from the latter cell-type.

The [^{3}H]NE transported into reserpinized, MAO-inhibited slices remains largely in a soluble form, not associated with synaptic vesicles, whose active amine transport mechanism is irreversibly inactivated by reserpine (41). The much higher resting efflux of [^{3}H]NE in these preparations, as compared to the one in untreated slices, is entirely compatible with a high concentration of soluble NE and with an impaired retention mechanism. The small electrically-induced, calcium-dependent release of NE seen in the present experiments, is also an indication of the low proportion of NE present in the vesicles, into which it is known to enter also by an ATP-independent, diffusional mechanism (36,37). On the other hand, the secretory mechanism as such, as judged by dopamine-β-hydroxylase release, is not affected by reserpine (41).

Finally, the lack of increase of the induced efflux peaks with voltages greater than 1.5 to 2 V, in the case of tyramine and of NE in reserpinized, MAO-inhibited slices, but not with NE, 5-HT, ACh, or DA in untreated slices (24,25,33, and our unpublished results), is interpreted as due to the less effective vesicular retention of the former amines, so that, as time proceeds, more of the amines are lost from the vesicles, and the fraction released does not continue to increase, in spite of the higher applied potentials. Entirely similar results have been reported regarding tyramine release from isolated perfused spleens (13).

In conclusion, [^{3}H]PEA was found to behave, under the present experimental conditions, as a nontransmitter substance, and a modulator role is tentatively assigned to it. Tyramine and octopamine, on the other hand, behave as transmitter substances. This implies that they may normally have a co-transmitter role in different central aminergic synapses.

ACKNOWLEDGMENTS

We are grateful to M. Ortiz for his fine technical assistance, to Martha Cuenca for secretarial help, to Dr. A. Gómez and to CIBA-Geigy for gifts of nialamide and reserpine, respectively, and, especially, to Dr. A. D. Mosnaim for his generous gift of [^{3}H]PEA, [^{3}H]tyramine, deprenyl, and pargyline, and for introducing us to the field of PEA.

REFERENCES

1. R. J. Baldessarini (1971). Release of aromatic amines from brain tissues of the rat in vitro. *J. Neurochem.*, 18:2509-2518.
2. R. J. Baldessarini and M. Vogt (1971). The uptake and subcellular distribution of aromatic amines in the brain of the rat. *J. Neurochem.*, 18:2519-2533.
3. R. J. Baldessarini and M. Vogt (1972). Regional release of aromatic amines from tissues of the rat brain in vitro. *J. Neurochem.*, 19:755-761.
4. A. A. Boulton and G. B. Baker (1974). The subcellular distribution of some monoamines following their intraventricular injection in the rat. *Can. J. Biochem.*, 52:288-293.
5. A. A. Boulton and G. B. Baker (1975). The subcellular distribution of β-phenylethylamine, p-tyramine and tryptamine in rat brain. *J. Neurochem.*, 25:477-481.
6. A. A. Boulton and P. H. Wu (1972). Biosynthesis of cerebral phenolic amines. I. In vivo formation of p-tyramine, octopamine and synephrine. *Can. J. Biochem.*, 50:261-267.
7. G. P. Burn and J. H. Burn (1961). Uptake of labelled noradrenaline by isolated atria. *Br. J. Pharmacol.*, 16:344-351.
8. A. Carlsson and B. Waldeck (1963). β-Hydroxylation of tyramine in vivo. *Acta Pharmacol. Toxicol. (Kbh.)*, 20:371-374.
9. C. R. Creveling and J. W. Daly (1971). Assay of enzymes of catecholamine biosynthesis and metabolism. In: *Methods of Biochemical Analysis, Suppl. Vol.: Analysis of Biogenic Amines and their Related Enzymes*, edited by D. Glick. Wiley-Interscience, New York, pp. 153-182.
10. D. A. Durden, S. R. Philips, and A. A. Boulton (1973). Identification and distribution of β-phenylethylamine in the rat. *Can. J. Biochem.*, 51:995-1002.
11. D. J. Edwards and K. Blau (1973). Phenethylamines in brain and liver of rats with experimentally induced phenylketonurialike characteristics. *Biochem. J.*, 132:95-100.

12. U. S. V. Euler and F. Lishajko (1960). Release of noradrenaline from adrenergic transmitter granules by tyramine. *Experientia*, 16:376-377.

13. J. E. Fischer, W. D. Horst, and I. J. Kopin (1965). β-Hydroxylated sympathomimetic amines as false neurotransmitters. *Br. J. Pharmacol.*, 24:477-484.

14. S. Friedman and S. Kaufamn (1965). 3,4-Dihidroxyphenylethylamine β-hydroxylase. Physical properties, copper content, and role of copper in the catalytic activity. *J. Biol. Chem.*, 240:4763-4773.

15. E. E. Inwang, A. D. Mosnaim, and H. C. Sabelli (1973). Isolation and characterization of phenylethylamine and phenylethanolamine from human brain. *J. Neurochem.*, 20:1469-1473.

16. R. Lindmar and E. Muscholl (1961). Die wirkung von cocain, guanethidin, reserpin, hexamethonium, tetracain und psicain auf die noradrenaline-freisetzung aus dem herzen. *Naunyn Schmiedeberg Arch. Exp. Pathol.*, 242:214-227.

17. A. D. Mosnaim and E. E. Inwang (1973). A spectrophotometric method for the quantification of 2-phenylethylamine in biological specimens. *Anal. Biochem.*, 54:561-577.

18. J. M. Musacchio, J. E. Fischer, and I. J. Kopin (1965). Effect of chronic sympathetic denervation on subcellular distribution of some sympathomimetic amines. *Biochem. Pharmacol.*, 14:898-900.

19. J. M. Musacchio and M. Goldstein (1963). Biosynthesis of norepinephrine and norsynephrine in the perfused rabbit heart. *Biochem. Pharmacol.*, 12:1061-1063.

20. J. M. Musacchio, I. J. Kopin, and V. K. Weise (1965). Subcellular distribution of some sympathomimetic amines and their β-hydroxylated derivatives in the rat heart. *J. Pharmacol. Exp. Ther.*, 148:22-28.

21. T. Nakajima, Y. Kakimoto, and I. Sano (1964). Formation of β-phenylethylamine in mammalian tissue and its effect on motor activity in the mouse. *J. Pharmacol. Exp. Ther.*, 143:319-325.

22. N. H. Neff, H.-Y. T. Yang, and J. A. Fuentes (1974). The use of selective monoamine oxidase inhibitor drugs to modify amine metabolism in brain. In: *Neuropsychopharmacology of Monoamines and their Regulatory Enzymes*, edited by E. Usdin. Raven Press, New York, pp. 49-57.

23. W. H. Oldendorf (1971). Brain uptake of radiolabeled amino acids, amines, and hexoses after arterial injection. *Am. J. Physiol.*, 221:1629-1639.

24. F. Orrego (1974). Identification of central transmitters with electrically stimulated brain slices. In: *Lipmann Symposium: Energy, Biosynthesis and Regulation in Molecular Biology*, edited by D. Richter. De Gruyter, Berlin, pp. 471-486.

25. F. Orrego, J. Jankelevich, L. Ceruti, and E. Ferrera (1974). Differential effects of electrical Stimulation on release of ^{3}H-noradrenaline and ^{14}C-α-aminoisobutyrate from brain slices. *Nature,* 251:55-57.

26. F. Orrego and R. Miranda (1976). Electrically induced release of (^{3}H)-GABA from neocortical thin slices. Effects of stimulus waveform and of amino-oxyacetic acid. *J. Neurochem.,* 26:1033-1038.

27. S. R. Philips, D. A. Durden, and A. A. Boulton (1974). Identification and distribution of p-tyramine in the rat. *Can. J. Biochem.,* 52:366-373.

28. J. J. Pisano, C. R. Creveling, and S. Udenfriend (1960). Enzymic conversion of p-tyramine to p-hydroxyphenylethanolamine (norsynephrine). *Biochim. Biophys. Acta,* 43:566-568.

29. J. M. Saavedra (1974). Enzymatic isotopic assay for and presence of β-phenylethylamine in brain. *J. Neurochem.,* 22:211-216.

30. H. C. Sabelli, A. D. Mosnaim, and A. J. Vazquez (1974). Phenylethylamine: Possible role in depression and antidepressive drug action. In: *Neurohumoral Coding of Brain Function,* edited by R. D. Myers and R. R. Drucker-Colin. Plenum, New York, pp. 331-357.

31. H. C. Sabelli and A. D. Mosnaim (1974). Phenylethylamine hypothesis of affective behavior. *Am. J. Psych.,* 131:695-699.

32. H. C. Sabelli, A. J. Vazquez, and D. Flavin (1975). Behavioral and electrophysiological effects of phenylethanolamine and 2-phenylethylamine. *Psychopharmacologia,* 42:117-125.

33. C. Saldate and F. Orrego (1975). Electrically induced release of (^{3}H)-5-hydroxytryptamine from neocortical slices in vitro: Influence of calcium but not of lithium ions. *Brain Res.,* 99:184-188.

34. H. J. Schümann and E. Weigmann (1960). Über den angriffspunkt der indirekten wirkung sympathicomimetischer amine. *Naunyn Schmiedeberg Arch. Exp. Path.,* 240:275-284.

35. R. P. Silkaitis and A. D. Mosnaim (1976). Pathways linking L-phenylalanine and 2-phenylethylamine with p-tyramine in rabbit brain. *Brain Res.,* 114:105-115.

36. T. A. Slotkin, R. M. Ferris, and N. Kirshner (1971). Compartmental analysis of amine storage in bovine adrenal medullary granules. *Mol. Pharmacol.,* 7:308-318.

37. T. A. Slotkin and N. Kirshner (1971). Uptake, storage, and distribution of amines in bovine adrenal medullary vesicles. *Mol. Pharmacol.,* 7:581-592.

38. S. R. Snodgrass (1974). Phenylalanine, brain phenethylamine and motor activity in the rat. *J. Pharm. Pharmacol.,* 26:931-936.

39. M. I. Steinberg and C. B. Smith (1970). Effects of desmethylimipramine and cocaine on the uptake, retention and metabolism of ^{3}H-tyramine in rat brain slices. *J. Pharmacol. Exp. Ther.*, 173:176-192.

40. M. I. Steinberg and C. B. Smith (1971). Uptake, retention and metabolism of ^{3}H-tyramine in rat atria. *J. Pharmacol. Exp. Ther.*, 176:139-148.

41. N. B. Thoa, G. F. Wooten, J. Axelrod, and I. J. Kopin (1975). On the mechanism of release of norepinephrine from sympathetic nerves induced by depolarizing agents and sympathomimetic drugs. *Mol. Pharmacol.*, 11:10-18.

42. S. Udenfriend and J. B. Wyngaarden (1956). Precursors of adrenal epinephrine and norepinephrine in vivo. *Biochim. Biophys. Acta*, 20:48-52.

43. O. H. Viveros, L. Arqueros, and N. Kirshner (1968). Release of catecholamines and dopamine-β-oxidase from the adrenal medulla. *Life Sci.*, 7:609-618.

44. P. H. Wu and A. A. Boulton (1974). Distribution, metabolism and disappearance of intraventricularly injected p-tyramine in the rat. *Can. J. Biochem.*, 52:374-381.

Chapter 9

EFFECTS OF PHENYLETHYLAMINES UPON SYNTHESIS OF BRAIN MONOAMINES

S. Robert Snodgrass and Norman J. Uretsky

Department of Neurology and Neuropathology
Harvard Medical School
and
Children's Hospital Medical Center
Boston, Massachusetts

I. INTRODUCTION

Many phenylethylamines can alter the synthesis of monoamine neurotransmitters (dopamine, DA; norepinephrine, NE; 5-hydroxytryptamine, 5-HT) in the mammalian nervous system. Amphetamine, for example, can release endogenous monoamine transmitters from nerve terminals (1-10). On repeated administration of amphetamine, monoamine levels are often found to be either normal or increased (11,12). This suggests that either the rate of synthesis increases or metabolism decreases. We have used PEA and related stimulants as tools to study the mechanisms by which tyrosine hydroxylase (TOH) activity is regulated. This enzyme is the rate-limiting enzyme in catecholamine biosynthesis in peripheral and central adrenergic neurons (13-16). Because of the importance of TOH in catecholamine biosynthesis, TOH regulation has been widely studied and reviewed (17-21). An important mechanism for the regulation of TOH activity appears to be feedback inhibition by catechols. When the transmitter is released from adrenergic neurons, this should produce a decrease in the amine concentration within some "critical pool" of catecholamines, thereby releasing TOH from feedback inhibition. Our studies support the view that the current theory requires modification in the future (21-23).

In addition to helping dissect out the various mechanisms for regulation of TOH, our studies of monoamine biosynthesis may be useful in categorizing stimulant drugs. These drugs have been grouped on the basis of their ability, 1) to stimulate adenyl cyclase in brain homogenates (24-25), 2) to produce behavioral changes despite inhibition of synthesis or storage of monoamines (26-28), 3) to displace or release stored amines (29-33), or to produce tachyphylaxis (34-36). These classifications may serve as theoretical bases to explain the many behavioral effects produced by this group of drugs. We have studied the effect of various amines on DA as well as 5-HT synthesis and have attempted to correlate in vivo and in vitro results.

II. ALPHA METHYL AMINO ACIDS AND THE FALSE TRANSMITTER HYPOTHESIS

Because of the historical importance of α-methyl DOPA (AMDOPA), first described as a decarboxylase inhibitor by Sourkes in 1954 (37), and first used clinically for blood pressure control in 1960 (38), we shall begin with a consideration of this compound. After the systemic administration of AMDOPA, α-methyl dopamine (AMDA) and α-methyl norepinephrine (AMNE) are found in central and peripheral nervous tissue (39-41). α-Methyl norepinephrine, the β-hydroxylated metabolite of AMDOPA, has been shown to satisfy many of the criteria for a neurotransmitter in postganglionic sympathetic neurons (30-31), and it has been accepted as a false transmitter. It can be stored in sites which normally contain NE and can be released from those sites by reserpine or by a depolarizing stimulus (42); AMNE can also stimulate alpha and beta adrenergic receptors, although its relative potency appears to be less than that of NE (43-46).

In the CNS, concentrations of AMDA, the DA analog formed from AMDOPA, are much higher in striatum than in other brain regions (47). After the systemic administration of AMDOPA, the time course of the accumulation and disappearance of AMDA in the striatum is similar to that for the fall and recovery of DA (47). Furthermore, the amount of AMDA accumulating in brain is similar to the amount of DA which is lost (39). Reserpine, a drug which inactivates storage sites, markedly reduces the retention of AMDA in the brain (48). These observations all suggest that the decline in DA which follows administration of AMDOPA is due to displacement of DA from its storage sites by AMDA. Displaced DA would then be metabolized by intraneuronal monoamine oxidase (MAO) (39). Changes in animal behavior observed after injections of AMDOPA may be related to decreased stimulation of post-synaptic DA receptors (39,49-51) since AMDA postulated to be released in place of DA appears to be a less potent agonist (24).

This simple explanation fails to consider known compensatory homeostatic mechanisms related to the nigrostriatal DA pathway. Neuroleptic drugs are believed to be antipsychotic because of their ability to block postsynaptic DA receptors in the forebrain (52,53). Nigrostriatal DA neurons increase their firing rate in response to blockade of synaptic transmission by neuroleptics (54). There is an increased DA turnover in the striatum (23,55), which would tend to reduce the receptor-blocking effect of the neuroleptic (56). Similarly, apomorphine, which acts directly on postsynaptic DA receptors (57), decreases the firing rate of nigrostriatal DA neurons and the rate of DA turnover (54,58). The effects observed with apomorphine and neuroleptics suggest that the decrease in DA receptor stimulation believed to occur after AMDOPA administration would evoke increases in the firing rate of nigrostriatal DA neurons and in DA turnover.

Such changes would tend to overcome the functional deficit produced by the false transmitter, AMDA. This suggests that the false transmitter hypothesis alone cannot account for the behavioral changes (sedation and decreased motor activity) observed after AMDOPA administration because increasing DA synthesis should be able to overcome the impairment of function resulting from decreased receptor stimulation and furthermore would displace the stored AMDA. The fact that marked behavioral changes are seen and that DA depletion is severe suggests that homeostatic mechanisms of the nigrostriatal DA system must be impaired after AMDOPA administration.

Our previous studies suggested that AMDOPA inhibits the synthesis of DA in the striatum (47). After administration of AMDOPA at a dose of 200 mg/kg, i.p., the concentration of DA and its metabolites, dihydroxyphenylacetic acid (DOPAC) and homovanillic acid (HVA) declined in parallel between 1 and 4 hr and slowly recovered. The parallel decline of both transmitter and metabolites is not consistent with release of transmitter from its storage sites and suggests synthesis inhibition. Furthermore, AMDOPA also decreased DA accumulation after a monoamine oxidase inhibitor (MAOI), pargyline. Finally, AMDOPA reduced the accumulation of [^{3}H]DA and its metabolites in the striatum 10 min after the injection of [^{3}H]tyrosine. These data agree with those of Dominic and Moore (41), who found that AMDOPA decreased the formation of radioactive DA from tyrosine in whole brain.

The decrease in synthesis could be the result of decreased precursor availability. Alternatively, AMDOPA and AMDA are both catechols and should decrease TOH activity by competing with the pteridine co-factor for binding sites on the enzyme. Finally, AMDOPA was introduced as a decarboxylase inhibitor but this mechanism cannot explain the effect of AMDOPA administration because more potent decarboxylase inhibitors produce neither the marked decrease in DA nor the behavioral changes (59,61).

We used small slices of rat striatum to further analyze the several possible mechanisms for the AMDOPA effect. Slices 0.25 mm in length were prepared with a McIlwain tissue chopper (62) and were then incubated with L-3, [5-^{3}H]tyrosine (ring labeled). As first established by Nagatsu et al. (14), formation of tritiated water is a good index of tyrosine hydroxylation. The measurement of [^{3}H]water by vacuum distillation permits quantitative study of tyrosine hydroxylation under conditions where the naturally occurring pteridine co-factors in the tissues are used and where tissue membranes are relatively intact. We also determined the accumulation of newly synthesized [^{3}H]DA in tissue and medium, as a check and to study release of transmitter. Although under some conditions the newly synthesized DA is metabolized, estimates of tyrosine hydroxylation by the two methods were in good agreement as reported by others (63). This agreement suggests that the release of stored DA by drugs might not be associated with metabolism of the transmitter by catabolic enzymes such as MAO.

Treatment of striatal slices with AMDOPA resulted in a decrease of both [^{3}H]water and [^{3}H]DA accumulation. These decreases occurred at tissue concentrations of AMDOPA similar to those found in vivo after the systemic administration of AMDOPA, 200 mg/kg, i.p. (47). A 100-μM concentration of AMDOPA decreased both indices of tyrosine hydroxylation by 90%, without altering either tyrosine uptake or tyrosine concentration in the slices. The similar decrease in both [^{3}H]water and [^{3}H]DA formation could not be due to decarboxylase inhibition and suggests inhibition of TOH. This effect was on the enzyme rather than on its precursor because there was no change in tyrosine specific activity. In additional experiments, the potent decarboxylase inhibitor, brocresine (NSD-1055), was added to the incubation medium and failed to diminish the ability of AMDOPA to inhibit TOH. Although AMDA can inhibit the enzyme, AMDOPA alone is sufficient to produce the effect when decarboxylation is prevented.

α-Methyl dopamine is a potent inhibitor of TOH, and can inhibit TOH activity at concentrations less than those found in vivo after the usual 200 mg/kg dose of AMDOPA (47). One reason for this high potency is that AMDA is actively taken up into DA nerve terminals by the DA membrane pump (64). We found that AMDA, when added to the incubation medium at a concentration of 0.3 μM, was concentrated to a tissue/medium ratio of 15.6.

TABLE 1

Effect of AMDA on the Release of $[^3H]$DA and Inhibition of $[^3H]H_2O$ in Slices of Rat Striatum[a]

AMDA Conc. in Medium (μM)	$[^3H]$DA in Medium (percent of total $[^3H]$amine)	$[^3H]H_2O$ Formation (percent inhibition)
0.0	3.43 ± 0.490	
0.3	5.88 ± 0.718	40.67 ± 5.388 (3)
0.6	9.15 ± 1.631	52.71 ± 4.210 (3)
1.2	19.05 ± 4.086	78.07 ± 3.406 (3)
10.0	66.98 ± 3.328	
100.0	85.67 ± 1.243	

[a]Striatal slices were treated as described for the release studies except that the initial incubation was in media that contained no radioactivity. The slices were washed three times, incubated for 20 min at 37°C, washed an additional time, and then added to 5 ml of medium containing $[^3H]$tyrosine (1.35 μCi/ml) and AMDA. The reaction was stopped in 30 min and $[^3H]H_2O$ accumulation was determined. Results are expressed as mean ± SEM, and the amount of $[^3H]$water formed in the control slices was 3.84 ± 0.331 (3). For comparison, data are presented from the release studies and calculated as $[^3H]$DA in medium as a percentage of total $[^3H]$amine. Data taken from Uretsky et al. (67).

Because the false transmitter hypothesis suggests that the major action of AMDOPA is to displace DA from striatal storage sites, we also studied the effects of both AMDOPA and AMDA on the release of radioactive DA from striatal slices in vitro (Table 1).

We used the method of Ziance and Rutledge (65), preloading slices with exogenous $[^3H]$DA, followed by washing and resuspension in fresh medium. α-Methyl dopamine was found to be a potent releasing agent, with an ED_{50} of approximately 5 μM, while AMDOPA had no releasing effect. The potency of AMDA for inhibiting TOH was greater than its potency for releasing exogenous $[^3H]$DA. These results suggest that inhibition of TOH is more important than release of stored transmitter in producing the decrease of striatal DA, which occurs after AMDOPA administration.

The ability of AMDOPA and AMDA to inhibit TOH is probably related to the catechol structure which they both possess, catechols being able to inhibit the binding of pteridine cofactor to the enzyme (66). However, this formulation cannot account for the ability of another α-methyl amino acid, α-methyl meta tyrosine (AMMT), to inhibit TOH. The decrease in striatal DA produced by α-methyl meta tyrosine (33) appears to be mediated through the effects of its decarboxylated metabolite, α-methyl metatyramine (AMMTA) (67). Neither AMMT nor AMMTA contain a catechol group, so we felt it important to test the possibility that they might act by TOH inhibition.

Our in vivo studies with these compounds yielded equivocal results. After the injection of AMMT, 100 mg/kg, i.p., striatal HVA rose by 41% at 1 hr and then fell progressively. To a smaller extent, DOPAC rose and also fell at later time periods. The rise in HVA and DOPAC at 1 hr is consistent with release of DA from storage sites, with a later fall in both acids being due to decreased synthesis. However, parallel studies indicated that AMMT injections decreased the conversion of [^{3}H]DA to [^{3}H]DOPAC, suggesting MAO I.

Studies in striatal slices showed that both AMMT and AMMTA decreased TOH (both in terms of [^{3}H]water formation and accumulation of [^{3}H]catechols) without altering specific activity of tissue tyrosine. The inhibition of TOH by AMMT could be reversed by increasing the concentration of tyrosine in the incubation medium, suggesting that the action of AMMT is to compete with tyrosine for sites on the enzyme. In homogenates of lysed striatal tissue, AMMTA could not inhibit TOH, suggesting that its effect on the enzyme was indirect and perhaps mediated by changes in a cytoplasmic constituent within the slice. α-Methyl metatyramine releases DA from striatal slices (67), and we assumed that the inhibitory effects of AMMTA upon synthesis might be mediated by release of endogenous DA. In contrast to AMDOPA, AMDA, and α-methyl paratyrosine (AMPT), the TOH inhibition resulting from AMMTA was incomplete and never exceeded 60% inhibition of control values. This partial inhibition could not be explained by a plateau in accumulation of AMMTA within slices, nor by a plateau in release of [^{3}H]DA by AMMTA. Why then is the inhibition of TOH by AMMTA never total? A portion of the TOH may be inaccessible to feedback inhibition by the DA which is released by AMMTA. It is possible that AMMTA might directly or indirectly alter the conformation of TOH enzyme molecules such that they were less susceptible to feedback inhibition. α-Methyl metatyramine was found to be more potent in inhibiting [^{3}H]water formation than in releasing either newly synthesized or stored DA from striatal slices. This observation suggests that a small amount of the DA released from storage sites is able to inhibit a relatively large fraction of TOH actively.

III. EFFECTS OF AMPHETAMINE ON TOH ACTIVITY IN STRIATAL SLICES

Amphetamine differs from both AMDA and AMMTA in that it contains no phenolic hydroxyl groups. As a consequence of this difference, amphetamine is more lipid soluble than the other two compounds and enters brain readily after systemic injection. In addition to its lipid solubility, there is evidence for a saturable transport system which transports amphetamine from blood to brain. It is noteworthy that this carrier system also appears to transport methylphenidate (MPD) and PEA into brain (68).

Amphetamine has been classified as an indirectly acting amine (57,69,70). In peripheral nerve, amphetamine effects are caused by release of NE from postganglionic sympathetic nerves (1). In the CNS, amphetamine has been shown to produce release of endogenous DA and NE (10,63,71-74). Many of the behavioral effects which result from amphetamine administration are believed to result from DA release in the forebrain (75-78). In spite of the strong evidence that amphetamine is a potent releaser of endogenous DA, amphetamine administration is often associated with normal or increased striatal DA concentrations (12,79-82). This observation suggests that DA synthesis is indeed increased, as has been reported by Costa et al. (83).

Most studies of DA synthesis in vitro, after systemic amphetamine administration, whether employing slices or synaptosomes, have reported that amphetamine decreases TOH activity (21,63,80,82,84). The mechanisms mediating stimulatory effects of amphetamine on TOH activity observed in vivo may be fragile and not preserved in these in vitro preparations. The decrease in TOH observed in vitro after systemic administration of amphetamine has been believed to result from inhibition of firing in the nigrostriatal DA pathway (58,63,82). This inhibition of neuronal firing rate is thought to be a compensatory response to the stimulation of DA receptors in the striatum (54), but recent evidence suggests that an additional factor may be release of endogenous DA from dendrites in the substantia nigra with activation of DA receptors located on neuronal perikarya (58,85,86).

The addition of amphetamine in vitro to the incubation medium of striatal synaptosomes results in an increase of striatal TOH activity (82,87). This is the opposite effect from that obtained with reserpine, which decreases TOH activity in vitro in synaptosomal preparations (87). It is believed that TOH activity is normally inhibited by a small pool of cytoplasmic catecholamine (17). Reserpine would increase the size of such a pool by release of DA into the cytoplasm after destruction of storage sites. Amphetamine can release striatal DA (88) and produce behavioral changes (26,27) even in reserpinized rats, so it appears that amphetamine can release DA from sites other than the normal storage sites. Release of cytoplasmic or nonstored DA should decrease the size of the inhibitory DA pool and thereby increase TOH activity.

Amphetamine has been shown to release newly synthesized DA from the striatum both in vivo and in vitro (88,89). The enzymes responsible for DA synthesis, TOH and DOPA decarboxylase, are cytoplasmic enzymes (57), so it is likely that newly synthesized DA is formed in the cytoplasm. New synthesis could be the main source of that inhibitory pool of DA which is postulated to exist in the cytoplasm (17,20). Administration of amphetamine in vitro is known to release newly synthesized DA, and this release could account for the stimulation seen of TOH activity.

However, if the release of newly synthesized DA is the basis for the amphetamine-induced increase in TOH activity, then blockade of DA synthesis should prevent the amphetamine-induced increase in TOH activity. We administered brocresine, a well-known decarboxylase inhibitor (90), and studied the effects of amphetamine on the synthesis of labeled catechols in striatal slices. Our results were generally similar to those obtained by others who have used synaptosomes (82,87). Amphetamine had a biphasic effect upon the synthesis of [^{3}H]catechols: Low concentrations of amphetamine increased TOH activity, with maximal stimulation resulting from an amphetamine concentration of approximately 1 μM. Higher concentrations of amphetamine had less stimulant effect, and 1 mM amphetamine produced significant inhibition of TOH activity.

The ability of amphetamine to stimulate TOH in the slice preparation depended upon the concentration of tyrosine in the medium, in addition to its relationship to drug concentration (Table 2).

Tyrosine concentrations in the medium below 1 μM prevented amphetamine stimulation, and maximal amphetamine stimulation (at the maximally effective drug concentration of 1 μM) was seen with 10 μM tyrosine concentrations in the medium. This relationship to medium tyrosine concentration could not be explained on the basis of differences in the ability of amphetamine to release newly synthesized DA: Amphetamine was shown to release [^{3}H]DA from the striatal slices at either 1 or 10 μM tyrosine concentrations. Neither could the effect be due to a decrease in medium tyrosine concentration when it was low, because the formation of [^{3}H]catechols was linear with time and amount of tissue at all concentrations of tyrosine which were studied.

In considering this effect, it has been reported that the ability of high potassium media or veratridine to increase DA synthesis does vary with the medium tyrosine concentration (91,92). Both of these factors increased the apparent K_m and V_{max} of tyrosine hydroxylation in vitro. Evidence was presented to indicate that veratridine inhibits the accumulation of [^{3}H]tyrosine into tissue from the medium (92). Such an effect upon precursor uptake would be more likely to alter TOH activity at low medium concentrations of tyrosine, where enzyme activity may be limited by substrate

TABLE 2

Effect of Different Concentrations of Tyrosine on the Enhanced Formation of $[^3H]$Catechols from $[^3H]$Tyrosine Produced by Amphetamine[a]

Initial Tyrosine Concentration (μM)	Total Catechols (nmoles/g/15 min) Normal Medium Control	Normal Medium Amphetamine	Normal Medium Percent Change	Brocresine (250 μM) Control	Brocresine (250 μM) Amphetamine	Brocresine (250 μM) Percent Change
0.25	—	—		0.612 ± 0.033 (3)	0.670 ± 0.025 (4)	+9
1.00	3.082 ± 0.190 (13)	3.702 ± 0.166 (12)	+20	2.192 ± 0.224 (6)	**3.144 ± 0.262 (5)	+43
10.00	2.070 ± 0.090 (21)	*6.160 ± 0.500 (20)	+196	2.220 ± 0.240 (11)	*7.990 ± 1.370 (10)	+259

*$P < 0.001$.

**$P < 0.05$.

[a]Striatal slices were incubated at 4°C in normal medium or medium containing amphetamine (10^{-6} M). In some experiments, brocresine (250 μM), a decarboxylase inhibitor, was also added. After the addition of $[^3H]$tyrosine, the slices were incubated at 37°C for 15 min. The amount of $[^3H]$catechols in both tissue and medium was determined. Each value is the mean ± SEM, and the number of determinations is given in the parenthesis.

TABLE 3

Effect of Amphetamine (1 μM) on the Accumulation of [^{3}H]Tyrosine into Striatal Slices

Treatment	Concentration of Tyrosine in Medium	2.5 min (nmol/g)	5 min (nmol/g)
Control	10	35.80 ± 3.87	30.49 ± 2.84
Amphetamine	10	40.05 ± 6.46	30.46 ± 2.60
Control	1	3.42 ± 0.27	4.90 ± 0.58
Amphetamine	1	3.93 ± 0.38	4.96 ± 0.28

[a]Striatal slices were incubated in normal medium or medium containing amphetamine, 1 μM for 5 min. [^{3}H]Tyrosine at a final concentration of either 1 or 10 μM, was added and the accumulation of [^{3}H]tyrosine was determined at 2.5 or 5 min. The reaction was stopped filtration. Blank values corresponding to the accumulation of [^{3}H]tyrosine into slices at 4°C were subtracted from values obtained from samples incubated at 37°C. Each value is the mean ± SEM of three to six determinations.

availability. At higher (saturating) concentrations of precursor, changes in tyrosine concentration should have much less effect upon TOH activity. We studied the effect on amphetamine upon uptake of [^{3}H]tyrosine into slices at both 1 and 10 μM tyrosine concentrations and found no inhibition of uptake (Table 3). Either the relationship of amphetamine stimulation of TOH and medium tyrosine concentration is unrelated to uptake, or an effect in uptake is not apparent because of the small percentage of striatal cells which are DA neurons (93).

Contrary to our predictions, the ability of amphetamine to stimulate TOH was not reduced by pre-incubation of tissue with brocresine. Brocresine actually increased the amphetamine stimulation of TOH. We, therefore, checked to be certain that brocresine did block DA synthesis. Brocresine at a concentration of 0.25 mM inhibited virtually all DA synthesis, and over 90% of the catechols formed were [^{3}H]DOPA. Therefore, DA synthesis may not be required for the stimulant effect of amphetamine on TOH activity.

A possible explanation for the failure of brocresine to prevent amphetamine stimulation of TOH is that amphetamine can release DOPA from the striatal slices. DOPA is a catechol, is known to inhibit the enzyme (14,22), and could be the agent of feedback inhibition in this circumstance. We checked this possibility by determining the amount of [^{3}H]DOPA in both slices and in medium with brocresine and the effect of amphetamine (1 μM) on the distribution of labeled DOPA. Amphetamine did increase the formation of [^{3}H]DOPA but did not alter the tissue/medium ratio of [^{3}H]DOPA: More than 85% of the [^{3}H]DOPA was found in the medium in both groups.

Another explanation for the failure of brocresine to prevent amphetamine stimulation of TOH is that the cytoplasmic inhibitory pool of DA is refilled from storage sites when brocresine is administered. Amphetamine would then release this DA and relieve feedback inhibition. Reserpinized rats (10 mg/kg, given 4 hr before killing) were used, and slices from these and control rats were incubated with brocresine, amphetamine, and [^{3}H]tyrosine. Although the striatal slices from reserpine-treated rats did show increased TOH activity in the presence of brocresine, amphetamine stimulation was still present when compared with slices from control rats (Table 4).

Despite the depletion of stored DA and the prevention of new DA synthesis, amphetamine stimulation of TOH was still present and was not even reduced in magnitude. This must mean that amphetamine stimulation need not depend upon release of transmitter and consequent alleviation of feedback inhibition. It also suggests a possible direct effect of amphetamine not mediated by endogenous transmitter. In this assertion, we assume that the double inhibition of DA storage and synthesis reduced the DA concentration in all transmitter pools within DA neurons (94).

Dibutyryl cyclic adenosine monophosphate (AMP) has been shown to stimulate TOH activity in striatal slices and synaptosomes (95-97). We have confirmed this result in our slice preparation, and we also find that phosphodiesterase inhibitors can increase TOH activity. Cyclic AMP has been shown to alter the kinetic properties of TOH in

TABLE 4

Effect of Pretreatment with Reserpine (10 mg/kg, i.p.) and Incubation with Brocresine on the Increased Rate of Formation of [^{3}H]Tyrosine Produced by Amphetamine (10^{-6} M)[a]

Treatment of Rats and Additions to Media	[^{3}H]Catechol (nmoles/g/15 min)
Control (10)	2.316 ± 0.148
Amphetamine (10)	8.284 ± 0.842**
Reserpine pretreatment (9)	3.254 ± 0.492*
Reserpine pretreatment + amphetamine (9)	9.240 ± 1.332***

[a]Rats were injected with reserpine 10 mg/kg, 4 hr before killing. Striatal slices were prepared as previously described and incubated for 15 min with [^{3}H]tyrosine, 10 μM and brocresine 0.25 mM in the presence and absence of amphetamine (10^{-6} M). Each value is the mean ± SEM. The number of determinations is given in the parenthesis. *0.1 < P < 0.05 when compared with the control value. **P < 0.001 when compared with the control value. ***P < 0.001 when compared with the value of slices from reserpine-pretreated rats not incubated with amphetamine.

supernatants prepared from striatum and hippocampus, lowering the K_m for the substrate tyrosine, as well as for the pteridine cofactor and increasing the apparent K_i for catecholamines (82,95). These data have been interpreted as suggesting that cyclic AMP alters the conformation of TOH enzyme molecules. This effect need not be a direct one and might be mediated by a protein kinase, as are many other effects of cyclic nucleotides (98). Protein kinase or a protein phosphorylated by it might then produce an allosteric activation of TOH.

Amphetamine stimulation of TOH activity might relate to activation of adenyl cyclase, resulting in increased cyclic AMP concentration at some critical site inside the neuron, causing the stimulation of TOH. One test of this hypothesis is to determine whether the amphetamine stimulation still occurs in slices exposed to a maximal concentration of cyclic AMP (3 mM in our system). As shown

TABLE 5

Effect of Amphetamine on the Formation of [^{3}H]Catechols in Striatal Slices in the Presence of Dibutyryl Cyclic AMP[a]

Medium	Total [^{3}H]Catechols (nmol/g/30 min)	Percent Change from Control
Experiment A		
Control	4.992 ± 0.399	
DB-CAMP	13.434 ± 0.874	169
Amphetamine	14.657 ± 0.296	194
DB-CAMP + amphetamine	21.928 ± 1.434	339
Experiment B		
Control	6.088 ± 0.290	
DB-CAMP	18.986 ± 1.218	211
Amphetamine	26.011 ± 1.161	327
DB-CAMP + amphetamine	33.631 ± 1.877	452

[a]Striatal slices were added to a medium containing dibutyryl cyclic AMP, 3 mM, and amphetamine, 1 μM, (Experiment A) or these compounds and brocresine, 250 μM (Experiment B). The samples were incubated for 30 min. Each value is the mean ± SEM of four determinations (Experiment A) or five determinations (Experiment B).

in Table 5, each treatment could increase TOH activity, but an additive effect was seen when both were present. Amphetamine stimulation of TOH activity was independent of that produced by cyclic AMP. Although isobutyl methylxanthine (IBMX), a phosphodiesterase inhibitor, could increase striatal TOH activity in slices, the effects produced by amphetamine were also additive with IBMX. Therefore, it is not likely that the stimulation of catechol formation in striatal slices by amphetamine is mediated through the production of increased cyclic AMP.

Media containing high concentrations of potassium can stimulate TOH activity in striatal slices, and this effect has been reported to be additive with the increase produced by dibutyryl cyclic AMP (97).

TABLE 6

Effect of Various Drugs on the Formation of [^{3}H]Catechols from 1 and 10 μM [^{3}H]Tyrosine[a]

Drug	Concentration	Tyrosine (10 μM) Percent Control	Tyrosine (1 μM) Percent Control
D-Amphetamine	10^{-7}	125 ± 6.0 (6)	84 ± 3.0 (3)
	10^{-6}	248 ± 25 (8)	98 ± 4.0 (6)
	10^{-5}	194 ± 19 (5)	59 ± 4.0 (2)
	10^{-4}	89 ± 7.0 (3)	64 ± 3.0 (3)
Phenethylamine	10^{-7}	109 ± 1.0 (3)	88 ± 5.0 (3)
	10^{-6}	152 ± 3.0 (3)	80 ± 2.0 (3)
	10^{-5}	217 ± 14 (3)	90 ± 1.0 (3)
Phenylethanolamine	10^{-6}	110 ± 5.0 (6)	91 ± 12 (3)
	10^{-5}	114 ± 9.0 (5)	105 ± 4.0 (3)
	10^{-4}	113 ± 3.0 (5)	100 ± 6.0 (3)
Fenfluramine	10^{-6}	114 ± 6.0 (3)	101 ± 9.0 (3)
	10^{-5}	126 ± 3.0 (3)	79 ± 4.0 (3)
	10^{-4}	61 ± 2.0 (3)	40 ± 1.0 (3)
Methylphenidate	10^{-6}	79 ± 8.0 (3)	112 ± 5.0 (3)
	10^{-5}	68 ± 5.0 (3)	115 ± 8.0 (3)
	10^{-4}	53 ± 5.0 (3)	92 ± 8.0 (3)
Potassium chloride	55×10^{-2}	+247 ± 9.0 (4)	72 ± 4.0 (4)
Cocaine	3×10^{-5}	86 ± 8.0 (5)	
Pargyline	10^{-4}	85 ± 7.0 (3)	

[a]Striatal slices (15 mg in weight) were preincubated with the drugs for 5 min at 37°C in 100% oxygen atmosphere. [^{3}H]Tyrosine was then added at a final concentration of 1 and 10 μM, and the incubation continued for an additional 30 min. [^{3}H]Catechols were estimated by adsorption and elution from alumina. Each value is the mean ± SEM, and the number of determinations are given in parenthesis. The control values for 10 μM tyrosine were 6.40 ± 0.18 (37); for 1 μM tyrosine 4.97 ± 0.19 (30); and for 0.25 μM tyrosine 1.18 ± 0.07 (6).

For this reason, we studied the effects of high potassium concentration and of amphetamine on TOH activity (Table 6). Amphetamine could not increase the activity of TOH beyond that produced by media containing increased potassium concentration. The increase produced by the presence of both treatments was no greater than that produced by either treatment alone. Stimulation of TOH by high potassium media has also been reported to depend upon medium concentrations of tyrosine (91) just as we have found with amphetamine. These studies suggest that amphetamine and media with increased potassium concentration use similar mechanisms to increase TOH activity.

We have studied the effects of some drugs related to amphetamine or believed to interact with DA nerve terminals upon TOH activity in slices. We studied each drug in the presence of two different medium concentrations of tyrosine, 1 and 10 μM (see Table 6).

Table 6 shows that PEA is most similar to amphetamine and high potassium in its effects upon TOH activity. Phenylethylamine increased the formation of [^{3}H]catechols in media containing 10 μM tyrosine, but not in 1 μM tyrosine media. Maximal stimulation of TOH activity was obtained with a drug (PEA) concentration of 10 μM. Phenylethanolamine (POLA), Fenfluramine (FFM), and methylphenidate (MPD) failed to increase TOH activity, and MPD treatment actually decreased TOH activity. This result is in contrast to the findings of Kuczenski et al. (87) who reported that MPD increased TOH activity in striatal synaptosomes. We have no explanation for this discrepancy.

Because of the very prominent relationship between medium concentrations of tyrosine and the stimulation of [^{3}H]catechol formation in striatal slices by several drugs, we investigated the possible effect of medium tyrosine concentration upon the effects produced by AMDA and AMMTA. The studies indicate that medium tyrosine concentration does not change the effect of AMDA on [^{3}H]catechol formation (Fig. 1). On the other hand, the inhibitory effect of AMMTA upon TOH was related to tyrosine concentration in the medium (Fig. 2).

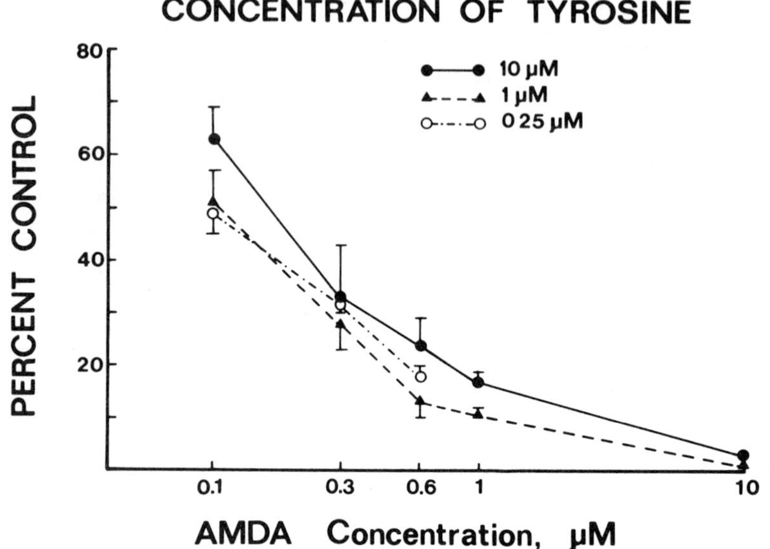

FIG. 1. Effect of AMDA on the formation of [^{3}H]catecholamines. Slices were preincubated with various concentrations of AMDA for 5 min at 37°C in 100% oxygen. [^{3}H]Tyrosine, at final concentrations of 0.25, 1, and 10 µM, was then added, and the slices were incubated for an additional 30 min. [^{3}H]Catechols were then determined. Results are expressed as a percentage of control values. Each value is the mean ± SEM of three experiments. Control values at different tyrosine concentrations are given in the legend to Table 6.

When medium tyrosine concentration was 0.25 µM, AMMTA inhibited TOH with maximal inhibition at a drug (AMMTA) concentration of 1 µM. When the concentration of tyrosine in the medium was increased slightly to 1 µM, 1 µM AMMTA decreased TOH as before, but higher drug concentrations resulted in a lessening of inhibition, with stimulation of TOH at drug concentrations of 10 and 100 µM. The inhibition of TOH by AMMTA was smaller for a medium tyrosine concentration of 10 µM, and when drug concentration was increased to 10 and 100 µM, no significant decrease in TOH activity was found. These studies show that AMMTA, which can produce behavioral stimulation (51,99-101), resembles amphetamine and PEA in that its effect upon the TOH activity of striatal slices depends upon the medium

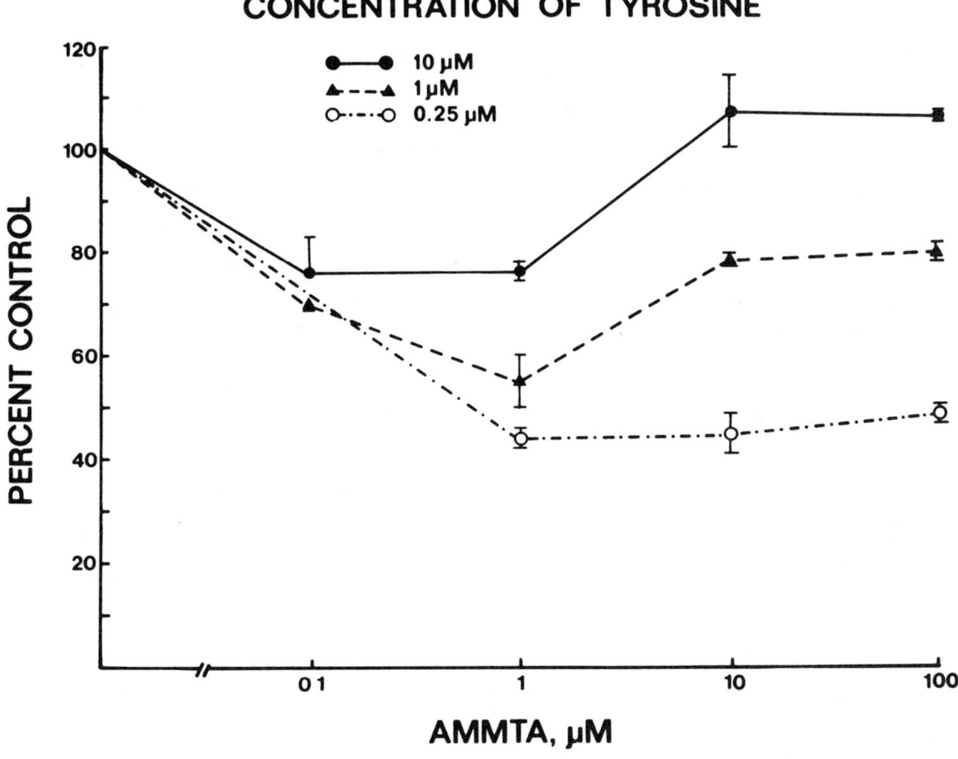

FIG. 2. Effect of AMMTA on the formation of [^{3}H]catechols from [^{3}H]tyrosine. Slices were pre-incubated with various concentrations of AMDA for 5 min at 37°C under oxygen. [^{3}H]Tyrosine, at final concentrations of 0.25, 1, and 10 μM was then added, and the slices were incubated for an additional 30 min. [^{3}H]Catechols were then determined. Results are expressed as a percentage of control values. Control values at different tyrosine concentrations are given in Table 6. Each value is the mean ± SEM of three to six experiments.

tyrosine concentration. Because it releases DA from storage sites into the cytoplasm, AMMTA may inhibit TOH, but this inhibitory effect is accentuated at low medium concentrations of tyrosine. With increasing tyrosine concentration in the medium, AMMTA may behave like amphetamine and lose its inhibitory effect upon TOH activity. The greatest degree of reversal is seen with 10 μM tyrosine concentrations.

IV. EFFECTS OF AMPHETAMINE AND PEA UPON 5-HT SYNTHESIS IN VIVO

Structural specificity of enzymes, receptors, and uptake sites is not absolute, and drugs which interact with catecholamine receptors, uptake sites, and enzymes often interact with 5-HT sites, making it difficult to determine which effect is responsible for their behavioral effects. For example, some drugs, including methiothepin (102-104), LSD (105-107), and some ergot alkaloids (108) appear to interact with both DA and 5-HT receptors, complicating the interpretation of their chemical and behavioral effects. Because of this structural "overlap" between catecholamine and 5-HT mechanisms, and because of the striking similarities between TOH and tryptophan hydroxylase (109-111), which are both pterin-dependent enzymes and share many kinetic features, as well as the ability to hydroxylate phenylalanine to some extent (111), we have also studied the effects of amphetamine, PEA, and POLA on 5-HT synthesis in vivo and also in slice preparations.

Systemic injection of amphetamine and PEA but not POLA increases brain tryptophan levels and 5-HT synthesis in some but not all brain regions. The data of Table 7 were obtained during the course of a study of regional metabolism and synthesis of 5-HT in the brain (112). Because the correlation between steady-state levels of 5-HT and 5-HIAA and rates of 5-HT synthesis is often imperfect, we also studied rates of 5-HT synthesis more directly by measuring the accumulation of 5-HT 30 min after the systemic injection of the MAOI pargyline (113,114) and in a smaller number of rats by the accumulation of 5-hydroxytryptophan (5-HTP) after decarboxylase inhibition with the drug RO 4-4602 (115-117). The latter method was also used because the use of MAOI to determine 5-HT turnover, suffers from the objection that compensatory mechanisms may be activated by increasing concentrations of 5-HT (118). Furthermore, pargyline often increases brain tryptophan (113), and brain tryptophan levels are a major determinant of 5-HT synthesis rates (119-121). Both techniques yielded similar results in our study, but our unpublished data indicates that accumulation of 5-HT after pargyline is not linear for more than 30 min in many brain regions.

As previously reported, amphetamine did increase tryptophan concentrations in some but not all brain regions and in some regions where no increase in 5-HT synthesis was detected (122). A similar regional increase was seen in PEA-treated rats, as shown by Table 7. Also of interest is the fact that propranolol, a β-adrenergic blocking agent, blocks the increase in brain tryptophan and the increased synthesis produced by acute administration of PEA or amphetamine (Table 7). It is possible that propranolol blocked the drug increase in 5-HT synthesis by a peripheral effect, preventing the increase in free plasma tryptophan, which is the most likely basis for the increased tryptophan levels seen after administration of PEA (132).

Haloperidol, in contrast to propanolol, blocked the increase in 5-HT synthesis, without changing the elevation of tissue tryptophan concentration produced by PEA. These results suggest that brain tryptophan concentration is only one of the factors regulating 5-HT synthesis (130). The heterogenity of brain tissue and the very small percentage of nerve endings which are 5-HT terminals all limit the significance of tissue tryptophan levels. Peripheral factors acting to alter plasma-free tryptophan (131), the ratio of plasma tryptophan to other neutral amino acids (120,121), or the entry of tryptophan into brain (122,132) are clearly important in regulating brain 5-HT synthesis. Several catecholamine-related drugs, such as apomorphine, appear to be able to change brain 5-HT synthesis (133-135). In vitro studies, where plasma tryptophan is not a factor, may permit clarification of central regulatory mechanisms and may help to establish whether 5-HT neurons have receptors for other transmitters [such as catecholamines (136,137)] whose activation directly alters neuronal 5-HT synthesis.

Chronic amphetamine treatment has been reported to increase whole brain concentrations of 5-HT as well as increasing the rate of synthesis of labeled 5-HT from intracisternally injected [^{3}H]tryptophan (138). Thus, the chronic administration of amphetamine for

TABLE 7

Regional Brain Concentrations of Tryptophan, 5-Hydroxyindoles, and 5-HT Synthesis Rate after Administration of Stimulant Drugs[a]

	Hypo	Str	PMB	NAS	Hcps	CCtx	Sp cord[†]
A. Controls (10)							
Trp (μg/g)	8.28 ± 0.73	5.52 ± 0.59	3.11 ± 0.25	5.47 ± 0.32	3.31 ± 0.19	2.86 ± 0.25	5.53 ± 0.66
5-HT (ng/g)	1399 ± 124	992 ± 73	766 ± 51	1395 ± 127	646 ± 45	449 ± 35	520 ± 33
5-HIAA (ng/g)	1121 ± 72	795 ± 68	888 ± 57	995 ± 98	675 ± 64	434 ± 44	334 ± 29
5-HT accumulation (ng/g/hr)[†]	479 ± 105	322 ± 73	348 ± 80	625 ± 191	290 ± 31	227 ± 43	143 ± 36
5-HTP accumulation (ng/g/hr)[†]	407 ± 68	359 ± 50	325 ± 37	558 ± 65	274 ± 50	208 ± 30	
Turnover time (hr)	2.92	2.86	2.20	2.22	2.22	1.98	3.64
B. Amphetamine 5 mg/kg/i.p., 1 hr before killing (5)							
Trp (μg/g)	8.46 ± 0.11	5.98 ± 0.50	4.56 ± 0.44*	5.66 ± 0.27	3.93 ± 0.46	4.43 ± 0.55*	5.73 ± 0.62
5-HT (ng/g)	1514 ± 0.16	866 ± 0.35	991 ± 80	1325 ± 115	516 ± 57*	340 ± 85*	473 ± 74
5-HIAA (ng/g)	1189 ± 113	918 ± 96	884 ± 101	1238 ± 801	801 ± 82*	516 ± 56	342 ± 62
5-HT accumulation (ng/g/hr)[†]	451 ± 108	410 ± 76	456 ± 90*	658 ± 159	377 ± 53*	348 ± 70*	170 ± 59
Turnover time (hr)	3.36	2.19	2.17	2.01	1.37	0.98	2.79
C. PEA 75 mg/kg/i.p., 1 hr before killing (4)							
Trp (μg/g)	9.01 ± 0.56	6.48 ± 0.70	3.86 ± 0.60	5.43 ± 0.76	4.60 ± 0.74	4.39 ± 0.23*	
5-HT (ng/g)	1305 ± 129	866 ± 67	832 ± 129	1506 ± 131	609 ± 116	427 ± 57	
5-HIAA (ng/g)	1186 ± 137	932 ± 92	1070 ± 77*	1250 ± 117	754 ± 85	544 ± 60	
5-HTP accumulation (ng/g/hr)[†]	563 ± 118	450 ± 41*	509 ± 93*	547 ± 104	420 ± 50*	291 ± 62*	
D. PEA 75 mg/kg and DL-propranolol 5 mg/kg (4)							
Trp (μg/g)	7.66 ± 0.91	5.51 ± 1.09	2.85 ± 0.73	5.16 ± 0.68	2.71 ± 0.44	3.19 ± 0.58	
5-HT (ng/g)	1200 ± 139	805 ± 117	538 ± 142	1441 ± 139	590 ± 73	369 ± 80	
5-HIAA (ng/g)	1058 ± 151	761 ± 100	733 ± 112	1140 ± 91	531 ± 68	502 ± 105	
5-HTP accumulation (ng/g/hr)[†]	511 ± 80	313 ± 91	354 ± 70	544 ± 60	259 ± 51	267 ± 45	

E. PEA 75 mg/kg and haloperidol 1 mg/kg, 1 hr (4)						
Trp (μg/g)	8.97 ± 0.99	6.61 ± 0.71	2.96 ± 0.66	5.98 ± 0.85	3.81 ± 0.94	3.90 ± 0.48
5-HT (ng/g)	1419 ± 161	991 ± 104	810 ± 71	1566 ± 188	491 ± 68	390 ± 52
5-HIAA (ng/g)	966 ± 115	722 ± 89	756 ± 41	987 ± 154	669 ± 74	356 ± 30
5-HTP accumulation (ng/g/hr)[†]	456 ± 70	295 ± 58	309 ± 61	519 ± 121	288 ± 44	196 ± 68
F. PEA 75 mg/kg and haloperidol 7.5 mg/kg, 1 hr (3)						
Trp (μg/g)	9.22 ± 1.21	5.30 ± 0.78	2.90 ± 0.58	4.87 ± 0.95	3.42 ± 0.81	3.31 ± 0.41
5-HT (ng/g)	1535 ± 181	940 ± 89	642 ± 125	1317 ± 165	511 ± 78	431 ± 89
5-HIAA (ng/g)	1201 ± 148	740 ± 108	837 ± 75	1159 ± 206	478 ± 65	229 ± 100

[a]All rats were kept in a 32°C chamber between the time of drug injection (saline injection for controls) and killing. Rats were decapitated, and their brains dissected on a chilled metal tray, following the procedure of Glowinski and Iversen (123) assisted in the case of the nucleus accumbens septi by reference to Konig and Klippel (124). Different animals were used for the spinal cord data than for the brain regions. After dissection, tissue was homogenized in 0.4 N perchloric acid (PCA) containing 2.5 mg/ml sodium metabisulfite and 1 mg/ml dithiothreitol. The homogenate was fractionated by a modification of the method of Atack and Lindquist (125), which involved initial extraction with butanol : petroleum ether (2 : 1) to obtain a fraction containing 5-HIAA, and passage of the aqueous phase through a 40 x 6 mm column of the resin, Amberlite CG-50, 100 to 200 mesh, Na form, buffered to pH 6.3, after adjustment of pH with KOH. The material not retained, together with 3 ml of 50 ethanol washings, was adjusted to pH 8.2 with Tris buffer containing sodium metabisulfite 5 mg/ml and thiourea 2 mg/ml as antioxidants, and passed through a column of Dowex 1 x 4, 200 to 400 mesh, formate form anion exchange resin. After washing with water, tryptophan and 5-HTP were eluted with 2 N formic acid in 50% ethanol. Tryptophan concentration was determined by fluorometric assay (126), and 5-HT and 5-HIAA fluorometric assays followed the method of Atack and Lindquist with minor modifications. All concentrations given are corrected for recovery, which was determined by adding radioactive 5-HT or 5-HIAA to homogenates. Trp recovery was calculated from internal standards of nonradioactive tryptophan.

Serotonin-accumulation studies involved the determination of 5-HT concentrations in controls and in rats 30 min after systemic injection of the MAO I Pargyline (114) at a dose of 100 mg/kg. Accumulation was calculated by subtracting the 5-HT values for control rats from those injected with pargyline and multiplying by 2. Accumulation of 5-HTP was measured in rats after injection of the decarboxylase inhibitor, RO4-4602, 800 mg/kg, i.p. After homogenization, 5-HTP was isolated on a Dowex 1 resin, in the same fashion used for isolation of tryptophan, after adjustment of pH to 8.2. Fluorometric assay followed the procedure of Atack and Lindquist. Turnover time was calculated as described by Curzon and Marsden (127) following the approach of Neff and Tozer (128) and can be summarized as the time interval in hours required to synthesize the amount of 5-HT present in the region under study.

(†) Four rats were used for this determination; (*) Difference from corresponding regional value in controls was significant at $P \leq 0.05$ by Dunnett's test (129). Abbreviations used: Hypo, hypothalamus; STR, striatum; PMB, pons-midbrain; NAS, nucleus accumbens septi; hcps, hippocampus; CCtx, cerebral cortex; sp cord, spinal cord; Trp, tryptophan; 5-HIAA, 5-hydroxyindoleacetic acid.

1 month increased both steady-state levels of 5-HT and rates of 5-HT synthesis. In Table 8, we present data comparing rats who received injections of PEA for 5 days with saline-injected controls. The data indicate increased steady-state levels of 5-HIAA with decreased 5-HT levels in the PEA group suggesting that 5-HT synthesis was not decreased. The table also shows that striatal DA and HVA were increased in rats receiving PEA for five days, these results differ from the results of Jackson and Smythe (140).

Table 7 indicates that acute administration of amphetamine produced large increases in 5-HT accumulation in the hippocampus and cerebral cortex. The cortex was of particular interest, having a relatively low concentration of 5-HT, but one that appeared to turn over more rapidly than the 5-HT store of regions with higher concentrations such as hypothalamus and nucleus accumbens septi. Not only did cerebral cortex have the shortest turnover time in the control rats, but it also had the greatest response to amphetamine in terms of shortening of the turnover time—amphetamine halved the cortical 5-HT turnover time.

Table 9 compares the effects of amphetamine and PEA on HVA concentrations in striatum and cerebral cortex. It was necessary to pool the cortical tissue from two rats (about 1.2-1.4 g) to obtain enough tissue for HVA determination, as the levels are so much less than those of striatum. Note that while amphetamine administration increased HVA in both striatum and cortex, only striatal HVA was increased by PEA. This indicates one way in which the biochemical effects of amphetamine differ from those produced by PEA. Since cortical DA terminals may be particularly important in emotional behavior and the effects of antipsychotic drugs (141,142), our observations may indicate that the two drugs will be found to differ in their behavioral effects.

TABLE 8

Chronic Administration of PEA[a]

	Hypothal	str	pmb	NAS	hcps	cctx
Saline Controls						
Trp (μg/g)	7.65 ± 0.84	5.31 ± 0.53	3.67 ± 0.45	4.88 ± 0.95	3.35 ± 0.67	3.06 ± 0.41
5-HT (ng/g)	1470 ± 158	871 ± 90	694 ± 70	1495 ± 205	568 ± 77	433 ± 38
5-HIAA (ng/g)	1075 ± 123	717 ± 56	897 ± 84	938 ± 151	629 ± 80	401 ± 62
PEA-injected rats						
Trp (μg/g)	8.06 ± 1.04	5.99 ± 0.83	4.00 ± 0.75	4.59 ± 1.05	3.12 ± 0.58	3.76 ± 0.61
5-HT (ng/g)	1377 ± 142	958 ± 127	642 ± 79	1639 ± 177	595 ± 101	458 ± 40
5-HIAA (ng/g)	1028 ± 168	845 ± 82	1025 ± 46	1020 ± 118	755 ± 49	533 ± 69
Saline controls (4)		Striatum		PEA rats (4)	Striatum	
DA (μg/g)		6.09 ± 0.53			5.86 ± 0.44	
HVA (ng/g)		594 ± 73			765 ± 58	

[a]Rats were given injections of saline or PEA (75 mg/kg, i.p.) once daily for 5 days and killed 12 hr after the last injection. Brains were dissected, homogenized, and analyzed as described in the legend to Tables 1 and 7. HVA determinations were by a modification of the method of Westerink and Korf (139) in which PCA homogenates were first extracted into butyl acetate : petroleum ether (3 : 1), were then back-extracted into Tris-HCl 50 mM solution, pH 8.0 containing 0.5 mg/ml sodium metabisulfite, before passing through the Sephadex G-10 columns. All concentrations given are corrected for recovery.

TABLE 9

Effects of Amphetamine and PEA upon HVA Concentrations[a]

	Controls (8)	Amphetamine (5 mg/kg) (8)	PEA (80 mg/kg) (8)
Striatal HVA (ng/g)	585 ± 57	791 ± 84	820 ± 61
Cortical HVA	17 ± 4	38 ± 9	20 ± 8

[a]Rats were killed 1 hr after a single injection of the drug indicated, controls receiving saline injections of identical volume. Striata of four rats were used for the determinations in each drug group, since the combined striata of one rat were sufficient for analysis, while cortical tissue from two rats was combined at the butyl acetate/ petroleum ether stage, the combined organic phases being extracted in the legend to Table 8. Values are corrected for recovery.

V. STUDIES ON 5-HT SYNTHESIS IN VITRO

Because amphetamine and other stimulants were found to alter brain 5-HT synthesis in vivo, in addition to changing TOH activity in striatal slices, we studied the effects of these drugs on 5-HT synthesis. Because of the similarities mentioned between the two synthetic enzymes, we suspected that a similar effect might be found (111,115,143,144). Because of the marked in vivo stimulation of cortical 5-HT synthesis by amphetamine, we studied slices of both striatum and cortex. The experimental methods were similar to those used for study of DA synthesis in striatal slices. Small (0.25-mm) slices were suspended in a HEPES buffered medium (NaCl 120 mM, KCl 4.8 mM, $CaCl_2$ 1.4 mM, $MgCl_2$ 1.0 mM, NaH_2PO_4 1.5 mM, HEPES 22 mM, D-glucose 7.5 mM, Na ascorbate 4 mM, adjusted to pH 7.3 with solid Tris base). This medium differs from the Krebs-Ringer buffer used for the catecholamine synthesis studies but gave higher control 5-HT synthesis rates. Amphetamine or another test drug was added at the start of the pre-incubation period, usually 10 min prior to the addition of labeled tryptophan. Several different incubation periods were studied. After homogenization in perchloric acid containing 2.5 mg/ml sodium metabisulfite and 1 mg/ml dithiothreitol, the 5-HT fraction was isolated by Amberlite CG-50 chromatography (see legend to Table 7). The HCl eluate from the Amberlite CG-50 columns was counted in a scintillation counter; in some experiments, fluorimetric assay of 5-HT content was done (125), and the tryptophan fraction, which did not stick at pH 6.5 to the Amberlite CG-50 column, was brought to pH 8.2 and added onto a column of Dowex 1 x 4. Tryptophan was eluted with a mixture of formic acid and methanol. Aliquots of this eluate were then used for both scintillation counting and fluorescent assay (126).

We studied the effects of amphetamine and PEA at several different concentrations of tryptophan in the medium. Figure 3 indicates that both drugs could stimulate 5-HT synthesis in both striatal and cortical slices and that this stimulation was related to medium concentrations of tryptophan. In striatal slices, both drugs inhibited synthesis at a medium tryptophan concentration of 0.25 μM, but stimulated synthesis at higher tryptophan concentrations. The figure also suggests a difference between striatal and cortical tissue, the effect of medium tryptophan concentration being greater for striatal than cortical slices. The stimulation obtained was never as great as that seen in the DA system. Other studies indicated that amphetamine and PEA did not alter tryptophan uptake by cortical or striatal slices at 5- or 10-min incubation times. It is interesting that these stimulant drugs which all increase brain tryptophan (122,132) in vivo can stimulate 5-HT synthesis in vitro without increasing either tryptophan uptake or slice content of tryptophan. At 0.25 μM tryptophan concentrations, where no stimulation of synthesis was demonstrable, amphetamine did increase both content of 5-HIAA (this was in slices plus medium combined) and the proportion of [^{3}H]5-HT found in the medium suggesting that release of transmitter can be seen under conditions where no stimulation of synthesis occurs.

Preliminary studies have suggested the following additional points: The presence of an MAOI (either pargyline or nialamide) markedly reduces and sometimes abolishes the amphetamine or PEA stimulation of [^{3}H]5-HT accumulation without significantly altering

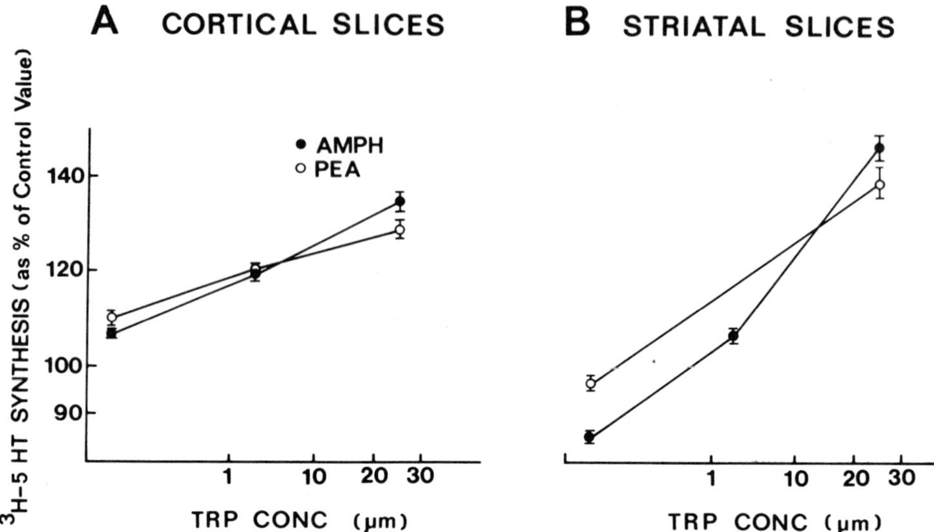

FIG. 3. After 10 min preincubation with the drug indicated (amphetamine at a concentration of 1.5 μM, PEA at 15 μM) slices of striatum or cerebral cortex were incubated with [^{3}H]tryptophan at the concentrations indicated on the abscissa. Flasks contained 25 mg wet wt. of tissue in a total of 4 ml HEPES buffer (see text). These flasks were incubated for 15 min slices and medium homogenized in PCA and [^{3}H]5-HT isolated by ion exchange chromatography on Amberlite CG-50 (Na^+ form) resins. Radioactivity in the eluate was corrected for other [^{3}H]amines by thin layer chromatography on silica gel using chloroform : methanol : glacial acetic acid (20 : 1 : 5) as the developing solvent and locating the amines by spraying with ethanolic p-dimethylaminocinnamaldehyde. Each point represents three or four different experiments, each experiment done in quadruplicate. The vertical bars indicate the standard error of the mean.

control accumulation (15 min incubation times, MAOI added at the same time as amphetamine, 10 min before addition of [^{3}H]tryptophan). This, of course, suggests that release of 5-HT from a cytoplasmic source may be able to alter synthesis (118). Feedback inhibition of 5-HT synthesis remains controversial (20,145-149), but may be more readily demonstrated in slices than in vivo. The stimulant effect is seen when the decarboxylase inhibitor, RO4-4602, is added at the same time as amphetamine, suggesting a possible effect of drug on the tryptophan hydroxylase step (130). Amphetamine also appears to promote efflux of endogenous tryptophan from tissue in vitro, but this is a weak effect and cannot explain the stimulation of 5-HT accumulation because the drugs also increase synthesis when it is expressed as a conversion index, taking into account specific activity of tissue tryptophan. Finally, we have been unable to demonstrate the effect in slices of pons, which are enriched in raphe nuclei and should contain 5-HT cell bodies but much fewer terminals than do the forebrain regions where we found the effect (57). Therefore, stimulation of synthesis may be a property of nerve terminals and not of perikarya. These observations all require further study, but they do coincide with the observations made in the DA system.

The study of 5-HT synthesis is technically complex for several reasons. First, there is no single chromatographic technique as useful as alumina absorption is for catechols. Second, it is necessary to correct the radioactivity found in the amine fraction for [^{3}H]tryptamine and occasionally other labeled substances which require checking by thin layer chromatography. Finally, use of a decarboxylase inhibitor results in the accumulation of a substance (5-HTP) which is separated from tryptophan only with difficulty, unlike the catecholamine system where DOPA is readily absorbed by alumina and tyrosine is not.

Because of these problems and the difficulties in isolating metabolites, the decarboxylase inhibition technique introduced by Carlsson is especially appealing. Its elimination of newly synthesized amine is an additional benefit. Since no measure of turnover or synthesis is above criticism, it is usually desirable to have two independent methods.

Serotonin neurons appear to be important in many brain functions and are closely interrelated chemically and functionally to catecholamine neurons (137,150). Both transmitters require the same co-factor for their synthesis. With increasing usage of both DOPA and 5-HTP in clinical medicine (151-155), the fact that 5-HTP can interact with catecholamine neurons and DOPA with 5-HT neurons is important in understanding the benefits and side effects of such treatments.

Previous studies have reported that the systemic administration of amphetamine to rats results in a depression of 5-HT synthesis measured in vitro (143,153). When the drug or its metabolites, p-hydroxyamphetamine or p-hydroxynorephedrine, were added to homogenates in vitro, no alteration of 5-HT synthesis was detected, although details, such as tryptophan concentrations, were not given. There are many possible reasons for the difference between our results and those of Knapp. Not only did we use slices rather than synaptosomes, but we have found many factors to influence the stimulatory effect of amphetamine and PEA: Not only precursor concentration in the medium, but time of incubation, duration of preincubation, and calcium content of the incubation medium all can alter the results (157). The fact that our effects are seen at low drug concentrations (less than 10 μM in both systems) suggests that the stimulation effect may be relevant to regulatory phenomena which occur in vivo as well.

In summary, we find that amphetamine and PEA appear to stimulate the synthesis of both DA and 5-HT measured in small brain slices, and that this effect persists in spite of decarboxylase inhibition. This indicates that newly synthesized amine is not responsible for the stimulation effect. In both systems, the stimulation effect is dependent upon the medium concentration of the precursor amino acid. This observation is similar to the effects reported for veratridine and high potassium concentrations on TOH activity in striatal synaptosomes (91,92). This suggests that the mechanism of the increased TOH and tryptophan hydroxylase activity produced by the depolarization of in vitro preparations with intact neuronal membranes may be similar to that produced by amphetamine and PEA.

Our results suggest that these stimulatory effects on transmitter synthesis are not mediated solely by release of transmitter amines. Since these drugs appear to lack effects on synthesis when studied in hypotonic homogenates and probably have no direct effects upon the enzymes, a cytoplasmic or membrane factor may mediate the effect. Cyclic nucleotides and calcium have been shown to alter the activity of TOH (95-97,158-161), and calcium has been reported to alter tryptophan hydroxylase activity (130,143).

Our studies suggest that cyclic nucleotides do not mediate the effects of amphetamine and PEA on TOH activity in striatal slices. Because of conflicting evidence about the effects of amphetamine on cyclic nucleotides in the brain (25,162,163), it is still possible that a relationship between cyclic nucleotides and enzyme stimulation may exist. It has been suggested that endogenous PEA may mediate the behavioral effects of amphetamine (164), but our work provides no evidence on this point.

ACKNOWLEDGMENTS

We thank A. Davis and L. Kamal for technical assistance, Dr. P. M. Beart and Dr. Reynold Spector for discussions and advice. This work is from the Children's Hospital Mental Retardation Center and was supported by N.I.H. grants NS 12368, Developmental Neurology program project grant NS-HD09704, and Mental Retardation Core Grant number 5P30-HD06276-05. We are especially grateful to Hoffman-LaRoche, Inc. (Dr. W. E. Scott) for donations of R04-4602, the CIBA-Geigy Corporation (Mr. C. A. Brownley) for supplies of methylphenidate, and to Lederle Laboratories (Dr. J. M. Smith) for supplies of brocresine.

REFERENCES

1. U. Trendelenburg (1963). Supersensitivity and subsensitivity to sympathomimetic amines. *Pharmacol. Rev.*, 15:225-276.

2. K. E. Moore (1964). The role of endogenous norepinephrine in the toxicity of D-amphetamine in isolated and aggregated mice. *J. Pharmacol. Exp. Ther.*, 144:45-51.

3. L. Stein (1964). Self-stimulation of the brain and the central stimulant action of amphetamine. *Fed. Proc.*, 23:836-850.

4. J. Glowinski and J. Axelrod (1965). The effect of drugs on the uptake, release, and metabolism of ^{3}H-norepinephrine in the rat brain. *J. Pharmacol. Exp. Ther.*, 149:43-50.

5. A. Weisman and B. K. Koe (1965). Behavioral effects of L-α-methyltyrosine, an inhibitor of tyrosine hydroxylase. *Life Sci.*, 4:1037-1048.

6. A. Randrup and I. Munkvad (1966). The role of catecholamines in the amphetamine excitatory response. *Nature*, 210:540.

7. L. C. F. Hanson (1967). Evidence that the central action of (+)-amphetamine is mediated via catecholamines. *Psychopharmacologia*, 10:289-297.

8. W. D. Reid (1970). Turnover rate of brain 5-hydroxytryptamine increased by d-amphetamine. *Br. J. Pharmacol.*, 40:483-491.

9. K. Fuxe and U. Ungerstedt (1970). Histochemical, biochemical, and functional studies on central monoamine neurons after acute and chronic amphetamine administration. In: *Amphetamines and Related Compounds*, edited by E. Costa and S. Garattini. Raven Press, New York, pp. 257-288.

10. A. J. Azzaro and C. O. Rutledge (1973). Selectivity of release of norepinephrine, dopamine and 5-hydroxytrptamine by amphetamine in various regions of the rat brain. *Biochem. Pharmacol.*, 22: 2801-2813.

11. A. Pletscher, G. Bartholini, H. Bruderer, W. O. Burkhard, and K. F. Gey (1964). Chlorinated arylalkylamines affecting the cerebral metabolism of 5-hydroxytryptamine. *J. Pharmacol. Exp. Ther.*, 145:344-350.

12. N. E. Anden, T. Magnusson, and G. Stock (1973). Effects of drugs influencing monoamine mechanisms on the increase in brain dopamine produced by axotomy or treatment with gamma hydroxybutyric acid. *Navryn Schmiedeberg's Arch. Pharmacol.*, 278:363-372.

13. M. Levitt, S. Spector, A. Sjoerdsma, and S. Udenfriend (1965). Elucidation of the Rate-Limiting step in norepinephrine biosynthesis in the perfused guinea pig heart. *J. Pharmacol. Exp. Ther.*, 148:1-8.

14. T. Nagatsu, M. Levitt, and S. Udenfriend (1964). Tyrosine hydroxylase: The initial step in norepinephrine biosynthesis. *J. Biol. Chem.*, 239:2910-2917.

15. E. Costa and N. H. Neff (1966). Isotopic and non-isotopic measurements of the rate of catecholamine biosynthesis. In: *Biochemistry and Pharmacology of the Basal Ganglia*, edited by E. Costa, L. J. Cote, and M. D. Yahr. Raven Press, New York, pp. 141-155.

16. A. Alouisi and N. Weiner (1966). The regulation of norepinephrine synthesis in sympathetic nerves: Effect of nerve stimulation, cocaine, and catecholamines. *Proc. Nat. Acad. Sci. USA*, 56:1491-1495.

17. N. Weiner (1970). Regulation of norepinephrine biosynthesis. *Ann. Rev. Pharmacol.*, 10:273-290.

18. W. Dairman (1973). Multiple measures of compensatory adaptation in catecholamine biosynthesis. In: *New Concepts in Neurotransmitter Regulation*, edited by A. J. Mandell. Plenum Press, New York, pp. 1-20.

19. N. Weiner, F. C. Borne, R. Bjun, G. Cloutier, and S. Z. Lauger (1973). Norepinephrine biosynthesis in relationship to neural activation. In: *New Concepts in Neurotransmitter Regulation*, edited by A. J. Mandell. Plenum Press, New York, pp. 89-103.

20. E. Costa and J. L. Meek (1974). Regulation of biosynthesis of catecholamines and serotonin in the CNS. *Ann. Rev. Pharmacol.*, 14:491-511.

21. R. Kuczenski (1975). Conformational adaptability of tyrosine hydroxylase in the regulation of striatal dopamine biosynthesis. In: *Neurobiological Mechanisms of Adaptation and Behavior*, edited by A. J. Mandell. Raven Press, New York, pp. 109-125.

22. J. R. Walters and R. H. Roth (1974). Dopaminergic neurons: Drug induced antagonism of the increase in tyrosine hydroxylase activity produced by cessation of impulse flow. *J. Pharmacol. Exp. Ther.*, 191:82-91.

23. L. E. Murrin and R. H. Roth (1976). Dopaminergic neurons: Effects of electrical stimulation on dopamine biosynthesis. *Mol. Pharmacol.*, 12:463-475.

24. R. J. Miller, A. S. Horn, L. L. Iversen, and R. M. Pinder (1974). Effects of dopamine-like drugs on rat striatal adenyl cyclase have implications for CNS dopamine receptor topography. *Nature*, 250:238-240.

25. J. A. Ferrendelli, D. A. Kiuscherf, and D. Kipnis (1970). Effects of amphetamine, chlorpromazine, and reserpine on cyclic GMP and cyclic AMP levels in mouse cerebellum. *Biochem. Biophys. Res. Commun.*, 41:1061-1067.

26. A. Weissman, B. K. Koe, and S. S. Tenen (1966). Antiamphetamine effects following inhibition of tyrosine hydroxylase. *J. Pharmacol. Exp. Ther.*, 151:339-352.

27. H. H. Wolf, D. E. Rollius, C. R. Rowland, and T. G. Reigle (1969). The importance of endogenous catecholamines in the activity of some CNS stimulants. *Int. J. Neuropharmacol.*, 8:319-328.

28. J. Scheel-Kruger (1971). Comparative studies of various amphetamine analogues demonstrating different interactions with the metabolism of the catecholamines of the brain. *Eur. J. Pharmacol.*, 14:47-59.

29. A. Giachetti and P. A. Shore (1966). Studies in vitro of amine uptake mechanisms in heart. *Biochem. Pharmacol.*, 15:607-614.

30. C. A. Stone and C. Porter (1966). Methyl DOPA and adrenergic nerve function. *Pharmacol. Rev.*, 18:569-575.

31. I. J. Kopin (1968). False adrenergic transmitters. *Ann. Rev. Pharmacol.*, 8:377-394.

32. G. P. Breese, I. Kopin, and J. K. Weise (1970). Effects of amphetamine derivatives on brain dopamine and noradrenaline. *Br. J. Pharmacol.*, 38:537-545.

33. R. L. Dorris and P. A. Shore (1971). Localization and persistance of metaraminol and α-methyl-m-tyramine in rat and rabbit brain. *J. Pharmacol. Exp. Ther.*, 179:10-14.

34. B. Bhagat, E. K. Gordon, and I. J. Kopin (1965). Norepinephrine synthesis and the pressor responses to tyramine in the spinal cat. *J. Pharmacol. Exp. Ther.*, 147:319-323.

35. J. N. Eble and A. D. Rudzik (1966). The blockade of the pressor response to tyramine by amphetamine in the reserpine treated dog. *J. Pharmacol. Exp. Ther.*, 153:62-69.

36. M. D. Day (1967). The lack of crossed tachyphylaxis between tyramine and some other indirectly acting sympathomimetic amines. *Br. J. Pharmacol. Chem.*, 30:631-643.

37. T. L. Sourkes (1954). Inhibition of dihydroxy-phenylalamine decarboxylase by derivatives of phenylalamine. *Arch. Biochem. Biophys.*, 51:444-456.

38. J. A. Oates, L. Gillespie, S. Udenfriend, and A. Sjoerdsuea (1960). Decarboxylase inhibition and blood pressure reduction by α-methyl 3,4-dihydroxy-D,L-phenylalamine. *Science*, 131: 1890-1892.

39. A. Carlsson and M. Lindquist (1962). In-vivo decarboxylation of α-methyl DOPA and α-methylmetatyrosine. *Acta Physiol. Scand.*, 54:87-94.

40. H. Schümann and H. Grobecker (1965). Ueber dre Werbung von α-methyl DOPA auf den brenzcatchinarmingehalt von meerschweinchenorganen. *Arch. Exp. Pathol. Pharmakol.*, 251:48-61.

41. J. A. Dominic and K. E. Moore (1971). Depression of behavior and the brain content of α-methylnorepinephrine and α-methyldopamine following the administration of α-methyl DOPA. *Neuropharmacology*, 10:33-44.

42. E. Muscholl and L. Maitre (1963). Release by sympathetic stimulation of α-methyl-noradrenaline stored in the heart after the administration of α-methyldopa. *Experientia*, 19:658-659.

43. M. Day and M. Rand (1964). Some observations on the pharmacology of α-methyl-dopa. *Br. J. Pharmacol.*, 22:72-86.

44. P. Holtz and D. Pahn (1967). On the pharmacology of α-methylated catecholamines and the mechanism of the antihypertensive action of α-methyldopa. *Life Sci.*, 1847-1857.

45. A. S. Horn (1975). Structure activity relations for neurotransmitter receptor agonists and antogonists. In: *Handbook of Psychopharmacology, Vol. 2*, edited by L. L. Iversen, S. D. Iversen, and S. H. Snyder. Plenum Press, New York, pp. 179-243.

46. J. C. Besse and R. F. Furchgott (1976). Dissociation constants and relative efficacies of agonists acting on alpha adrenergic receptors in rabbit aorta. *J. Pharmacol. Exp. Ther.*, 197:66-78.

47. N. J. Uretsky (1974). Effect of α-methyldopa on the metabolism of dopamine in the striatium of the rat. *J. Pharmacol. Exp. Ther.*, 189:359-369.

48. N. J. Uretsky and L. S. Seiden (1969). Effect of α-methyl dopa on the reserpine induced suppression of motor activity and the conditioned avoidance response. *J. Pharmacol. Exp. Ther.*, 168: 153-162.

49. M. Day and M. Rand (1963). A hypothesis for the mode of action of α-methyl DOPA in relieving hypertension. *J. Pharm. Pharmacol.*, 15:221-224.

50. O. S. Ray and L. W. Bivens (1965). Behavioral effects of L-alpha methyldopa. *Life Sci.*, 4:823-826.

51. J. M. van Rossum (1963). Mechanism of the central stimulant action of α-methyl-meta tyrosine. *Psychopharmacologia*, 4:271-280.

52. L. L. Iversen (1975). Dopamine receptors in the brain. *Science*, 188:1084-1089.

53. I. Creese, D. R. Burt, and S. H. Snyder (1976). Dopamine receptor binding predicts clinical and pharmacological potencies of anti-schizophrenic drugs. *Science*, 192:481-483.

54. B. S. Bunney, J. R. Walters, R. H. Roth, and G. K. Aghajanian (1973). Dopaminergic neurons: effect of antipsychotic drugs and amphetamine on single cell activity. *J. Pharmacol. Exp. Ther.*, 185:560-571.

55. R. H. Roth, J. R. Walters, and V. H. Morgenroth (1974). Effects of alterations in impulse flow on transmitter metabolism in central dopaminergic neurons. *Adv. Biochem. Psychopharmacol.*, 12: 369-384.

56. P. A. Shore and R. L. Dorris (1975). On a prime role for newly synthesized dopamine in striatal function. *Eur. J. Pharmacol.*, 30:315-318.

57. J. R. Cooper, F. E. Bloom, and R. H. Roth (1974). *The Biochemical Basis of Neuropharmacology*, 2nd ed. Oxford University Press, New York.

58. P. M. Groves, C. J. Wilson, S. J. Young, and G. V. Rebec (1975). Self-inhibition by dopaminergic neurons. *Science*, 190:522-529.

59. C. C. Porter (1971). Aromatic amino acid decarboxylase inhibitors. *Fed. Proc.*, 30:871-876.

60. B. B. Brodie, R. Kuntzman, C. W. Hirsch, and E. Costa (1962). Effects of decarboxylase inhibition on the biosynthesis of brain monoamines. *Life Sci.*, 3:81-84.

61. R. J. Levine and A. Sjoerdsma (1964). Dissociation of the decarboxylase inhibiting and norepinephrine depleting effects of α-methyl DOPA, α-ethyl DOPA, 4-bromo-3-hydroxybenzyloxyamine and related substances. *J. Pharmacol. Exp. Ther.*, 146:42-47.

62. H. McIlwain and R. Rodnight (1962). *Practical Neurochemistry*. J. and A. Churchill, London.

63. M. J. Besson, A. Cheramy, and J. Glowinski (1971). Effects of some psychotropic drugs on dopamine synthesis in rat striatum. *J. Pharmacol. Exp. Ther.*, 177:196-205.

64. A. S. Horn (1973). Structure-activity relationships for the inhibition of catecholamine uptake into synaptosomes from noradrenaline and dopamine neurones in brain homogenates. *Br. J. Pharmacol.*, 47:332-335.

65. R. J. Ziance, A. J. Azzaro, and C. O. Rutledge (1972). Characteristics of amphetamine-induced release of norepinephrine from rat cerebral cortex in vitro. *J. Pharmacol. Exp. Ther.*, 182: 284-294.

66. S. Udenfriend, P. Zaltzman-Nirenberg, and T. Nagatsu (1965). Inhibitors of purified beef adrenal tyrosine hydroxylase. *Biochem. Pharmacol.*, 14:837-845.

67. N. J. Uretsky, G. J. Chase, and A. V. Lorenzo (1975). Effect of α-methyl DOPA on dopamine synthesis and release in rat striatum in vitro. *J. Pharmacol. Exp. Ther.*, 193:73-87.

68. W. M. Pardridge and J. D. Connor (1973). Saturable transport of amphetamine across the blood-brain barrier. *Experientia*, 29:302-304.

69. U. Trendelenburg, A. Muskus, W. W. Fleming, and B. G. A. de la Sierra (1962). Modification by reserpine of the action of sympathomimetic amines in spinal cats: A classification of sympathomimetic amines. *J. Pharmacol. Exp. Ther.*, 138:170-180.

70. I. R. Innes and M. Nickerson (1975). Norepinephrine, epinephrine, and the sympathomimetic amines. In: *The Pharmacological Basis of Therapeutics*, 5th ed., edited by L. S. Goodman and A. Gilman. Macmillan, New York, pp. 477-513.

71. G. M. McKenzie and J. C. Szerb (1968). The effect of dihydroxyphenylalanine, pheniprazine and d-amphetamine on the in vivo release of dopamine from the caudate nucleus. *J. Pharmacol. Exp. Ther.*, 162:302-308.

72. L. Stein and C. D. Wise (1969). Release of norepinephrine from hypothalamus and amygdala by rewarding medial forebrain bundle stimulation and by amphetamine. *J. Comp. Physiol. Psychol.*, 67:189-198.

73. L. A. Carr and K. E. Moore (1970). Effects of amphetamine on the contents of norepinephrine and its metabolites in the effluent of the perfused cerebral ventricles of the cat. *Biochem. Pharmacol.*, 19:2361-2374.

74. P. F. Von Voigtlander and K. E. Moore (1973). Involvement of nigro-striatal neurons in the in vivo release of dopamine by amphetamine, amantadine, and tyramine. *J. Pharmacol. Exp. Ther.*, 184:542-552.

75. I. Creese and S. D. Iversen (1974). The role of forebrain dopamine systems in amphetamine induced stereotyped behavior in the rat. *Psychopharmacologia*, 39:345-357.

76. I. M. Asher and G. K. Aghajanian (1974). 6-Hydroxydopamine lesions of olfactory tubercles and caudate nuclei: Effect on amphetamine-induced stereotyped behavior in rats. *Brain Res.*, 82:1-12.

77. D. C. Roberts, A. P. Zis, and H. C. Fibiger (1975). Ascending catecholamine pathways and amphetamine-induced locomotor activity: Importance of dopamine and apparent non-involvement of norepinephrine fibers. *Brain Res.*, 93:441-454.

78. P. H. Kelly, P. W. Seviour, and S. D. Iversen (1975). Amphetamine and apomorphine responses in the rat following 6-OHDA lesions of the nucleus accumbens septi and corpus striatum. *Brain Res.*, 94:507-522.

79. C. B. Smith (1965). Effects of d-amphetamine upon brain amine content and locomotor activity in mice. *J. Pharmacol. Exp. Ther.*, 147:96-102.

80. L. Y. Koda and J. W. Gibb (1973). Adrenal and striatal tyrosine hydroxylase activity and methamphetamine. *J. Pharmacol. Exp. Ther.*, 185:42-48.

81. R. H. Roth, J. Walters, and G. Aghajanian (1973). Effect of impulse flow on the release and synthesis of dopamine in the striatum. In: *Frontiers in Catecholamine Research*, edited by E. Usdin and S. H. Snyder. Pergamon Press, Oxford, pp. 567-574.

82. J. E. Harris, R. J. Baldessarini, and R. H. Roth (1975). Amphetamine-induced inhibition of tyrosine hydroxylation in homogenates of rat corpus striatum. *Neuropharmacology*, 14:457-471.

83. E. Costa, A. Gropetti, and M. K. Naimzada (1972). Effects of amphetamine on the turnover rate of brain catecholamines and motor activity. *Br. J. Pharmacol.*, 44:742-751.

84. J. E. Harris and R. J. Baldessarini (1973). Amphetamine induced inhibition of tyrosine hydroxylation in homogenates of rat corpus striatum. *J. Pharm. Pharmacol.*, 25:348-351.

85. B. S. Bunney and G. K. Aghajanian (1974). Electrophysiological effects of amphetamine on dopaminergic neurons. *Biochem. Pharmacol.*, supplement 2:792-797.

86. L. B. Geffen, T. M. Jessell, A. C. Cuello, and L. L. Iversen (1976). Release of dopamine from dendrites in rat substantia nigra. *Nature*, 260:258-260.

87. R. Kuczenski and D. S. Segal (1975). Differential effects of d- and l-amphetamine and methylphenidate on rat striatal dopamine biosynthesis. *Eur. J. Pharmacol.*, 30:244-251.

88. C. C. Chieuh and K. E. Moore (1975). d-Amphetamine-induced release of "newly synthesized" and "stored" dopamine from the caudate nucleus in vivo. *J. Pharmacol. Exp. Ther.*, 192:642-653.

89. M. J. Besson, A. Cheramy, P. Feltz, and J. Glowinski (1969). Release of newly synthesized dopamine from dopamine containing terminals in the striatum of the rat. *Proc. Nat. Acad. Sci. USA*, 62:741-748.

90. E. Hansson, R. M. Fleming, and W. G. Clark (1964). Effect of some benzylhydrazines and benzyloxyamines on DOPA and 5-hydroxytryptophan decarboxylase in vivo. *Neuropharmacology*, 3:177-188.

91. G. Bustos, M. J. Kuhar, and R. H. Roth (1972). Effect of gamma-hydroxybutyrate and gamma-butyrolactone on dopamine synthesis and reuptake by rat striatum. *Biochem. Pharmacol.*, 21:2649-2652.

92. R. L. Patrick, T. E. Snyder, and J. D. Barchas (1975). Regulation of dopamine synthesis in rat brain striatal synaptosomes. *Mol. Pharmacol.*, 11:621-631.

93. T. Hokfelt, G. Jonsson, and P. Lidbrink (1970). Electron microscopic identification of monoamine nerve ending particles in rat brain homogenates. *Brain Res.*, 22:147-151.

94. M. Doteuchi, C. Wang, and E. Costa (1974). Compartmentation of dopamine in rat striatum. *Mol. Pharmacol.*, 10:225-234.

95. J. E. Harris, R. J. Baldessarini, V. H. Morgenroth, and R. H. Roth (1975). Activation by cyclic 3':5' adenosine monophosphate of tyrosine hydroxylase in rat brain. *Proc. Nat. Acad. Sci. USA*, 72:789-793.

96. B. Ebstein, C. Roberge, K. Tabachnik, and M. Goldstein (1974). The effect of dopamine and of apomorphine on di-butyryl-cyclic AMP induced stimulation of synaptosomal tyrosine hydroxylase. *J. Pharm. Pharmacol.*, 26:975-977.

97. C. Roberge, B. Ebstein, and M. Goldstein (1974). Stimulation of tyrosine hydroxylase activity by dibutyryl cyclic AMP in synaptosomal preparations. *Fed. Proc.*, 33:521.

98. P. Greengard (1976). Possible role for cyclic nucleotides and phosphorylated membrane proteins in post synaptic actions of neurotransmitters. *Nature*, 260:101-108.

99. P. L. Carlton (1963). Behavioral stimulation due to α-methyl-meta-tyrosine. *Nature*, 200:586-587.

100. A. Weissman and B. K. Koe (1967). Contrasting locomotor effects of catecholamine releasers and tyrosine hydroxylase inhibitors in MAO-treated mice. *Psychopharmacologia*, 11:282-286.

101. M. K. Shellenberger (1971). Effects of -methyl-m-tyrosine on spontaneous and caudate-induced electroencephalogram activity and regional catecholamines in cat brain. *J. Pharmacol. Exp. Ther.*, 177:481-490.

102. M. Jaefre, A. Ruch-Monachon, and W. Haefely (1974). Methods for assessing the interaction of agents with 5-hydroxytryptamine neurons and receptors in the brain. *Adv. Biochem. Psychopharmacol.*, 10:121-134.

103. K. G. Lloyd and G. Bartholini (1974). The effect of methiothepin on cerebral monoamine neurons. *Adv. Biochem. Psychopharmacol.*, 10:305-309.

104. J. P. Bennett and S. H. Snyder (1976). Serotonin and lysergic acid diethylamide binding in rat brain membranes: Relationship to post-synaptic serotonin receptors. *Mol. Pharmacol.*, 12: 373-389.

105. M. DaPrada, A. Saner, W. P. Burkhard, G. Bartholini, and A. Pletscher (1975). Lysergic acid diethylamide: Evidence for stimulation of cerebral dopamine receptors. *Brain Res.*, 94: 67-73.

106. K. Von Hungen, S. Roberts, and D. F. Hill (1975). Interactions between lysergic acid diethylamide and dopamine sensitive adenyl cyclase systems in rat brain. *Brain Res.*, 94:57-66.

107. I. Creese, D. R. Burt, and S. H. Snyder (1975). The dopamine receptor: Differential binding of D-LSD and related agents to agonist and antagonist states. *Life Sci.*, 17:1715-1720.

108. H. Corrodi, L. O. Farnebo, K. Fuxe, and B. Hamberger (1975). Effect of ergot drugs on central 5-hydroxytryptamine neurons: Evidence for 5-hydroxytryptamine release or 5-hydroxytryptamine receptor stimulation. *Eur. J. Pharmacol.*, 30:172-181.

109. P. A. Friedman, A. H. Kappelman, and S. Kaufman (1972). Partial purification and characterization of tryptophan-5-hydroxylase. *J. Biol. Chem.*, 247:4165-5173.

110. S. Kaufman and D. B. Fisher (1974). Pterin-requiring amino acid hydroxylases. In: *Molecular Mechanisms of Oxygen Activation*, edited by O. Hayaishi. Academic Press, New York, pp. 191-237.

111. J. H. Tong and S. Kaufman (1975). Tryptophan hydroxylase: Purification and some properties of the enzyme from rabbit hindbrain. *J. Biol. Chem.*, 250:4152-4158.

112. S. R. Snodgrass and N. J. Uretsky (1976). Regional studies of 5-hydroxytryptamine turnover in the rat brain. Submitted for publication to *Neuropharmacology*.

113. Y. Morot-Gaudry, M. Hamon, S. Bougoin, J. P. Ley, and J. Glowinski (1974). Estimation of the rate of 5-HT synthesis in the mouse brain by various methods. *Naunyn Schmiedeberg's Arch. Pharmacol.*, 282:223-238.

114. N. H. Neff, R. C. Lin, S. H. Ngai, and E. Costa (1969). Turnover rate measurements of brain serotonin in unanesthetized rats. *Adv. Biochem. Psychopharmacol.*, 1:91-109.

115. A. Carlsson, W. Kehr, M. Lindquist, T. Magnusson, and C. V. Atack (1972). Regulation of monoamine metabolism in the central nervous system. *Pharmacol. Rev.*, 24:371-384.

116. S. Bourgoin, J. P. Ternaux, A. Boireau, F. Hery, and M. Hamon (1975). Effects of halothane and nitrous oxide anesthesia on 5-HT turnover in the rat brain. *Naunyn Schmiedeberg's Arch. Pharmacol.*, 288:109-121.

117. J. L. Comenares, R. J. Wurtman, and J. D. Fernstrom (1975). Effects of the ingestion of a carbohydrate-fat meal on the levels and synthesis of 5-hydroxyindoles in various regions of the rat central nervous system. *J. Neurochem.*, 25:825-829.

118. M. Hamon, S. Bourgoin, and J. Glowinski (1973). Feedback regulation of 5-HT synthesis in rat striatal slices. *J. Neurochem.*, 20:1727-1745.

119. H. Green, S. M. Greenberg, R. W. Erickson, J. L. Sawyer, and R. J. Ellison (1962). Effect of dietary phenylalanine and tryptophan on rat brain amine levels. *J. Pharmacol. Exp. Ther.*, 136:174-178.

120. J. D. Fernstrom and R. J. Wurtman (1971). Brain serotonin content: Physiological dependence on plasma tryptophan levels. *Science*, 173:149-151.

121. R. J. Wurtman and J. D. Fernstrom (1972). L-tryptophan, L-tyrosine, and the control of brain monoamine biosynthesis. In: *Perspectives in Neuropharmacology*, edited by S. H. Snyder. Oxford University Press, New York, pp. 143-166.

122. A. Tagliamonte, P. Tagliamonte, J. Perez-Cruet, S. Stern, and G. L. Gessa (1971). Effect of psychotropic drugs on tryptophan concentration in the rat brain. *J. Pharmacol. Exp. Ther.*, 177:475-480.

123. J. Glowinski and L. L. Iversen (1966). Regional studies of catecholamines in the rat brain—I. *J. Neurochem.*, 13:655-669.

124. J. F. R. Konig and R. A. Klippel (1970). *The Rat Brain*. Robert E. Krieger Publishing Co., Huntington, N.Y.

125. C. Atack and M. Lindquist (1973). Conjoint native and orthopthaldialdehyde-condensate assays for the fluorimetric determination of 5-hydroxyindoles in brain. *Naunyn Schmiedeberg's Arch. Pharmacol.*, 279:267-284.

126. W. D. Denckla and H. K. Dewey (1967). The determination of tryptophan in plasma, liver and urine. *J. Lab. Clin. Med.*, 69:160-169.

127. G. Curzon and C. A. Marsden (1975). Metabolism of tryptophan load in the hypothalamus and other brain regions. *J. Neurochem.*, 25:251-256.

128. N. H. Neff and T. N. Tozer (1968). In vivo measurement of brain serotonin turnover. *Adv. Pharmacol.*, 6A:97-109.

129. C. W. Dunnett (1955). A multiple comparison procedure for comparing several treatments with a control. *J. Am. Stat. Assn.*, 50:1096-1121.

130. M. Hamon, S. Bourgoin, F. Hery, J. P. Ternaux, and J. Glowinski (1976). In vitro and in vivo activation of soluble tryptophan hydroxylase from rat brain stem. *Nature*, 260:61-63.

131. P. J. Knott and G. Curzon (1972). Free tryptophan in plasma and brain tryptophan metabolism. *Nature*, 239:452-453.

132. D. A. Brase and H. H. Loh (1976). The increase in brain tryptophan caused by amphetamine-like drugs: Correlation with an increase in body temperature. *Life Sci.*, 18:115-121.

133. M. Grabowska (1975). The influence of apomorphine on brain serotonin turnover rate. *Pharmacol. Biochem. Behav.*, 3:589-591.

134. L. Rochette and J. Bralet (1976). Effect of the norepinephrine receptor stimulating agent clonidine on the turnover of 5-hydroxytryptamine in some areas of the rat brain. *J. Neural Trans.*, 37:259-267.

135. J. Scheel-Kruger and E. Hasselager (1974). Studies of various amphetamines, apomorphine and clonidine on body temperature and brain 5-hydroxytryptamine metabolism in rats. *Psychopharmacologia*, 36:189-202.

136. C. Blondaux, A. Judge, F. Sordet, G. Chouvet, M. Jouvet, and J. F. Pujol (1973). Modification du metabolisme de la serotonine cerebrale induite chez le rat par administration de 6-hydroxydopamine. *Brain Res.*, 50:101-114.

137. R. Samanin and S. Garattini (1975). The serotonergic system in the brain and its possible functional connections with other aminergic systems. *Life Sci.*, 17:1201-1210.

138. J. L. Diaz and M. O. Huttunen (1972). Altered metabolism of serotonin in the brain of the rat after chronic ingestion of d-amphetamine. *Psychopharmacologia*, 23:305-312.

139. B. H. C. Westerink and J. Korf (1975). Determination of Nanogram Amounts of Homovanillic Acid in the Nervous System with a rapid semiautomated method. *Biochem. Med.*, 12:106-115.

140. S. M. Jackson and S. B. Smythe (1974). The effect of β-phenethylamine on noradrenaline and dopamine turnover in rat brain. *J. Pharm. Pharmacol.*, 26:456-458.

141. S. H. Snyder (1972). Catecholamines in the brain as mediators of amphetamine psychosis. *Arch. Gen. Psychiat.*, 27:169-179.

142. M. Trabucchi, S. Govoui, G. C. Tonou, and P. F. Spano (1976). Localization of dopamine receptors in rat cerebral cortex. *J. Pharm. Pharmacol.*, 28:244-245.

143. S. Knapp, A. J. Mandell, and M. A. Geyer (1974). Effects of amphetamine on regional tryptophan hydroxylase activity and synaptosomal conversion of tryptophan to 5-HT in rat brain. *J. Pharmacol. Exp. Ther.*, 189:676-689.

144. S. Knapp, A. J. Mandell, and W. P. Bullard (1975). Calcium activation of brain tryptophan hydroxylase. *Life Sci.*, 16: 1583-1594.

145. J. L. Meek and K. Fuxe (1971). Serotonin accumulation after monoamine oxidase inhibition. *Biochem. Pharmacol.*, 20:693-706.

146. M. Hamon and J. Glowinski (1974). Regulation of serotonin synthesis. *Life Sci.*, 15:1533-1548.

147. D. G. Grahame-Smith (1971). Studies on vivo on the relationship between brain tryptophan, brain 5-HT synthesis and hyperactivity in rats treated with a monoamine oxidase inhibitor and tryptophan. *J. Neurochem.*, 18:1053-1066.

148. J. B. Macon, L. Sokoloff, and J. Glowinski (1971). Feed back control of rat brain 5-hydroxytryptamine synthesis. *J. Neurochem.*, 18:323-331.

149. S. A. Millard, E. Costa, and E. M. Gal (1972). On the control of brain serotonin turnover rate by end product inhibition. *Brain Res.*, 40:545-551.

150. G. R. Breese, B. R. Cooper, and A. S. Hollister (1975). Involvement of brain monoamines in the stimulant and paradoxical inhibitory effects of methylphenidate. *Psychopharmacologia*, 44:5-10.

151. G. C. Cotzias, P. S. Papavasiliou, and R. Gellene (1969). Modification of Parkinsonism: Chronic treatment with L-Dopa. *N. Eng. J. Med.*, 280:337-345.

152. F. Lhermitte, R. Marteau, and C. Degos (1972). Analyse pharmacologique d'un nouveau cas de myoclonies d' intention et d' action postanoxiques. *Rev. Neurol.*, 126:107-114.

153. C. D. Marsden (1975). The neuropharmacology of abnormal involuntary movement disorders. In: *Modern Trends in Neurology, Vol. 6*, edited by D. Williams. Butterworths, London, pp. 41-166.

154. M. H. Van Woert and V. H. Sethy (1975). Therapy of intention myoclonus with L-5-hydroxytryptophan and a peripheral decarboxylase inhibitor. *Neurology*, 25:135-140.

155. A. Barbeau (1976). Six years of high level levodopa therapy in severely akinetic Parkinsonian patients. *Arch. Neurol.*, 33:333-338.

156. S. Knapp (1975). Differential drug effects on brain serotonergic systems. In: *Neurobiological Mechanisms of Adaptation and Behavior*, edited by A. J. Mandell. Raven, New York, pp. 155-168.

157. N. J. Uretsky and S. R. Snodgrass (1977). Amphetamine stimulation of tryosine hydroxylation in rat striatal slices. *J. Pharmacol. Exp. Ther.* (in press).

158. M. Goldstein, T. Backstrow, Y. Cji, and R. Frenkel (1970). The effects of Ca^{++} ion on the ^{14}C-catecholamine biosynthesis from ^{14}C-tyrosine in slices from the striatum of rats. *Life Sci.,* 9(pt.I):919-924.

159. B. Anagnoste, C. Shirron, E. Ruedinan, and M. Goldstein (1974). Effect of dibutyryl cyclic adenosine monophosphase on ^{14}C-dopamine biosynthesis in rat brain striatal slices. *J. Pharmacol. Exp. Ther.,* 191:370-376.

160. V. H. Morgenroth, M. C. Boadle-Biber, and R. H. Roth (1975). Activation of tyrosine hydroxylase from central noradenergic neurons by calcium. *Mol. Pharmacol.,* 12:41-48.

161. V. H. Morgenroth, M. C. Boadle-Biber, and R. H. Roth (1976). Dopaminergic neurons: Activation of tyrosine hydroxylase by a calcium chelator. *Mol. Pharmacol.,* 12:41-48.

162. S. W. Gumulka, V. Dinnendalil, H. D. Peters, and P. S. Schonhofer (1976). Effects of dopaminergic stimulants on cyclic nucleotide levels in mouse brain in vivo. *Naunyn Schmiedeberg's Arch. Pharmacol.,* 293:75-80.

163. A. Carenzi, S. Govoni, P. F. Spano, and M. Trabucchi (1976). On the cyclic nucleotide involvement in rat striatal function. *Pharmacol. Res. Commun.,* 8:143-147.

164. R. L. Borison, A. D. Mosnaim, and H. C. Sabelli (1975). Brain 2-phenylethylamine as a major mediator for the central actions of amphetamine and methylphenidate. *Life Sci.,* 17:1331-1344.

Chapter 10

COMPARATIVE PHYSICOCHEMISTRY OF PHENYLETHYLAMINE TARGETS

E. Albert Zeller*

Department of Biochemistry
Northwestern University Medical School
Chicago, Illinois

---

*Current affiliation: National Institute of Environmental Health Sciences, Research Triangle Park, North Carolina.

I. INTRODUCTION

β-Phenylethylamine (PEA) is known to interact with several cellular components. Our knowledge of the forces binding it to various receptors, however, is severely limited. Since its mode of interaction with its receptors is essential to our understanding of its biofunction, I attempted to fill out some of the major informational gaps. This was done by analyzing kinetic and specificity data secured in this and other laboratories. The results of this endeavor served as a basis for comparison of the action of PEA with that of some of its hydroxylated derivatives.

The formation of the receptor-amine complex is followed by a second phase, consisting of chemical changes inflicted upon the substrate, platelet aggregation, or biogenic amines release. The forces involved in the second phase will not be discussed here unless it is necessary to separate them from those acting in the first step. This separation is not always feasible, because the binding energy between substrate and receptor can be utilized to bring about an increase in the reaction rate as recently summarized by Jencks (1).

Some data collected from the literature pertaining to the aromatic amino acid decarboxylase (AAD) have been included here. This

*Joint project with E. C. Rossi.

enzyme catalyzes the production of PEA and its active site offers a welcome contrast to the PEA receptors.

II. FORMATION OF PEA AS CATALYZED BY AROMATIC AMINO ACID DECARBOXYLASE

Table 1 (2) shows that L-phenylalanine, in contrast to L-DOPA, is rather slowly degraded in the presence of AAD (aromatic-L-amino-acid carboxylase, 1.4.28). If phenylalanine were as good a substrate as DOPA, most of the nutritionally supplied phenylalanine would be destroyed by the action of AAD. This statement is based on clinical observations: When several grams of L-DOPA are given to Parkinsonian patients a day, they are so rapidly metabolized by the combined action of AAD, monoamine oxidase (MAO), and other enzymes that only a few milligrams of dopamine (DA) will reach their intended target, namely, the nigrostriatal region.

Why is there such a wide span between the enzymic decarboxylation rates of phenylalanine and DOPA? A partial answer is given in the following sections.

TABLE 1

Kinetic Constants for the Degradation of Aromatic Amino Acids[a]

	K_m (mM)	V_{max} (μg amines mg^{-1} hr^{-1})
L-Phenylalanine	20	104
L-Tyrosine	13	34
L-meta-Tyrosine		2000
L-DOPA	0.4	6400
L-Tryptophan	3	220
5-Hydroxytryptophan	0.02	1000

[a]Kinetic constants characterizing the decarboxylation of L-phenylalanine and some derivatives. A 10-fold purified guinea pig kidney preparation served as the enzyme source (2).

A. Estimates of Free Energy Changes

The Michaelis constant of steady-state relationship, in its simplest form, is as follows:

$$K_m \simeq \frac{k_1 + k_{cat}}{k_{+1}} \qquad (I)$$

If $k_{cat} << k_{-1}$, then K_m converges toward K_s, the latter being the dissociation constant of the enzyme-substrate complex. In such a case, we can use $1/K_m$ as a basis for the computation of the free energy of binding. If K_m and K_s are substantially different, $K_s < K_m$, the free energies calculated represent minimal values. For the enzyme AAD, no information regarding the relative values of K_m and K_s appears to be available.

B. Hydrophobic Bonding

The enzyme catalyzes the degradation of aromatic acids, but not of smaller aliphatic amino acids. The aromatic ring, therefore, seems to be a part of the nonreactive moiety which is essential for binding the substrates to the active site by hydrophobic forces. From K_m values for the decarboxylation of phenylalanine, we arrive at ΔG = 2.3 kcal. If we make the not too far-fetched assumption that the substrate's aralkyl residue will be bound to similar moiety of the active site, we would arrive at 2 kcal for ring-to-ring binding (3) and 2 x 0.8 kcal for the binding of the two methylene residues. Either $K_m >> K_s$ or else the hydrophobic attachment of the benzene ring to another aromatic ring of the active site may be the main force binding in the enzyme-substrate complex formation.

C. The Role of the Catechol Moiety on the Substrate Binding

The introduction of a para-hydroxylic group into phenylalanine causes a three-fold decrease in reaction rate (Table 1), while m-tyrosine is decarboxylated 20 times faster than phenylalanine. This compares with a 60-fold rate increase when we replace phenylalanine by DOPA.

Apparently, the meta-hydroxyl moiety is the important part of the catechol residue with respect to reaction rate. In trying to analyze this observation, we have, unfortunately, to do it without the K_m value for meta-tyrosine degradation. The semiquantitative evaluation of the free energies of binding of the various phenylalanine derivatives supports the concept that *one* hydroxylic group is involved in the enzyme-substrate complex formation. The increment of free energy binding accompanying the introduction of the catechol moiety is 2.4 kcal. While this value is too small for hydrogen bonding, it may be the resultant of two opposite trends: an increase of binding energy through hydrogen bonding and a reduction of hydrophobicity by ring hydroxylation. If a similar relationship between log K_m and partition coefficient π would hold as seen in the MAO reaction (Section III.C), then a loss of 1 to 2 kcal would follow the appearance of a catechol moiety. In adding this energy to the 2.4 kcal, we would reach the range of hydrogen bonding. With both hydroxylic groups attached to the active site by hydrogen bonding, the Michaelis constant would be in the range of 10^{-7} to 10^{-9}, far below the actual value of 4×10^{-4}.

D. Effect of Hydroxylic Groups on Electron Distribution

Hydroxylation in meta-position not only produces a stronger binding of the substrate but also a sharp increase in the reaction rate. Is this behavior due to a reduction of the electron density at the α-carbon atom? In considering the (average) Hammett data for our compounds (Table 2), one would expect that after meta-substitution a slight increase in electron density at the α-carbon atom and a reduction in reaction rate would follow, while the opposite would be true after 4- and 3,4-substitution. Obviously, the data do not support this concept. Thus, after elimination of this conjecture, we tentatively conclude that the 3-hydroxy group, possibly by hydrogen bonding, helps to bind the molecule to the active site and to orient it in such a way that part of the increase of binding

TABLE 2

Average Hammett and Hansch Constants for Aromatic Hydroxylic Groups[a]

	3-OH	4-OH	3,4-OH
σ	+ 0.121	- 0.370	- 0.249
π	- 0.61	- 0.49	- 1.10

[a]Taken from Hansch et al. (4).

energy can be utilized to accelerate the reaction. This could be done by stabilizing the transition state of the interaction between amino group and pyridoxal phosphate. Recently, a quantitative approach to this problem has been made by observing marked changes of the enzyme-substrate interaction when a critical dimension of the enzyme-substrate complex was altered by less than 0.2 Å (5).

E. In Vivo Degradation of L-Phenylalanine

The decarboxylation of phenylalanine appears to be a very minor metabolic pathway. This, however, is not always true. Since the in vivo concentration of phenylalanine (e.g., 10^{-4} M in human blood plasma) is way below the concentration defined by K_m (2×10^{-2}), one can expect the concentration-rate relationship in vivo to be close to linear. Thus, with an increase in the phenylalanine concentration, the decarboxylation rate could rise perceptibly. This conclusion indeed is substantiated by experimental data: Rats fed with large doses of phenylalanine doubled the PEA content in their blood (6). In patients suffering from phenylketonuria, the excretion of PEA multiplied several hundredfold, provided that MAO inhibitors (MAOI) prevented the degradation of PEA. Without inhibitor, there were only small amounts of PEA, the same as found in healthy persons (7). This displays the great efficiency of MAO in the elimination of PEA. Under these circumstances the decarboxylation of phenylalanine becomes a major metabolic route (7).

III. ON THE BINDING OF SUBSTRATES TO THE ACTIVE SITE OF MAO

The forces governing the binding of substrates (and inhibitors) to the active site of MAO (amine : oxygen oxidoreductase (deaminating) (flavin-containing) 1.4.3.4) are not fully explored as yet, although the line of investigation goes back to 1959 (8). On the basis of the results obtained with a large number of substrates and hydrazine inhibitors, it was then suggested that the substrates are bound to an aromatic amino acid in the active site, for example, to phenylalanine. Shortly afterwards, hydrophobic forces were held responsible for the binding of the nonpolar moieties (9). If we assume the amino group to be in the plane of the aromatic ring of the hypothetical phenylalanine, the distance is 5.2 Å from the center of the ring to the receptor. On the basis of quantitative structure-activity studies on MAOI, Johnson recently assigned the value of 5.25 Å to this distance (10). Another important step was made by McEwen et al. who concluded that the nonprotonated form was the molecular species acted upon by MAO (11). But the exact nature of the binding of the amino residue to the active site still remained unanswered.

A. Phenylethylamine as a Substrate of MAO

The structure of PEA, with only one polar group present, simplified the task to gain an overall view of the various types of forces involved in the enzyme-substrate complex formation. McEwen's group provided us with a firm basis for computations of free energy changes. a) They showed K_m essentially to be the same as K_s (dissociation constant of the enzyme-substrate complex) by demonstrating that K_i for one substrate serving as competitive inhibitor of a second substrate was equal to its K_m (11). b) When they tested various substrates at several pH levels, the K_m values remained the same provided that the concentration of the nonprotonated amines served as a basis for the calculations ($\tilde{K}$). A few pertinent data are given in Table 3, including free energies computed on the basis of K_m and K_i values. The average free energy of binding of PEA to MAO is 10.5 kcal. If

TABLE 3

List of K_m, K_i, and $\Delta F°$ Values for the System MAO-Phenylethylamine and Related Amines[a]

Substrate or Inhibitor	Enzyme Source	$\tilde{K}_m$ (μM)	$\Delta\tilde{G}$ (kcal)	$\tilde{K}_i$ (μM)	$\Delta\tilde{G}$ (kcal)
Benzylamine	I	5.1	7.3	4.8	7.3
	II	2.3	7.8		
N-Methylbenzylamine	I			4.1	7.3
N-Dimethylbenzylamine	I			56	5.9
Phenylethylamine	I	0.10	9.7	0.095	9.7
	III	0.01	10.9		
	IV			0.024	10.2
	V	0.02	10.6		
β-Phenylethanolamine	I	5.1	7.3	5.1	7.3
	III	4.9	7.2		
Tyramine	I	0.4	8.8	0.4	8.8
	III	0.5	8.7		
	IV	1.6	8.0	1.6	8.0
	V	0.3	9.0		
	VI	1.0	8.3		
	VII	1.0	8.3		
meta-Tyramine	VI	0.8	8.4		
Octopamine	I	31	6.2	31	6.2
	III	2.4	7.7		
Dopamine	IV	0.7	8.5	1.9	7.9
	III	0.95	8.3		
Norepinephrine	VII	2.5	7.7		

[a] K_m and K_i values are calculated on the basis of the concentration of nonprotonated amines. Enzyme preparations: I, purified human liver MAO (11); II, purified beef liver MAO (12); III, purified beef liver MAO (13); IV, human brain mitochondria (14); V, washed and sonicated human platelets (15); VI, albino mouse liver mitochondria; VII, homogenates of rabbit tissues.

we subtract from this figure approximately 3.6 kcal for the binding of the phenylethyl moiety to a similar nonpolar group in the active site (Section II.B), 7 kcal are left for binding the rest of the molecule. This value is within the range of hydrogen bonding. It is, however, rather high for such a bond involving nitrogen. A suggestion for elimination of this difficulty may come from the fact that MAO easily attacks monomethylated derivatives. Table 3 indicates the binding energy to be about the same for benzylamine and its monomethyl derivative. The introduction of a second methyl residue, though, is accompanied by a marked depression of the binding energy We visualize, therefore, the existence of two bonds, one with the nitrogen serving as a hydrogen donor. Some time ago it was suggested that the amino group may be hooked up with peptide bonds (8).

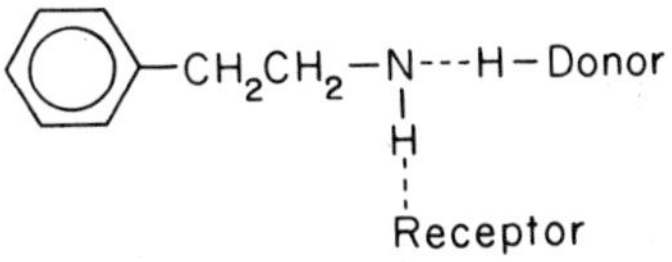

$$C_6H_5-CH_2CH_2-N(-H\cdots Receptor)\cdots H-Donor \qquad (II)$$

A recent study of Hiramatsu et al. produced evidence for the existence of two "essential" imidazole rings participating in the binding of the substrate molecule to the active site (17), which could serve as hydrogen-donor and -acceptor.

B. Effects of β-Hydroxylic Residues on the Free Energy of Binding

Aralkyl alcohols appear to occupy essentially the same spot on the active surface of MAO as the corresponding aralkylamines. This conclusion is based on the following facts: a) aliphatic and aralkyl alcohols compete with amines for MAO. For purified human liver, MAO and β-phenylethanolamine, K_i = 0.3 mM and ΔG = 4.8 kcal (11): Thus, the alcoholic group, according to our conjecture, seems to be able to participate in hydrogen bonding in lieu of the amino residue. b) When either benzylamine or benzyl alcohol is anchored to the active site by this hydrogen bond, their aromatic rings are unable to reach the aromatic center 5.25 Å away. In the following diagram, two positions of the benzyl residues are indicated, without any consideration as to what the *real* conformation may be. Accordingly, a

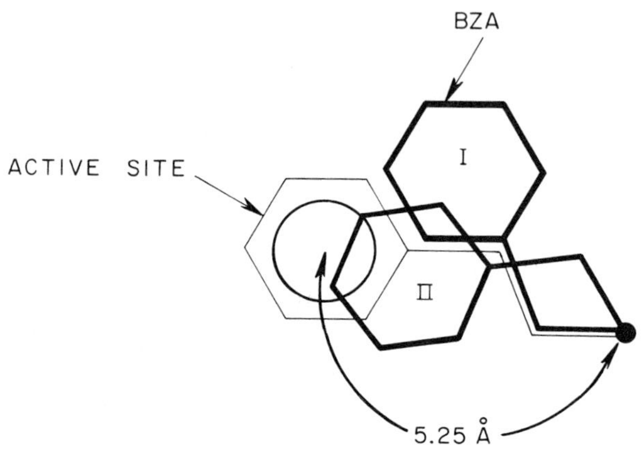

(III)

substantial loss of hydrophobic bonding is to be expected. The pair PEA and benzylamine differs by $\Delta G = 2.4$ kcal with respect to their binding energies (Table 3).

We consider, therefore, that β-phenylethanolamine may react in two ways with the active site, either by leading with the amino or the hydroxylic group. Thus, a competition between the two polar groups appears to take place. When the hydroxylic residue is involved in the bonding, the free energy of binding should be similar to that of benzylamine; the data of Table 3 are in agreement with this concept.

Based on the same scheme used in the diagram III, these enzyme-substrate complexes are given schematically by the following sequence of formulae: the aromatic rings (thin lines) indicate the hydrophobic area of MAO, the small circles the *locus* of electron transfer from the α-carbon of the substrate to the flavin moiety of MAO; the hydrogen attached to the α-carbon of the substrate and to be removed by the enzymic process, is indicated as well as the *locus* of the hydrogen bonding (triangle).

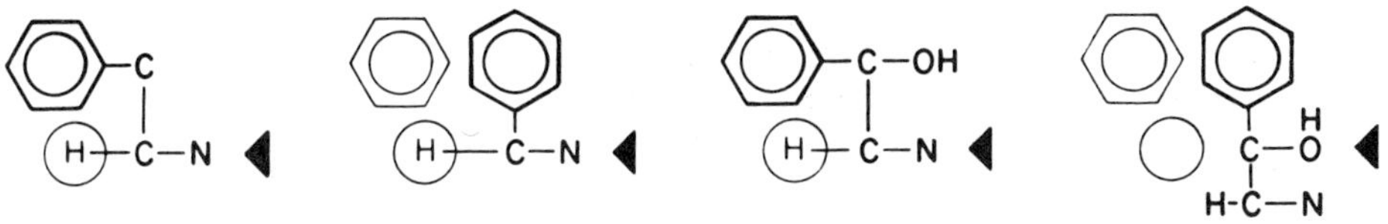

A PEA | B BZA | C β-PETA | D β-PETA

(TYPE B) (IV)

After conversion of PEA, tyramine, and DA to β-phenylethanolamine, octopamine, and norepinephrine (NE), respectively, the maximal velocities dropped down to 12 to 40% from those measured for the parent substances. The lowest value was found for the pair DA-NE when these compounds were degraded in presence of goldfish brain (unpublished). The β-hydroxylic group could be considered as a protective device against the inactivation of NE and epinephrine by type B MAO.

With this model IV, we can interpret these observations: When β-phenylethanolamine is oriented according to IVD, the substrate's α-carbon is too far removed from the electrophilic residue (EPR) of the flavin nucleus (presumably position IVA) to serve as an optimal electron pair donor, and the rate reduction induced by β-hydroxylation would, therefore, be due to the relative occurrence of complexes IVC and D.

This concept of nonoptimal orientation of the substrate molecule on the active site has been useful for the interpretation of many enzymological phenomena (1). These particular complexes have been called nonproductive (18) or dystopic (19). The latter term emerged in the course of extensive studies of the substrate pattern of MAO (20) and of the closely related flavin enzyme L-amino acid oxidase of snake venoms (5).

C. Role of β-Hydroxylic Groups in Determining MAO Specificity with Respect to Types A and B

β-Phenylethylamine is the classical substrate of the type B, while NE and epinephrine are mainly attacked by the type A. Why does not PEA fit into the active site of the A-enzyme, which can accommodate molecules which display the same backbone as PEA but which are larger than the latter? To find some answers to this question, we compared the kinetics of the degradation of PEA and β-phenylethanolamine by types A and B MAO. We observed that β-phenylethanolamine, in contrast to PEA, is easily degraded by the A-type present in mitochondria of rat liver and human placenta. These data suggest that the EPR of MAO is localized in a slightly different position within the two types. In the crude scheme outlined in diagram V, the orientation of PEA and β-phenylethanolamine are given. The former does not reach the EPR with its α-carbon, while the latter does. In using the terminology already cited, one can describe the situation in the following way: The β-hydroxylic group pulls the substrate molecule into a productive (eutopic) position in the active site of type A, but into a nonproductive (dystopic) attachment in type B. β-Phenylethylamine forms a dystopic complex with type A and (eutopic) complex with type B. The problem of the chemical differences between the two types of MAO has been discussed recently by Ekstedt and Oreland (21).

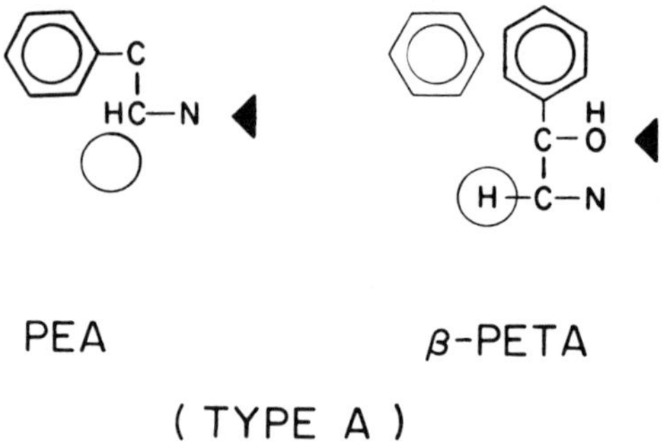

C
HC—N
H
C—O
H—C—N

PEA β-PETA

(TYPE A) (V)

D. The Effect of Phenolic Groups on Binding Energy

In contrast to the AAD reaction, ring hydroxylation of PEA only slightly changes the reaction rates (V_{max}) of MAO. However, the free energy of binding is consistently reduced upon hydroxylation. We explored the nature of this energy loss by applying Hansch's relationship to the analysis of our kinetic data. When the logarithm of dissociation constant of a given complex is plotted against π, which is the logarithm of the partition coefficient of a given substance in the system octanol-water, we often obtain a straight line if hydrophobic forces are predominant in the binding process. If it is true that $K_m \simeq K_s$, then K_m may serve as a dissociation constant of the MAO-substrate complex. In determining the kinetic constants for the degradation of benzylamine and 13 derivatives by partially purified beef liver MAO and in plotting log K_m versus π, we obtained a linear relationship with a slope of -0.43. From this line, we derive ΔG = 0.4 kcal that the appearance of the hydroxylic group in benzylamine reduces the free energy of binding by 0.4 kcal. Unfortunately, similar data are not available for analogous PEA derivatives. The scant and heterogenous material found in the literature, however, suggests that the slope of the Hansch relationship is considerably more negative for PEA derivatives than that for benzylamines. Data from many laboratories indicate $\Delta\tilde{G}$ = 8.6 kcal for tyramine (range 8.3 to 8.8), while the corresponding value for PEA is 9.7 kcal (Table 3). Thus, one phenolic group reduces the binding energy by 1.9 kcal. This result is compatible with a model outlined in Section II.B. The geometry of the enzyme-substrate complex is such that the hydrophobic areas of MAO and substrate share a much larger field when the substrate is a PEA derivative than when it is one of the benzylamines. Changes in the hydrophobicity of the former compounds must, therefore, have a much greater impact to free energy of binding than of the latter benzylamines.

In summary, the effect of ring hydroxylation on K_m appears to be due in part to a reduction of hydrophobicity of the substrate benzene ring.

E. Phenylethylamine and MAO: Inhibition by Competition and by High Substrate Concentrations

Two enzymic phenomena are probably related to the high affinity of the hydrophobic area of PEA for the active site of MAO.

1. PEA as an Efficient Competitive Inhibitor

To inhibit by 50% the degradation of tyramine by rat liver mitochondria, the tyramine : PEA ratio has to be 50 : 1, only one-fiftieth of PEA is required to produce 50% inhibition. By this competition, it may be able to protect other amines from destruction. The inhibition appears to be not purely competitive and contains other inhibitory components (13,14).

2. Inhibition by High Concentration

After reaching a maximum rate with increasing substrate concentration, the velocity goes down beyond a given concentration. In human platelets, 2×10^{-5} M PEA already induces inhibition (13). Inhibition by high concentrations is found only occasionally with other substrates. If we have two moieties in a substrate molecule which bind to the enzyme with similar free energies, two substrate molecules can bind to the active site and thus prevent each other from being degraded. When one moiety, for example the amino group, is first bound to its MAO counterpart, at high concentrations a second molecule may be attached to the aromatic MAO moiety before the first substrate molecule can do the same. This model was first proposed by Haldane (22) and further developed in this laboratory for the case of diamine oxidase (23) and acetyl cholinesterase (24).

With intact rat liver mitochondria, the inhibition becomes apparent at 10^{-5} M, whereas in mitochondria sonicated for 2 min the concentration had to be above 10^{-4} M to produce the same effect, showing that release of the enzyme from the mitochondrial membrane alters its physicochemical properties, including K_m values. Similar observations were made with beef liver mitochondria. Purified MAO, therefore, is to some degree an artifact. This is one of the reasons why we use homogenates and "intact" mitochondria when we want to study certain physicochemical properties of this enzyme that may be directly relevant to its biofunction.

IV. EFFECT OF PHENYLETHYLAMINE ON EPINEPHRINE-INDUCED PLATELET AGGREGATION (Joint Project with E. C. Rossi)

It is a fascinating task to find out whether and to what extent the same forces are involved in the binding of biogenic amines to the active site of enzymes or to aminergic receptors. As an example for the latter type, the binding of epinephrine (EPI) to human platelets is discussed here. As a consequence of the addition of EPI to platelet rich plasma (PRP), aggregation occurs. During this process, light transmission is sharply reduced, which can be followed photometrically and recorded electronically. While extensive work has been done in the analysis of this phenomenon (25), few systematic studies have been carried out regarding the agonist and antagonist specificity of this reaction. A series of experiments were undertaken to explore some of the forces which produce the binding of EPI and related compounds to the platelets (26,27). Again, PEA served as a valuable tool in the investigation of this process.

A. Establishment of Overall Specificity of Epinephrine-induced Platelet Aggregation with Phenylethylamine

From the results obtained in several laboratories it is well-established that EPI-induced aggregation is a biphasic one: At first adenosine phosphate (ADP) is released which in turn produces aggregation (25). We investigated the specificity of the first phase by adding PEA to the measuring system containing PRP at pH 7.4 before EPI was administered. β-Phenylethylamine prevented the aggregation induced by 1.0 μM EPI, but not of the aggregation initiated by 1.5 μM ADP or 0.4 μg/ml collagen. This was the first time that the EPI-initiated process could be separated from the two other ones by an inhibitor. It also strengthened the hypothesis of the existence of a receptor relatively specific for EPI.

B. Quantitative Evaluation of Epinephrine-induced Aggregation

To raise this study to a quantitative level, we used the following approach: When EPI is added to the PRP, a sigmoidal curve is recorded by the aggregometer. The instrument measures changes in light transparency which are due to aggregation. The tracings have a sigmoidal form with an inflection point at 50% aggregation (28). We measured the time interval between the administration of EPI and the appearance of the inflection point. The intervals increase with decreasing EPI concentration and can be considered as reciprocals of reaction rates of the aggregation process. In plotting these against the reciprocals of the EPI concentration, a linear relationship emerged which permitted us to determine the dissociation content of the receptor-EPI complex (K_d). For the platelets of five healthy young adults, we arrived at $K_d = 1.2 \pm 0.4$ μM. Although we have no direct evidence at the present time, we consider for pragmatic reasons that the nonprotonated form of EPI is the reactive species (see Section III). Thus, we gain $\tilde{K}_d = 0.005$ μM, $\Delta G = 8.1$ kcal, and $\Delta\tilde{G} = 11.4$ kcal.

When PEA, which is not an aggregation agonist, is added with EPI to our measuring system, the appearance of the EPI-induced curves is delayed. Therefore, PEA appears to be an inhibitor of EPI aggregation. The double-reciprocal plot gave a straight line with the same intercept on the 1/y axis as the EPI line. β-Phenylethylamine, therefore, seemed to act as a competitive inhibitor of EPI. The computation yields $K_i = 5.9 \pm 2.3$ μM, $\tilde{K}_i = 24$ nm, $\Delta G = 8.1$ kcal, and $\Delta\tilde{G} = 11.4$ kcal. The latter value is close to 9.7 to 10.9 kcal calculated for the PEA-MAO system. From these data we draw the tentative conclusion that PEA is bound to the platelet receptor in a way which is similar to the active site of MAO (Section II.A). The values for G_{EPI} and G_{PEA} are also fairly close.

C. The Role of Hydroxylic Groups in the Initiation of Aggregation

β-Phenylethylamine is almost as strongly bound to the platelet as EPI, but in contrast to the latter, it is incapable of initiating aggregation. Which of the substituents of EPI are then responsible for the action of EPI? To find the answer to this question, we tested a considerable number of derivatives of PEA as aggregation agonists and EPI antagonists. Three compounds only, and all three of them catecholamines, turned out to be able to initiate platelet aggregation. They are given in order of increasing efficiency.

EPI > NE >> DA

When only one hydroxylic group is present, as in tyramine and metatyramine or octopamine (p-hydroxyphenylethanolamine), no aggregation can be induced. The same is true if one of the phenolic groups is methylated as in metanephrine and normetanephrine. Apparently, the catechol moiety is essential for the action on aggregation, presumably for the release of ADP.

D. Substituted Arylalkylamines as Inhibitors of Epinephrine-induced Aggregation

Among arylalkylamines, we found many antagonists with a wide range of inhibitory power. The sequence of activities is the following:

PEA > Phenylpropylamine, phenylbutylamine > benzylamine

The biological activity of PEA displays marked sensitivity to relatively small structural changes (Table 4). Phenylephrine which is identical with EPI except for the lack of the para-phenolic group, is the most powerful antagonist. In contrast, octopamine, which differs from NE only by the omission of the meta-hydroxylic residue, exhibits very weak activity. Epinephrine seems to fit tightly into its receptor because bulky residues prevent significant agonistic and inhibitory action. When the N-methyl residue in EPI is replaced by a N-isopropyl moiety, isoproterenol is obtained which is almost devoid of action. Methylation of the meta-hydroxyls of EPI and NE

TABLE 4

Effect of Hydroxylation on Inhibitory Power of Phenylethylamine on EPI-induced Aggregation

Amine > PEA	Amine ≃ PEA	Amine < PEA	Amine << PEA
Phenylephrine	β-Phenylethanolamine	Phenylpropylamine	Benzylamine
		Phenylbutylamine	D,L-Amphetamine
		Tyramine	Octopamine
		Dopamine	meta-Tyramine
		Metanephrine	Normetanephrine
			Isoproterenol

[a]These symbols represent relative inhibitory power of aralkylamines in comparison to PEA effects

also abolishes most of the antagonistic power suggesting that the methoxy group prevents snug fitting of the molecule into the receptor slot. There is, however, another possibility to be considered: The methoxy residue, in contrast to the hydroxy group, cannot serve as a donor in hydrogen bonding. This latter concept is supported by the fact that phenylephrine is a strong inhibitor. Thus, the meta-hydroxy residue may participate in two functions of the EPI action: a) binding to the receptor and b) involvement of ADP release. With more kinetic data available, a more precise picture of the role of various parts of EPI the molecule in the biofunction of EPI will emerge.

V. SUMMARY

This study obviously is in an early stage. Its main points are summarized in Table 5. Although the conclusions are fraught with question marks, they point toward similarities and dissimilarities between the various targets of PEA.

This work was undertaken as a preparation for a similar analysis of the amine-releasing action of PEA. We believe that this is a major biofunction of PEA (29). An example has been discussed in the preceding paper (13), namely, the possible role of PEA in the response of the rabbit iris to inhibitors of MAO types A and B. However, more quantitative investigations regarding the specificity of amine-release phenomena are needed before this new field can be successfully treated.

There is one particular gap in our discussion which we consider more serious than most other ones. It pertains to the proposition that the intrinsic binding energy that results from the noncovalent interaction of a specific substrate with the active site of the enzyme is considerably larger than is believed. An important part of this binding energy may be utilized to provide the driving force for catalysis, so that the observed binding energy represents only what is left over after this utilization (1). It is an important task of the future to extend this enzymological concept to nonenzymic systems and to find out whether in aminergic receptors the binding energy (as calculated in this paper) represents only a fraction of the intrinsic binding energy.

ACKNOWLEDGMENTS

This study was in part supported by research grants from the National Institute of Mental Health, Bethesda, Maryland (MH20020). I thank M. S. Bieber for her great help in the preparation of the manuscript.

REFERENCES

1. W. P. Jencks (1975). *Adv. Enzymol.*, 43:219.
2. W. Lovenberg, H. Weissbach, and S. Udenfriend (1962). *J. Biol. Chem.*, 237:89.
3. C. Tanford (1962). *J. Am. Chem. Soc.*, 84:4240.
4. C. Hansch, A. Leo, S. H. Unger, K. H. Kim, D. Nikaitani, and E. J. Lien (1973). *J. Med. Chem.*, 16:1207.
5. E. A. Zeller, L. M. Clauss, and J. T. Ohlsson (1974). *Helv. Chim. Acta*, 57:2406.
6. J. C. David, W. Dairman, and S. Udenfriend (1974). *Arch. Biochem. Biophys.*, 160:561.

7. J. A. Oates, P. Z. Nirenberg, J. B. Jepson, A. Sjoerdsma, and S. Udenfriend (1963). *Proc. Soc. Exp. Biol. Med.*, 112:1078.

8. E. A. Zeller (1959). *Pharmacol. Rev.*, 11:387.

9. E. A. Zeller (1960). *Experientia*, 16:399.

10. C. L. Johnson (1976). *J. Med. Chem.*, 19:600.

11. C. M. McEwen, Jr., G. Sasaki, and W. R. Lenz, Jr. (1968). *J. Biol. Chem.*, 243:5217.

12. J. C. Miller (1968). Dissertation, Northwestern University, Evanston, Ill.

13. E. A. Zeller, A. D. Mosnaim, R. L. Borison, and S. V. Huprikar (1976). *Papers, Satellite Symposium on: New First and Second Messengers in Nervous Tissues.* Raven Press, New York.

14. H. L. White and J. C. Wu (1975). *J. Neurochem.*, 25:21.

15. E. A. Zeller, B. Boshes, J. Arbit, M. Bieber, E. R. Blonsky, M. Dolkart, and S. V. Huprikar (1976). *J. Neural Transm.*, in press.

16. L. P. Bausher (1976). *Invest. Ophthalmol.*, 15:529.

17. A. Hiramatsu, S. Tsurushiin, and K. T. Yasunobu (1975). *Eur. J. Biochem.*, 57:587.

18. C. Niemann (1964). *Science*, 143:1287.

19. E. A. Zeller (1963). *Ann. N.Y. Acad. Sci.*, 107:811.

20. E. A. Zeller (1963). *Biochem. Z.*, 339:13.

21. B. Ekstedt and L. Oreland (1976). *Biochem. Pharmacol.*, 25:119.

22. J. B. S. Haldane (1930). *Enzymes.* Longmans, Green and Co., London, p. 84.

23. E. A. Zeller (1940). *Helv. Chim. Acta*, 23:1418.

24. E. A. Zeller and A. Bissegger (1943). *Helv. Chim. Acta*, 26:1619.

25. J. F. Mustard and M. A. Packham (1970). *Pharmacol. Rev.*, 22:97.

26. E. C. Rossi and E. A. Zeller (1976). *Fed. Proc.*, 35:732.

27. E. C. Rossi and E. A. Zeller (1976). *Abstracts, Central Soc.*, in press.

28. E. C. Rossi and G. Louis (1975). *J. Lab. Clin. Med.*, 85:300.

29. L. Full, H. Grobecker, and J. Jonnson (1967). *Eur. J. Pharmacol.*, 2:202-207.

Part B

BEHAVIORAL STUDIES

Chapter 11

STEREOTYPED BEHAVIOR IN RATS INDUCED BY PHENYLETHYLAMINE, DEPENDENCE ON DOPAMINE AND NORADRENALINE, AND POSSIBLE RELATION TO PSYCHOSES?

Claus Braestrup and Axel Randrup

The Research Laboratory, Department E
St. Hans Mental Hospital
Roskilde, Denmark

I. INTRODUCTION

Pharmacological, biochemical, and behavioral investigations of amphetamine in animal experiments have been carried out in our laboratory for several years. These investigations were originally inspired by reports from the human clinic about an amphetamine psychosis resembling certain forms of schizophrenia, especially the paranoid forms, so much that these psychoses were misdiagnosed as genuine schizophrenia. It was believed that fundamental knowledge of the mechanism of action of amphetamine might be relevant to the understanding of the pathogenesis of schizophrenia and the action of antipsychotic drugs (4,54,55,78,92).

A massive body of evidence now corroborates the link between the action of amphetamine and dopamine (DA) transmission in the brain and further between hyperfunction of central DA transmission and schizophrenia.

On this background, it seems reasonable to consider the behavioral actions and pharmacological mechanisms of β-phenylethylamine (PEA). β-Phenylethylamine is chemically very closely related to amphetamine (α-methyl-β-phenylethylamine) but in contrast to amphetamine, PEA is synthesized within the living organism, presumably by decarboxylation of the naturally occurring amino acid l-phenylalanine. β-Phenylethylamine may, therefore, be anticipated to have effects in undrugged organisms. The purpose of this presentation is to review and extend the behavioral, biochemical, and pharmacological similarities between exogenous PEA and the well-described drug amphetamine in relation to stereotyped behavior. Stereotyped features are important characteristics of the amphetamine psychosis, and also patients suffering from the endogenous psychoses schizophrenia, mania, and depression exhibit stereotyped behaviors and stereotyped thinking (75).

II. AMPHETAMINE, BRAIN DOPAMINE, AND STEREOTYPED BEHAVIOR

This topic has been reviewed several times recently (15,70,76,81,83), and it now seems generally agreed that amphetamine elicits stereotyped behavior by an action on brain DA. Only pertinent parts of the extensive evidence shall be mentioned here.

The behavior described as stereotyped is seen very clearly in rats treated with 10 mg/kg D-amphetamine sulphate s.c. About 1 hr after the injection, the behavior develops into an extremely stereotyped pattern (i.e., it lacks variation and is apparently aimless), the behavior now consists of only continuous sniffing, licking, or biting of the cage floor. The rats sit in crouched postures and usually press their bodies against the cage wall. Normal activities such as grooming, eating, rearing, and forward locomotion are absent.

The stereotyped behavior induced by amphetamine can be inhibited by α-methylDOPA and completely prevented by α-methyltyrosine, which effectively inhibits the synthesis of the catecholamines DA and noradrenaline (36,69,72,77,84,86,94). This result shows that the catecholamines are involved in the stereotyped behavior and strongly indicates that amphetamine acts indirectly via release of catecholamines. The involvement of catecholamines is further supported by selective lesions of central catecholamine neurones with 6-hydroxydopamine (6-OH-DA). 6-Hydroxydopamine was injected into substantia nigra, which contains the cell bodies connected to more than 70% of the DA in brain nerve terminals, and this treatment completely prevented the amphetamine-induced stereotyped behavior (20,91). The aforementioned experiments do not point directly to DA rather than noradrenaline as mediator, but indirectly it seems that noradrenaline is not a prerequisite for the development of the stereotyped feature of behavior. It was found that diethyldithiocarbamate (DDC), the related drugs disulfiram, and FLA-63, which all inhibit the synthesis of noradrenaline but not DA, do not prevent the amphetamine-induced stereotypies (61,63,80, Braestrup, unpublished). Drugs which block the α- and β-adrenoceptors for noradrenaline or adrenaline are also unable to inhibit the stereotypies (37,48,49,79). These results exclude noradrenaline on a pharmacological basis, and experiments with microinjection of 6-OH-DA in the ventral and medial noradrenaline ascending axon bundles, effecting an extensive and selective destruction of central noradrenaline terminals without inhibition of the amphetamine-stereotypy (20, Braestrup, unpublished), further point to DA as mediator of stereotyped behavior.

The amphetamine stereotypies are antagonized by small doses of almost all antipsychotic drugs, which, as a common feature, possess the ability to block dopamine receptors [see review by Randrup and Munkvad (76)].

Furthermore, both amphetamine and DA can provoke stereotyped behavior when injected into the DA-rich area corpus striatum, and injection of antipsychotic drugs or lesions in this area can inhibit the stereotyped behavior (18,21,29,31,67). Also other DA-rich structures in the forebrain, such as tuberculum olfactorium, globus pallidus, and nucleus accumbens, may be involved (2,3,15,17,19,22).

III. PEA-INDUCED STEREOTYPED BEHAVIOR

The behavioral effect of β-phenylethylamine-HCl is rather weak and short lasting in rats when the drug is injected alone. We have found that 20 to 40 mg/kg PEA-HCl injected s.c. to male Wistar rats about 200 g in weight produces mostly some alerting effect for less than 1 hr. The rats are lying like sphinxes in their wire mesh cages almost without moving, but awake and with weak sniffing and slow head bobbing (head moving up and down). Intraperitoneal injection of PEA produces stronger behavioral effects, and 40 mg/kg PEA injected i.p. causes piloerection within few minutes and some increase in locomotion and rearing. The peak drug effect is observed between 10 and 25 min after injection where the rats sniff continuously in the air. Eighty milligrams per kilogram PEA i.p. produces a short lasting stimulation with high intensity. Already 10 min after injection the rats assume crouched postures and perform continuous sniffing for about 35 min. The sniffing period resembles the sniffing induced by s.c. injected PEA in combination with MAO inhibition (see discussion that follows). The periods before and after sniffing behavior differ from the pre- and post-sniffing phases of lower doses of D-amphetamine in that locomotion and rearing are absent after the high doses of PEA injected i.p. Our laboratory and others have previously described amphetaminelike stereotyped stimulation in rats by 60 mg/kg PEA-HCl i.v. (74) or 50 to 100 mg/kg i.m. (30,53; see also Jackson (47)).

To exclude the action of potentially active oxidized metabolites of PEA (43) and to prolong and potentiate the behavioral action, we combined PEA with monoamine oxidase inhibitors (MAOI). In rats pretreated with specific inhibitor of PEA oxydation (1-deprenyl, 8 mg/kg, s.c., 5 hr), we observed the following very characteristic and reproducible effect of 40 mg/kg s.c. PEA-HCl: After 5 min, the rats showed a weak increase in locomotor and rearing activities covering the whole wire mesh cage (floor 21 x 27 cm, 17 cm high). This behavior lasted until about 20 min after PEA injection, the locomotion and rearing then decreased in intensity and the rats eventually assumed a hunched back posture with continuous sniffing to a small area of the floor or sometimes the lower wall of the cage. While sniffing at the floor the head was turned from side to side, still with the nose near the bars of the cage. This behavior, which looked very much like that seen after lower doses of D-amphetamine-sulphate (3-5 mg/kg, s.c.), was not interrupted by locomotion, rearing, eating, drinking, grooming, or quietness until about 80 min after PEA injection, where the continuous sniffing wore off and was replaced by the reoccurrence of locomotion, rearing, and other varied normal activity. The period of sniffing was completely and clearly stereotyped, all other normal behaviors were absent and no interruptions were noted. The sniffing behavior was performed compulsively, extremely intensely, and persistently. Noise or tactile stimuli did not interrupt the behavior, even lifting the rats by their tails did not interrupt sniffing until the head was removed from the bars of the cage. No signs of aggressive behavior was observed in the sniffing phase, neither in response to objects nor to other rats.

Induction of amphetaminelike behavior by the combination of PEA and MAOI have also been reported by others (30,31,62, Jackson, personal communication).

A. Differential Effect of Inhibitors of Types A and B Monoamine Oxidase

Phenylethylamine, like the catecholamines, are degraded in the brain by monoamine oxidases (MAO) (6). It appears that DA and noradrenaline are degraded by one type of MAO, type A (8,34,35,50,51,88,99), while PEA is oxidized by another type of MAO, type B (98), to phenylacetic acid (6,60). Figure 1 shows that the inhibition of type B MAO strongly potentiates and prolongs the behavioral action of PEA, while the strong and selective inhibitor of type A MAO (clorgyline) does not potentiate or prolong the PEA effect. This result clearly shows that it is not the inhibition of catecholamine degradation by l-deprenyl that potentiates PEA, but rather the inhibition of the oxidation of PEA itself. The stereotyped sniffing produced by a low dose of amphetamine, which is not a substrate for MAO (6), was not potentiated by clorgyline (8 mg/kg, s.c., 5 hr before 2.5 mg/kg D-amphetamine-sulphate, four rats) or l-deprenyl (8 mg/kg, 5 hr before amphetamine, four rats). This result shows that the accumulation of PEA in the CNS after amphetamine (7) is probably not mediating the stereotyped behavior induced by amphetamine (7,25) since l-deprenyl then should be expected either to exhibit amphetaminelike effects by itself or to increase the amphetamine effect strongly. Inhibition of type B MAO by l-deprenyl in doses below 100 mg/kg does not induce behavioral activation (8).

B. l-Phenylalanine

All our attempts to produce the PEA-induced stereotyped behavior by administration of its precursor l-phenylalanine have been fruitless. l-Phenylalanine 1500 mg/kg perorally or 333 mg/kg subcutaneously do

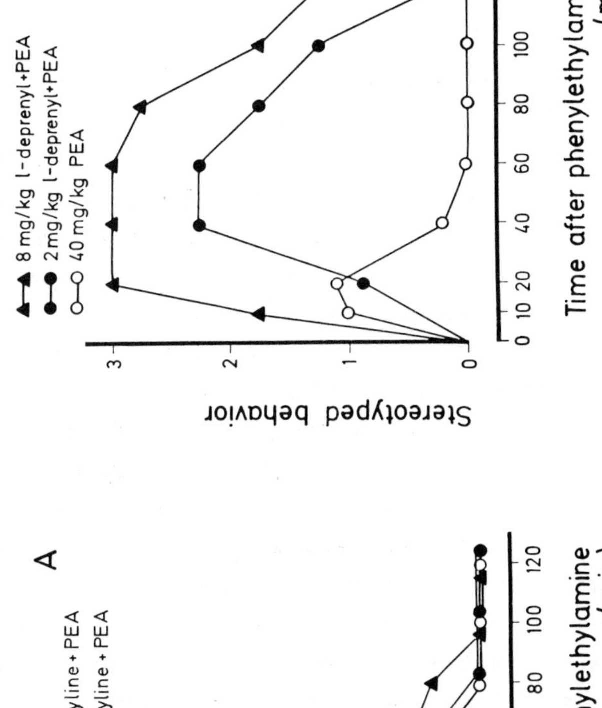

FIG. 1. Stereotyped behavior induced in groups of five rats by 40 mg/kg s.c. of β-phenylethylamine-HCl in combination with clorgyline (a type A MAO inhibitor) or l-deprenyl (a type B MAO inhibitor). 1: occasional sniffing and slow head movements up and down, and/or increased locomotor activity; 2: continuous sniffing, covering the whole cage and increased rearing and locomotion; 3: continuous sniffing at the bottom of the cage with head movements from side to side. No locomotion or rearing. Clorgyline or l-deprenyl alone (2 or 8 mg/kg) never exceeded a mean score of 0.2. (See (8).)

not induce stereotyped sniffing or other activations in our rats. Not even pretreatment 5 hr before l-phenylalanine with an inhibitor of type A MAO (clorgyline, 10 mg/kg, s.c.), type B MAO (l-deprenyl, 10 mg/kg, s.c.), or both produced stereotyped sniffing; only alertness and weak discontinuous sniffing with head bobbing were observed after the combination of phenylalanine and MAOI. The remarkably weak stimulating effect of l-phenylalanine in rats has also been noticed for locomotor activity (87).

C. Inhibition of PEA-stereotyped Behavior by Antidopamine Drugs

The only common characteristic of antipsychotic drugs is their unique ability to antagonize DA receptors in the brain (see Section II). Low doses of two most specific DA-receptor blocking antipsychotic drugs, haloperidol and spiramide (0.2 mg/kg, i.p., 1 hr before PEA), completely blocked the stereotyped sniffing behavior of PEA (40 mg/kg, s.c.; l-deprenyl 8 mg/kg, 5 hr before). This result strongly indicates that the stereotyped behavior of PEA is mediated via DA receptors, probably via release of DA (see discussion that follows).

The antidepressant drug desmethylimipramine (20 mg/kg, s.c., 1 hr before PEA) does not change the PEA-induced stereotypy (40 mg/kg, s.c.; neither with nor without pretreatment with l-deprenyl). Neither do the β-adrenoceptor antagonist propranolol (5 mg/kg, s.c., 0.5 hr) nor the serotonin (5-HT) antagonist methergoline (0.5 mg/kg, s.c., 2 hr). α-Adrenoceptor blocking drugs do not inhibit the stereotyped behavior but they induced profound changes (see Section V).

IV. INTERACTION OF PEA WITH RESERPINE AND α-METHYLTYROSINE

Although reserpine depletes the endogenous stores of biogenic amines, including the stores of DA, it does not inhibit the central stimulant action, including stereotyped behavior, induced by amphetamine (31, 52,72,84,86). For several years, this finding propagated the notion that the excitatory action of amphetamine was due to a direct action

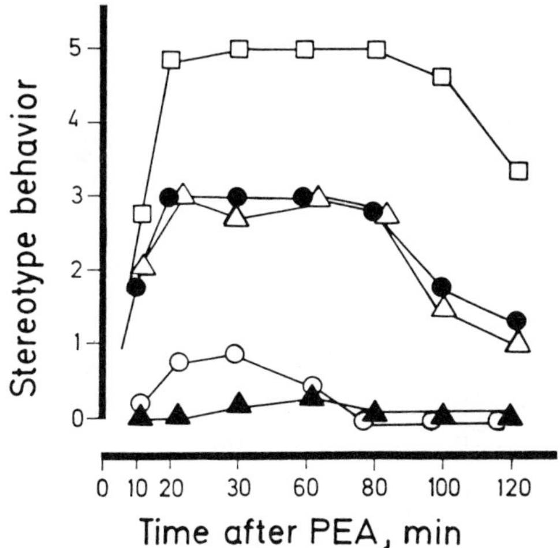

FIG. 2. Effects of reserpine and α-methyltyrosine on the stereotyped behavior induced in groups of four to five rats by 40 mg/kg s.c. phenylethylamine-HCl (all rats pretreated with 8 mg/kg l-deprenyl). Pretreatments: closed circles, saline (controls); open triangles, α-methyltyrosine methylester (H 44/68), 100 mg/kg, i.p., 0.5 hr before PEA; closed triangles, H 44/68, 250 mg/kg, i.p., 5 hr before PEA; open squares, reserpine, 7.5 mg/kg, s.c., 18 hr before PEA; open circles, reserpine (7.5 mg/kg, 18 hr) plus H 44/68, 25 mg/kg, 0.5 hr before PEA. Rating: 1 to 3 as Fig. 1; 4 as score 3 plus occasional licking or gnawing; 5, continuous licking or gnawing.

on receptors rather than being mediated via biogenic amines. The issue was settled in favor of an indirect, DA-mediated action of amphetamine by use of synthesis inhibitors (see Section II). The explanation for the noninhibitory action of reserpine on amphetamine lies in the fact that the synthesis of DA continues after reserpine even though the stores are severely depleted (1,5,71, Table 3). The noradrenaline synthesis is strongly inhibited (11).

β-Phenylethylamine clearly belongs to the reserpine-resistant group of amphetamine congeners, like also methylamphetamine, phenmetrazine, and l-amphetamine (14,86). As shown in Fig. 2, a high dose of reserpine has no inhibitory action on the PEA-induced stereotyped behavior in agreement with previous studies (30,31,47). On the contrary, the behavior appears in the pronounced gnawing form of stereotyped behavior, a feature never observed after PEA alone. The persistent effect of PEA in reserpinized rats indicates that this drug can somehow protect DA, which is currently synthesized during reserpine treatment, from rapid intraneuronal degradation and make this DA available for release.

The indirect mechanism of action for the PEA-induced stereotyped sniffing is clearly shown in Fig. 2. When the synthesis of DA and noradrenaline is inhibited by a high dose of α-methyltyrosine, which also depletes DA and noradrenaline from the brain, then the stimulation induced by PEA is completely prevented. The possibility that α-methyltyrosine inhibits the formation of a metabolite of PEA with a direct action is not likely, since the injection of α-methyltyrosine only 0.5 hr before PEA, when the synthesis inhibitory actions are highly developed, but when the catecholamines are not yet severely depleted (89), will not prevent the PEA-induced excitation (Fig. 2). Also d-amphetamine-induced stereotypy was less susceptible to inhibition by α-methyltyrosine when administered at short time intervals (69, Braestrup, in preparation).

The indirect action of PEA already indicated is further substantiated by the finding that only very small doses of α-methyltyrosine are needed to strongly inhibit the PEA-stereotypy in reserpinized rats (Fig. 2).

In contrast to the indirect mechanism of action for PEA-induced stereotyped behavior, studies on other parameters have indicated direct effect of PEA on central neurons (33), on turning behavior after unilateral DA lesion (23), and on locomotor activity (27,40, 45,46). Low doses of PEA in combination with MAOI, however, appear to affect locomotor activity by an indirect mechanism (43).

V. NORADRENERGIC INFLUENCE ON PEA BEHAVIOR

β-Phenylethylamine injected alone or in combination with MAOI never induce licking or gnawing behavior in our rats as we observe after other amphetaminelike drugs (i.e., D-amphetamine, methamphetamine, phenmetrazine, methylphenidate, pipradrol, and amfonelic acid (NCA)) (86) or the direct DA-receptor stimulant drug apomorphine (24). Licking or gnawing was not even observed after an almost lethal dose of 160 mg/kg s.c. PEA-HCl in combination with l-deprenyl.

Several authors have considered the transition from continuous, stereotyped sniffing to licking, biting, or gnawing as an indication of increased dopaminergic stimulation, but results with PEA indicate that the change in behavior from continuous sniffing to licking or gnawing may be inhibited by increased noradrenergic stimulation. This effect is shown when a variety of different drugs with the common property to reduce central noradrenaline transmission are combined with PEA. In all instances, that is, after reserpine [which depletes noradrenaline and inhibits its synthesis (11)], phenoxybenzamine (which blocks α-adrenoceptors), FLA-63 (which inhibits the synthesis of noradrenaline), or clonidine [which may inhibit release of noradrenaline (12)], PEA will induce the normally observed sniffing together with occasional or continuous licking or biting/gnawing (Table 1). This result indicates that PEA, besides a releasing action on brain DA, possesses a strong releasing action on brain noradrenaline, an action which inhibits the development of the extreme signs of stereotyped behavior in rats, that is, the licking, biting, and gnawing components. The biochemical findings presented in Section VI substantiate this conclusion.

Besides the pharmacological evidence presented here, experiments with lesions in the ascending noradrenaline neurons strongly point to a noradrenergic component in the PEA-induced sniffing behavior. The lesions described in Fig. 3 induced a severe depletion of noradrenaline in the CNS. The major noradrenaline metabolite MOPEG was decreased to about 50% of the level in sham-operated animals, and the uptake of noradrenaline was decreased to 30% of control in the occipital cortex, while DA, judged from the level of homovanillic acid (HVA), and 5-HT, judged from the level of 5-hydroxyindoleacetic acid was not affected. As shown in Fig. 3, PEA induced licking or gnawing in almost all animals tested with the described lesion. The

TABLE 1

Induction of Gnawing or Licking by Phenylethylamine After Pretreatment with Drugs Modifying Noradrenergic Function

Pretreatment[a]	Dose (mg/kg)	Time Before PEA Injection (hr)	Numbers of Gnawing or Licking/Numbers Used	Percent Possible Response
Saline			1/28	3
Reserpine s.c.[b]	7.5	3	4/5	80
Reserpine s.c.	7.5	20	9/9	100
Phenoxybenzamine	20	5	5/9	55
Clonidine	0.5	1	13/15	87
Phenoxybenzamine + clonidine	20 0.5	2 1	4/4	100
FLA-63	2 x 30	5 and 18	7/10	70

[a]All groups, except the one receiving reserpine 3 hr before PEA, received 8 mg/kg l-deprenyl s.c. (Knoll, Budapest) 5 hr before 40 mg/kg PEA-HCl s.c. plus the indicated pretreatment. The group pretreated with saline exhibited intense sniffing, starting 10 min after PEA and lasting 80 to 100 min.

[b]This group received only reserpine and 80 mg/kg phenylethylamine-HCl i.p.

[See Mogilnicka and Braestrup (65).]

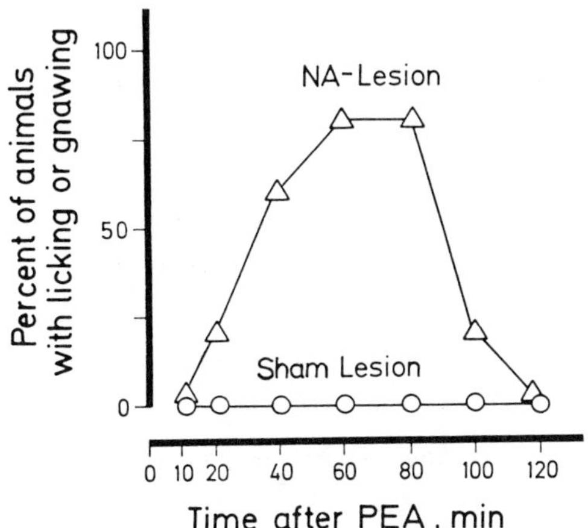

FIG. 3. Induction of licking or gnawing by 40 mg/kg s.c. phenylethylamine-HCl (8 mg/kg l-deprenyl 5 hr before) in rats with lesions in ascending noradrenaline neurons. Open triangles, five male Wistar rats weighing 180 g were injected with 8 μg 6-hydroxydopamine (6-OH-DA) at 0.4 mm anterior to the interaural line, 1.3 mm lateral to the midline and 1.9 mm and 2.9 mm below the horizontal zero plane (58). Total injection of 32 μg 6-OH-DA-Br to each rat. Open circles, five male Wistar rats with sham operations without injections. Fourteen days after the operation 8 mg/kg l-deprenyl was injected 5 hr before 40 mg/kg PEA-HCl s.c., and the rats were observed continuously in wire mesh cages without food and water. All rats exhibited sniffing, which in the lesioned animals was interrupted by occasional or continuous licking or gnawing (Braestrup, in preparation).

effects of lesions in central noradrenaline neurons on the stereotyped sniffing produced by low doses of apomorphine (0.25 mg/kg, s.c.) or D-amphetamine sulphate (5 mg/kg) were qualitatively the same though less pronounced (Braestrup, in preparation).

VI. BIOCHEMICAL INTERACTIONS BETWEEN PEA AND THE CATECHOLAMINES

The interaction between PEA and brain catecholamines can be demonstrated in several biochemical measures. In this respect there appears to be only quantitative differences between PEA and amphetamine.

The brain levels of noradrenaline are strongly depleted by PEA (30,41,44,53), while the effect on endogenous DA is biphasic with an increase 1 hr after a single injection and a decrease after repeated injections (30). Similar biphasic results can be obtained with amphetamine (38, see review by Welch and Welch (95)).

β-Phenylethylamine potentiates the depletions of central noradrenaline induced by α-methyltyrosine in schedules where PEA alone does not affect the level of noradrenaline (42). In contrast, the depletion of DA was not enhanced by PEA in the α-methyltyrosine test (42), but as shown by Papeschi et al. (69), the results of turnover studies using synthesis inhibition with α-methyltyrosine are highly dependent on the order of drug administration and the time and dosage schedules used. Amphetamine has thus induced increased, unchanged, or decreased disappearance of DA after α-methyltyrosine (16,32,59,69,96).

β-Phenylethylamine is a moderately potent inhibitor of noradrenaline uptake in the rat heart, being about six times less potent than D-amphetamine (39). We have found PEA to be about five times less potent than D-amphetamine as inhibitor of [^{3}H]DA uptake into crude synaptosomes of rat corpus striatum (Table 2). The uptake inhibitory activity of PEA on dopaminergic terminals cannot account for the production of stereotyped sniffing (see Braestrup and Scheel-Krüger (14)).

TABLE 2

Inhibition of Dopamine Uptake in Corpus Striatum by D-Amphetamine and Phenylethylamine[a]

	IC_{50} Molar Concentration
Amphetamine	2.8×10^{-7}
Phenylethylamine	1.5×10^{-6}

[a]Crude synaptosomes were prepared from rat corpus striatum according to Kuhar (57) and incubated for 3 min at 37°C with 10^{-7} M [^{3}H]DA according to Tuomisto et al. (90). The concentration causing 50% inhibition of uptake (IC_{50}) was estimated from 10 single analyses.

β-Phenylethylamine induces a profound increase in the level of HVA in the rat brain (Table 3). The increase in this major methylated DA metabolite in the CNS of rats strongly indicates that PEA releases DA from presynaptic terminals to the synaptic cleft in agreement with the pharmacological studies depicted in Section IV.

The proposed indirect mechanism of action of PEA on brain DA in reserpinized rats (Section IV) is in accordance with the biochemical results. The high levels of HVA and DOPAC in reserpinized rats (Table 3) show that the synthesis of DA is not impaired though the DA storage capacity is destroyed. The decrease in HVA and especially in the intraneuronally formed metabolite (DOPAC) induced by PEA in reserpinized rats (Table 3) show that PEA can influence the DA transmission system without an intact storage capacity. The decrease in both HVA and DOPAC may indicate a feedback inhibition of DA synthesis.

The major noradrenaline metabolite in the CNS of rats, conjugated MOPEG (10,13,68,85), which is believed to reflect central noradrenergic activity (9,12,13,56,93), is strongly increased by PEA (Table 3). Also tritium-labeled MOPEG, formed after intraventricular injection of the noradrenaline precursor tyrosine, is strongly increased by PEA (68). The PEA-induced increase in MOPEG is more pronounced than the increase induced by amphetamine, and this difference may be the biochemical indication for a very active release of noradrenaline induced by PEA. The pronounced release of noradrenaline by PEA is probably the factor that restricts the behavior in rats to continuous sniffing.

VII. SUMMARY AND DISCUSSION

The presented data and previous results strongly indicate that the stereotyped feature of the behavior induced by PEA is definitely dependent on the availability of endogenous DA for release. In this respect, PEA is amphetaminelike and belongs to the reserpine-resistant group of amphetamine congeners.

Besides the DA-releasing action, PEA appears to have a strong central noradrenaline-releasing action, and this ability may account

TABLE 3

The Level of Homovanillic Acid (HVA), 3,4-Dihydroxyphenylacetic Acid (DOPAC) and Total 3-Methoxy-4-hydroxyphenylglycol (MOPEG) in the Whole Rat Brain After Pretreatment with D-Amphetamine-sulfate, β-Phenylethylamine-HCl and Apomorphine[a]

Treatment	Time Before Decapitation	Percent of Vehicle (or vehicle + 1-deprenyl)		
		HVA	DOPAC	Total MOPEG
PEA, 80 mg/kg, i.p.	1 hr	290^{***} ± 10 (4)	114 ± 8 (4)	227^{***} ± 24 (5)
PEA, 40 mg/kg, s.c. + 1-deprenyl 2 mg/kg	2 hr 7 hr	152^{***} ± 6 (5)	89.5^{*} ± 1.3 (5)	218^{***} ± 13 (5)
D-Amphetamine, 10 mg/kg, s.c.	2 hr	183^{***} ± 6 (5)	62.4^{***} ± 5.8 (5)	154^{**} ± 10 (5)
Apomorphine, 0.5 mg/kg, s.c.	1 hr	66.6^{***} ± 2.5 (5)	71.3^{**} ± 3 (5)	120^{*} ± 7 (6)
PEA, 40 mg/kg, s.c. + 1-deprenyl 2 mg/kg + reserpine 7.5 mg/kg	2 hr 7 hr 19 hr	$166.7^{\dagger}$ ± 5.0 (4)	$72.1^{\ddagger}$ ± 2.7 (4)	22.5^{***} ± 0.6 (4)
Reserpine, 7.5 mg/kg	19 hr	204.5^{***} ± 18 (4)	141.5^{**} ± 8 (4)	17.4^{***} ± 1.4 (4)

[a]All values in percent of appropriate controls, 1-deprenyl alone had no effect on the metabolites. Controls: HVA, 76.6 ± 3.1 ng/g, n = (24); DOPAC, 97.9 ± 4.8 ng/g, n = (20); MOPEG, 64.8 ± 2.5 ng/g, n = (17) all corrected for recovery. (HVA and DOPAC were estimated according to Braestrup et al. (8); total MOPEG was estimated according to Braestrup (10). (*) $P < 0.05$, (**) $P < 0.01$, and (***) $P < 0.001$: versus controls. (†) $P < 0.05$ and (‡) $P < 0.001$: versus reserpine.

for the specific expression of the stereotyped behavior as compulsory sniffing.

A clinical perspective of the presented evidence can be found in the DA hypothesis of schizophrenia and of mania and depression (73,75). Scanty results have proposed that the level of PEA is profoundly increased in the urine of schizophrenic patients (28), and this finding, in combination with the finding that MAO may be decreased in chronic schizophrenic patients (64,66), and indicates that at least some forms of schizophrenia may be dependent on central dopaminergic hyperactivity caused by PEA-induced DA release. This highly interesting possibility should be followed by controlled studies on the PEA level in urine and plasma from schizophrenic patients. For such experiments, highly specific anlytical procedures, such as combined gas chromatography/mass spectrometry (97), are advantageous.

β-Phenylethylamine has also been implicated in the etiology of depression (26,82). The decreased levels of PEA found in the urine of depressed patients (see elsewhere this volume) may reflect decreased central concentrations and decreased central dopaminergic transmission. This hypothesized chain of events is consonant with the theory of DA involvement in depression (75). It should, however, be reckoned that the physiological significance of endogenous PEA is not yet settled.

ACKNOWLEDGMENT

This work has been aided by a grant from Statens laegevidenskabelige Forskningsråd, Copenhagen.

REFERENCES

1. N.-E. Andén, B.-E. Roos, and B. Werdinius (1964). Effects of chlorpromazine, haloperidol and reserpine on the levels of phenolic acids in rabbit corpus striatum. *Life Sci.*, 3:149-158.

2. E. B. Arushanyan and Yu. A. Belozertsev (1971). Amphetamine action on caudate reactions of solitary neurons in the sensorimotor cortex of cats. *Farmakol. Toksikol.*, 34:263-268.

3. E. Baum, P. Etevenon, M.-C. Piarroux, P. Simon, and J. R. Boissier (1971). Modifications comportementales et pharmacologiques obtenues chez le rat après lésion bilatérale de la substance noire. *J. Pharmacol.*, 2:423-434.

4. W. Belart (1942). Pathogenetisches und Therapeutisches aus Pervitinversuchen bei Schizophrenie. *Schweiz. Med. Wschr.*, 72:41-43.

5. M.-J. Besson, A. Cheramy, C. Gauchy, and J. Musacchio (1973). Effects of some psychotropic drugs on tyrosine hydroxylase activity in different structures of the rat brain. *Eur. J. Pharmacol.*, 22:181-186.

6. H. Blaschko (1952). Amine oxidase and amine metabolism. *Pharmacol. Rev.*, 4:415-458.

7. R. L. Borison, A. D. Mosnaim, and H. C. Sabelli (1975). Brain 2-phenylethylamine as a major mediator for the central actions of amphetamine and methylphenidate. *Life Sci.*, 17:1331-1344.

8. C. Braestrup, H. Andersen, and A. Randrup (1975). The monoamine oxidase B inhibitor deprenyl potentiates phenylethylamine behavior in rats without inhibition of catecholamine metabolite formation. *Eur. J. Pharmacol.*, 34:181-187.

9. C. Braestrup (1974). Effects of phenoxybenzamine, aceperone and clonidine on the level of 3-methoxy-4-hydroxyphenylglycol (MOPEG) in rat brain. *J. Pharm. Pharmacol.*, 26:139-141.

10. C. Braestrup (1973). Identification of free and conjugated 3-methoxy-4-hydroxyphenylglycol (MOPEG) in rat brain by gas chromatography and mass fragmentography. *Anal. Biochem.*, 55:420-431.

11. C. Braestrup and M. Nielsen (1975). Intra- and extraneuronal formation of the two major noradrenaline metabolites in the CNS of rats. *J. Pharm. Pharmacol.*, 27:413-419.

12. C. Braestrup and M. Nielsen (1976). Regulation in the central norepinephrine neurotransmission induced in vivo by α-adrenoceptor active drugs. *J. Pharmacol. Exp. Ther.*, 198:596-608.

13. C. Braestrup, M. Nielsen, and J. Scheel-Krüger (1974). Accumulation and disappearance of noradrenaline and its major metabolites synthesized from intraventricularly injected ^{3}H-dopamine in the rat brain. *J. Neurochem.*, 23:569-578.

14. C. Braestrup and J. Scheel-Krüger (1976). Methylphenidate-like effects of the new antidepressant drug nomifensine (HOE 984). *Eur. J. Pharmacol.*, 38:305-312.

15. A. R. Cools (1973). The caudate nucleus and neurochemical control of behaviour. The function of dopamine and serotonin in the caput nuclei caudati of cats. Thesis. Universitet Nijmegen, Holland.

16. H. Corrodi, K. Fuxe, and T. Hökfelt (1967). The effect of some psychoactive drugs on central monoamine neurons. *Eur. J. Pharmacol.*, 1:363-368.

17. B. Costall and R. J. Naylor (1974). Extrapyramidal and mesolimbic involvement with the stereotypic activity of D- and L-amphetamine. *Eur. J. Pharmacol.*, 25:121-129.

18. B. Costall, R. Naylor, and J. E. Olley (1972). Stereotypic and anticataleptic activities of amphetamine after intracerebral injections. *Eur. J. Pharmacol.*, 18:83-94.

19. B. Costall and R. J. Naylor (1975). The behavioural effects of dopamine applied intracerebrally to areas of the mesolimbic system. *Eur. J. Pharmacol.*, 32:87-92.

20. J. Creese and S. D. Iversen (1975). The pharmacological and anatomical substrates of the amphetamine response in the rat. *Brain Res.*, 83:419-436.

21. T. J. Crow (1972). A map of the rat mesencephalon for electrical self-stimulation. *Brain Res.*, 36:265-273.

22. I. Divac (1972). Drug-induced syndromes in rats with large, chronic lesions in the corpus striatum. *Psychopharmacologia (Berl.)*, 27:171-178.

23. D. J. Edwards and S. M. Antelman (1975). Phenylethylamine: Interaction with dopaminergic neurons. Lecture. American Society for Neurochemistry.

24. A. M. Ernst (1965). Relation between the action of dopamine and apomorphine and their O-methylated derivatives upon the CNS. *Psychopharmacologia (Berl.)*, 7:391-399.

25. E. Fischer and B. Heller (1974). Methyl-amphetamine and phenethylamine. *Arzneim.-Forsch.*, 24:956.

26. E. Fischer, B. Heller, and A. H. Miró (1968). β-Phenylethylamine in human urine. *Arzneim.-Forsch.*, 11:1486.

27. E. Fischer, R. I. Ludmer, and H. C. Sabelli (1967). The antagonism of phenylethylamine to catecholamines on mouse motor activity. *Acta Physiol. Lat. Am.*, 17:15-21.

28. E. Fischer, H. Spatz, J. M. Saavedra, H. Reggiani, A. H. Miró, and B. Heller (1972). Urinary elimination of phenethylamine. *Biol. Psychiat.*, 5:139-147.

29. R. Fog, A. Randrup, and H. Pakkenberg (1970). Lesions in corpus striatum and cortex of rat brains and the effect on pharmacologically induced stereotyped, aggressive and cataleptic behaviour. *Psychopharmacologia (Berl.)*, 18:346-356.

30. K. Fuxe, H. Grobecker, and J. Jonsson (1967). The effect of β-phenylethylamine on central and peripheral monoamine-containing neurons. *Eur. J. Pharmacol.*, 2:202-207.

31. K. Fuxe and U. Ungerstedt (1970). Histochemical, biochemical and functional studies on central monoamine neurons after acute and chronic amphetamine administration. In: *Amphetamines and Related Compounds*, edited by E. Costa, and S. Garattini. Raven Press, New York, pp. 257-288.

32. H. J. Gerhards, A. Carenzi, and E. Costa (1974). Effect of nomifensine on motor activity, dopamine turnover rate and cyclic 3',5'-adenosine. *Naunyn Schmiedeberg's Arch. Pharmacol.*, 286: 49-63.

33. W. J. Giardina, W. A. Pedemonte, and H. C. Sabelli (1973). Iontophoretic study of the effects of norepinephrine and 2-phenylethylamine on single cortical neurons. *Life Sci.*, 12:153-161.

34. C. Goridis and N. H. Neff (1971). Evidence for a specific monoamine oxidase associated with sympathetic nerves. *Neuro pharmacology*, 10:557-564.

35. C. Goridis and N. H. Neff (1971). Monoamine oxidase in sympathetic nerves: A transmitter specific enzyme type. *Br. J. Pharmacol.*, 43:814-818.

36. L. C. F. Hanson (1967). Evidence that the central action of (+)-amphetamine is mediated via catecholamines. *Psychopharmacologia (Berl.)*, 10:289-297.

37. Z. S. Herman (1967). Influence of some psychotropic and adrenergic blocking agents upon amphetamine stereotyped behaviour in white rats. *Psychopharmacologia (Berl.)*, 11:136-142.

38. R. J. Hitzemann, H. H. Loh, and E. F. Domino (1971). Effect of paramethoxyamphetamine on catecholamine metabolism in the mouse brain. *Life Sci.*, 10:1087-1095.

39. L. L. Iversen (1965). The inhibition of noradrenaline uptake by drugs. *Adv. Drug Res.*, 2:5-45.

40. D. M. Jackson (1975). β-phenylethylamine and locomotor activity in mice: Interaction with catecholaminergic neurones and receptors. *Arzneim.-Forsch.*, 25:622-626.

41. D. M. Jackson and D. B. Smythe (1973). The distribution of β-phenylethylamine in discrete regions of the rat brain and its effect on brain noradrenaline, dopamine and 5-hydroxytryptamine levels. *Neuropharmacology*, 12:663-668.

42. D. M. Jackson and D. B. Smythe (1974). The effect of β-phenethylamine on noradrenaline and dopamine turnover in rat brain. *J. Pharm. Pharmacol.*, 26:456-458.

43. D. M. Jackson (1975). Some further observations of the effect of β-phenethylamine on locomotor activity in mice. *J. Pharm. Pharmacol.*, 27:278-280.

44. D. M. Jackson (1971). The effect of β-phenethylamine on noradrenaline concentrations in guinea-pig brain. *J. Pharm. Pharmacol.*, 23:623-624.

45. D. M. Jackson (1972). The effect of β-phenethylamine upon spontaneous motor activity in mice: A dual effect on locomotor activity. *J. Pharm. Pharmacol.*, 24:383-389.

46. D. M. Jackson (1974). The involvement of noradrenergic systems in the locomotor activity stimulation in mice produced by β-phenethylamine. *J. Pharm. Pharmacol.*, 26:651-654.

47. D. M. Jackson (1977). β-Phenylethylamine: Studies on the mechanism of its stimulant effects. *Phenylethylamine: Biological Mechanisms and Clinical Aspects*, edited by A. D. Mosnaim and M. E. Wolf. Marcel Dekker, New York. (This volume.)

48. P. Janssen, C. Niemegeers, and K. Schellekens (1965). Is it possible to predict the clinical effects of neuroleptic drugs (major tranquillizers) from animal data? Part I: "Neuroleptic activity spectra" for rats. *Arzneim.-Forsch.*, 15:104-117.

49. P. Janssen, C. Niemegeers, K. Schellekens, and F. Lenaerts (1967). Is it possible to predict the clinical effects of neuroleptic drugs (major tranquillizers) from animal data? Part IV: An improved experimental design for measuring the inhibitory effects of neuroleptic drugs on amphetamine- or apomorphine-induced "chewing" and "agitation" in rats. *Arzneim.-Forsch.*, 17:841-854.

50. B. Jarrot (1971). Occurrence and properties of monoamine oxidase in adrenergic neurons. *J. Neurochem.*, 18:7-16.

51. J. P. Johnston (1968). Some observations upon a new inhibitor of monoamine oxidase in brain tissue. *Biochem. Pharmacol.*, 17:1285-1297.

52. W. Jonas and J. Scheel-Krüger (1969). Amphetamine induced stereotyped behaviour correlated with the accumulation of O-methylated dopamine. *Arch. Int. Pharmacodyn. Ther.*, 177: 379-389.

53. J. Jonsson, H. Grobecker, and P. Holtz (1966). Effect of β-phenylethylamine on content and subcellular distribution of norepinephrine in rat heart and brain. *Life Sci.*, 5: 2235-2246.

54. O. J. Kalant (1966). *The Amphetamines. Toxicity and Addiction.* Brookside Monograph No. 5. University of Toronto Press, Toronto.

55. P. Karli (1960). Troubles du comportement induits chez le rat par l'amphetamine et le phénidylate (Ritaline). *Arch. Int. Pharmacodyn. Ther.*, 122:344-351.

56. J. Korf, G. Aghajanian, and R. Roth (1973). Stimulation and destruction of the locus coeruleus: Opposite effects on 3-methoxy-4-hydroxyphenylglycol sulfate levels in the rat cerebral cortex. *Eur. J. Pharmacol.*, 21:305-310.

57. M. Kuhar, R. Roth, and G. Aghajanian (1972). Synaptosomes from forebrain of rats with midbrain raphe lesions: Selective reduction of serotonin uptake. *J. Pharmacol. Exp. Ther.*, 181:36-45.

58. J. König and R. Klippel (1963). *The Rat Brain. A Stereotaxic Atlas of the Forebrain and Lower Parts of the Brain Stem.* Williams and Wilkins, Baltimore.

59. J. M. Littleton (1967). The interaction of dexamphetamine with inhibitors of noradrenaline biosynthesis in rat brain in vivo. *J. Pharm. Pharmacol.*, 19:414-415.

60. U. P. Madubuike, A. D. Mosnaim, and H. C. Sabelli (1974). Brain phenylacetic acid, a major metabolite of 2-phenylethylamine in rabbit brain. Abstract. *XXVI International Congress of Physiological Sciences*, New Delhi, October 20-26.

61. J. Maj and E. Przegaliński (1967). Disulfiram and some effects of amphetamine in mice and rats. *J. Pharm. Pharmacol.*, 19:341-342.

62. P. Mantegazza and M. Riva (1963). Amphetamine-like activity of β-phenethylamine after a monoamine oxidase inhibitor in vivo. *J. Pharm. Pharmacol.*, 15:472-478.

63. O. Mayer and V. Eybl (1971). The effect of diethyldithiocarbamate on amphetamine-induced behaviour in rats. *J. Pharm. Pharmacol.*, 23:894-896.

64. Y. Meltzer and S. Stahl (1974). Platelet monoamine oxidase activity and substrate preferences in schizophrenic patients. *Res. Commun. Chem. Pathol. Pharmacol.*, 7:419-431.

65. E. Mogilnicka and C. Braestrup (1976). Noradrenergic influence on the stereotyped behaviour induced by amphetamine, phenylethylamine and apomorphine. *J. Pharm. Pharmacol.*, 28:253-255.

66. D. L. Murphy and R. J. Wyatt (1972). Reduced monoamine oxidase activity in blood platelets from schizophrenic patients. *Nature*, 238:225-226.

67. R. J. Naylor and J. E. Olley (1972). Modification of the behavioural changes induced by amphetamine in the rat by lesions in the caudate nucleus, the caudate-putamen and globus pallidus. *Neuropharmacology*, 11:91-99.

68. M. Nielsen (1976). Estimation of noradrenaline and its metabolites synthesized from ^{3}H-tyrosine in the rat brain. *J. Neurochem.*, 27:493-500.

69. R. Papeschi (1975). Behavioral and biochemical interaction between AMT and (+)-amphetamine: Relevance to the identification of the functional pool of brain catecholamines. *Psychopharmacologia*, 45:21-28.

70. V. Pedersen and A. V. Christensen (1972). Antagonism of methylphenidate induced stereotyped gnawing in mice. *Acta Pharmacol. Toxicol.*, 31:488-496.

71. T. Persson and B. Waldeck (1969). The interaction between different metabolic pathways of catecholamines in the brain studied by means of ^{3}H-DOPA. *Acta Pharmacol. Toxicol.*, 27:225-236.

72. R. Quinton and G. Halliwell (1963). Effects of α-methyl DOPA and DOPA on the amphetamine excitatory response in reserpinized rats. *Nature,* 200:178-179.

73. A. Randrup and I. Munkvad (1968). Behavioural stereotypies induced by pharmacological agents. *Pharmakopsychiat. Neuro-Psychopharmakol.,* 1:18-26.

74. A. Randrup and I. Munkvad (1966). DOPA and other naturally occurring substances as causes of stereotypy and rage in rats. *Acta Psychiat. Scand.,* suppl. 191 (ad vol. 42):193-199.

75. A. Randrup, I. Munkvad, R. Fog, J. Gerlach, L. Molander, B. Kjellberg, and J. Scheel-Krüger (1975). Mania, depression, and brain dopamine. In: *Current Developments in Psychopharmacology, Vol. II,* chapter VII, edited by W. B. Essmann and L. Valzelli. Spectrum Publications, Inc., New York, pp. 205-248.

76. A. Randrup and I. Munkvad (1974). Pharmacology and physiology of stereotyped behavior. *J. Psychiat. Res.,* 11:1-10.

77. A. Randrup and I. Munkvad (1966). Role of catecholamines in the amphetamine excitatory response. *Nature,* 211:540.

78. A. Randrup and I. Munkvad (1967). Stereotyped activities produced by amphetamine in several animal species and man. *Psychopharmacologia (Berl.),* 11:300-310.

79. A. Randrup, I. Munkvad, and P. Udsen (1963). Adrenergic mechanisms and amphetamine induced abnormal behaviour. *Acta Pharmacol. Toxicol.,* 20:145-157.

80. A. Randrup and J. Scheel-Krüger (1966). Diethyldithiocarbamate and amphetamine stereotype behaviour. *J. Pharm. Pharmacol.,* 18:752.

81. A. Randrup, I. Munkvad, and J. Scheel-Krüger (1973). Mechanisms by which amphetamines produce stereotypy, aggression and other behavioural effects. In: *Psychopharmacology, Sexual Disorders and Drug Abuse,* edited by T. Ban, J. Boissier, G. Gessa, H. Heimann, L. Hollister, H. Lehmann, I. Munkvad, H. Steinberg, F. Sulser, A. Sundwall and O. Vinař. North-Holland Publishing Company, Amsterdam-London. Avicenum, Czechoslovak Medical Press, Prague, pp. 659-673.

82. H. Sabelli and D. Mosnaim (1974). Phenylethylamine hypothesis of affective behavior. *Am. J. Psychiat.,* 131:695-699.

83. A. C. Sayers (1972). An investigation into hyperkinesia and akinesia induced by certain stimulant drugs. Thesis. University of Aston in Birmingham.

84. A. C. Sayers and S. L. Handley (1973). A study of the role of catecholamines in the response to various central stimulants. *Eur. J. Pharmacol.,* 23:47-55.

85. S. M. Schanberg, J. J. Schildkraut, G. R. Breese, and I. J. Kopin (1968). Metabolism of normetanephrine-H^3 in rat brain—Identification of conjugated 3-methoxy-4-hydroxyphenylglycol as the major metabolite. *Biochem. Pharmacol.*, 17:247-254.

86. J. Scheel-Krüger (1971). Comparative studies of various amphetamine analogues demonstrating different interactions with the metabolism of the catecholamines in the brain. *Eur. J. Pharmacol.*, 14:47-59.

87. S. R. Snodgrass (1974). Phenylalanine, brain phenethylamine and motor activity in the rat. *J. Pharm. Pharmacol.*, 26:931-936.

88. R. F. Squires (1972). Multiple forms of monoamine oxidase in intact mitochondria as characterized by selective inhibitors and thermal stability: A comparison of eight mammalian species. *Adv. Biochem. Psychopharmacol.*, 5:355-370.

89. A. M. Thierry, G. Blanc, and J. Glowinski (1971). Dopamine-norepinephrine another regulatory step of norepinephrine synthesis in central noradrenergic neurons. *Eur. J. Pharmacol.*, 14:303-307.

90. L. Tuomisto, J. Tuomisto, and E. Smissman (1974). Dopamine uptake in striatal and hypothalamic synaptosomes: Conformational selectivity of the inhibitor. *Eur. J. Pharmacol.*, 25:351-361.

91. U. Ungerstedt, A. Avemo, E. Avemo, T. Ljungberg, and C. Ranje (1973). Animal models of Parkinsonism. In: *Advances in Neurology, Vol. 3,* edited by D. Calne. Raven Press, New York, pp. 257-271.

92. H. Utena (1961). Behavior and neurochemistry. Special type of model psychosis: A chronic methamphetamine intoxication in man and animal. *Brain and Nerve,* 13:687-696.

93. D. S. Walter and D. Eccleston (1973). Increase of noradrenaline metabolism following electrical stimulation of the locus coeruleus in the rat. *J. Neurochem.*, 21:281-289.

94. A. Weissman, B. K. Koe, and S. S. Tenen (1966). Antiamphetamine effects following inhibition of tyrosine hydroxylase. *J. Pharmacol. Exp. Ther.*, 151:339-352.

95. B. L. Welch and A. S. Welch (1970). Control of brain catecholamines and serotonin during acute stress and after d-amphetamine by natural inhibition of monoamine oxidase: An hypothesis. In: *Amphetamines and Related Compounds,* edited by E. Costa and S. Garattini. Raven Press, New York, pp. 415-445.

96. B. L. Welch and A. S. Welch (1967). Stimulus-dependent antagonism of the β-methyltyrosine-induced lowering of brain catecholamines by (+)-amphetamine in intact mice. *J. Pharm. Pharmacol.*, 19:841-843.

97. J. Willner, H. F. Lefevre, and E. Costa (1974). Assay by multiple ion detection of phenylethylamine and phenylethanolamine in rat brain. *J. Neurochem.*, 23:857-859.

98. H.-Y. Yang and N. H. Neff (1973). β-Phenylethylamine: A specific substrate for type B monoamine oxidase of brain. *J. Pharmacol. Exp. Ther.*, 187:365-371.

99. H.-Y. Yang and N. H. Neff (1974). The monoamine oxidases of brain: Selective inhibition with drugs and the consequences for the metabolism of the biogenic amines. *J. Pharmacol. Exp. Ther.*, 189:733-740.

Chapter 12

BEHAVIORAL EFFECTS OF PHENYLETHYLAMINE— NEUROCHEMICAL CORRELATES

Beng T. Ho

Texas Research Institute of Mental Sciences
The University of Texas Health Science Center at Houston
Houston, Texas

I. INTRODUCTION

The existence of β-phenylethylamine (PEA), the simplest endogenous biogenic amine in structure, has been demonstrated in tissues of a number of species (11,12,32,40,41,44,45,47,50,55). β-Phenylethylamine has also been reported to be present in human urine and blood (1,3,15, 16,25,33,39,42), as well as in the human brain (23). It has been postulated that PEA may function as a stimulant and perhaps as a neuromodulator (52). The levels of PEA change in certain forms of mental disorders. High excretion of PEA in the urine was reported in the phenylketonuric (35,42) and during the manic phase of manic depressive psychosis (17). However, the opposite was found in the urine of patients with endogenous depression (3,15,17,24,39). The concentration of PEA was normal or elevated in the urine of schizophrenic and alcoholic patients (15,17). The brain PEA content in mice (40), rabbits (51), and rats (17) is increased by treatment with antidepressants such as imipramine, inhibitors of monoamine oxidase (MAO), and after electroshock; it is reduced by reserpine. Studies of human PEA levels and the effects of antidepressants on PEA in animals lead to the PEA hypothesis of affective disorders (52, 53).

Intramuscular or i.p. injection of PEA, with or without pretreatment of an inhibitor of MAO, produces significant depletion of brain catecholamines (CA) in rats (18,31,34) and in guinea pigs (26). Concentrations of 5-hydroxytryptamine (5-HT) in the midbrain and cortex of rat brains are also reduced by PEA (31). Questions then arise as to how PEA exerts its behavior effects: Is the action of PEA mediated by the change in concentrations of CA or 5-HT in the brain, or rather, does PEA produce a direct action at either the receptors of CA and 5-HT or its own specific PEA receptor?

II. NEUROCHEMICAL CORRELATION OF PEA-INDUCED BEHAVIOR

β-Phenylethylamine crosses the blood-brain barrier (10,41,43) and exerts stimulation of motor activity in mice (14,41). In animals pretreated with an inhibitor of MAO, PEA induces amphetaminelike effects including an increase in exploratory and spontaneous activity of mice (14,36,41,50), activation of EEG in rats (41), reduction of the amplitude of the slow negative wave of visual evoked responses (VER) in unanesthetized rabbits (49,50), anorexia in rats and dogs, hyperthermia in mice and rats, and exhibition of difference in lethality between isolated and aggregated mice (36). In view of these similarities of effects between PEA and amphetamine, it has been suggested that, like amphetamine, the stimulatory effect of PEA is attributed at least in part to the release of norepinephrine (NE) (18). Data presented against this hypothesis are the observed depressant, rather than stimulatory, effect of CA when CA were administered intraventricularly or their amino acid precursor DOPA given systemically, and the antagonism of epinephrine or DOPA-induced reduction of motor activity of mice by PEA plus a monoamine oxidase inhibitor (MAOI) and also by methamphetamine (14). Further differentiation between PEA and CA is provided by the finding that while PEA reduces the amplitude of the slow negative wave of VER (49,50), DOPA augments the response (50). Evidence for a difference between the central receptors for PEA and for CA has been provided by Giardina et al. (19), who demonstrated effects of PEA different from those of NE on the spontaneous spike rate of certain cortical neuron cells. Stein (54) proposed that the stimulatory effect of amphetamine is mediated by the release of a naturally occurring amine, and Saavedra and Fischer (48) suggested that the amine may be PEA. Because of the similar action observed between PEA and amphetamine, Sabelli et al. (52) speculated that amphetamine and its derivatives induce behavioral stimulation either by releasing endogenous PEA or by mimicking the action of PEA on specific PEA receptors.

β-Phenylethylamine is also found to antagonize the depressant effects of 5-HT and tryptamine on locomotor activity of mice (15,48).

Jackson (27,29) utilized a number of neurochemical agents in an attempt to establish the relationship of PEA-induced behavior with catecholaminergic neurons. It was found that i.p. injection of PEA to QS strain mice at doses higher than 75 mg/kg produced a biphasic increase of locomotor activity which peaked at about 5 and 30 min post-injection. This biphasic type of activity was not reported with amphetamine. Pretreatment with iproniazid lowered the doses of PEA required for the rise in locomotor activity to 10 or 25 mg/kg. According to Jackson, the peak at the shorter interval was related to the indirect effect of PEA mediated by the release of CA, since the motor activity between 0 and 20 min was blocked by diethyldithiocarbamate, disulfiram, perphenazine, and α-methyl-*p*-tyrosine (AMT), but not by reserpine (27,29). This first phase of activity, therefore, resembles that produced by amphetamine. The activity at the later time interval was thought to be attributed to a direct action of PEA on central DA receptors (27). The conclusion was based on an interpretation of the lack of effects of the aforementioned neurochemical agents on PEA-induced stimulation during the second phase (20-60 min) (27), as well as the blockade of this phase of activity by the DA receptor blockers, haloperidol and pimozide (29).

III. CURRENT INVESTIGATION

While i.p. administration of PEA produces an amphetaminelike increase of spontaneous locomotor activity in mice pretreated with iproniazid, no significant effect can be observed in rats (22,36). This species difference in response to PEA is interesting and is currently under investigation. In our laboratory, PEA-induced behavior is produced by i.v. injection to Sprague-Dawley rats pretreated 24 hr with 100 mg/kg of pargyline or iproniazid i.p. (Fig. 1). The locomotor activity was measured with an electronic motility meter (Motron-Produkter, Stockholm, Sweden, Model 40Fc): The horizontal movement of animals was sensed by a series of 40 photocells activated by an incandescent lamp mounted 1.0 m above a 20 x 16-cm cage, and the vertical movement was detected by a series of five photocells mounted horizontally about 10 cm above the cage. Individual animals were placed in the cage for 30 min to acclimatize to their surroundings before the administration of drugs. Accumulated counts over a period of 60 min were recorded, and results are expressed as percent of the saline control.

It was found that an i.v. dose of 0.75 mg/kg of d-amphetamine produces an increase in the horizontal movement of rats comparable to that induced by i.v. administration of 10 mg/kg of PEA in pargyline-pretreated animals (Fig. 1). These two doses of d-amphetamine and PEA were then chosen to study the efforts of various monoaminergic receptor blockers on behavioral stimulation induced by PEA or d-amphetamine.

An obvious difference exists between d-amphetamine and PEA in their effects on the vertical movement of rats. d-Amphetamine produces twice the magnitude of increase in the vertical movement as that in the horizontal movement (Fig. 2). Contrarily, PEA plus pargyline exerts no effect on the vertical movement (Fig. 2).

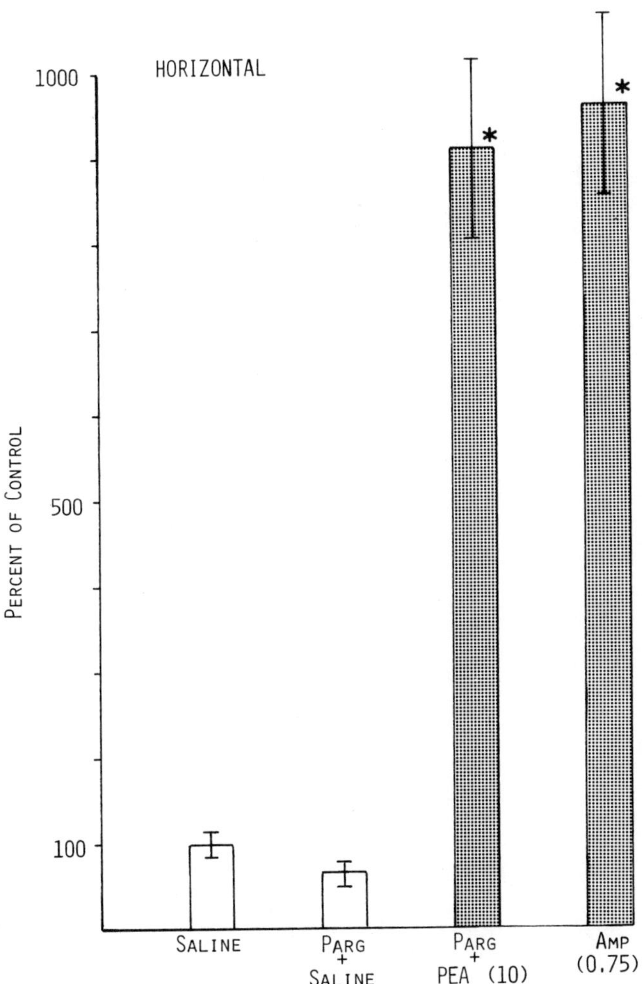

FIG. 1. Increase of horizontal movement over 60 min induced by d-amphetamine sulfate (Amp, 0.75 mg/kg, i.v.) and by PEA hydrochloride (PEA, 10 mg/kg, i.v.) in rats pretreated 24 hr with pargyline hydrochloride (Parg, 100 mg/kg, i.p.). The vertical lines represent the range of percent from five animals. *$P < 0.001$ different from the saline control.

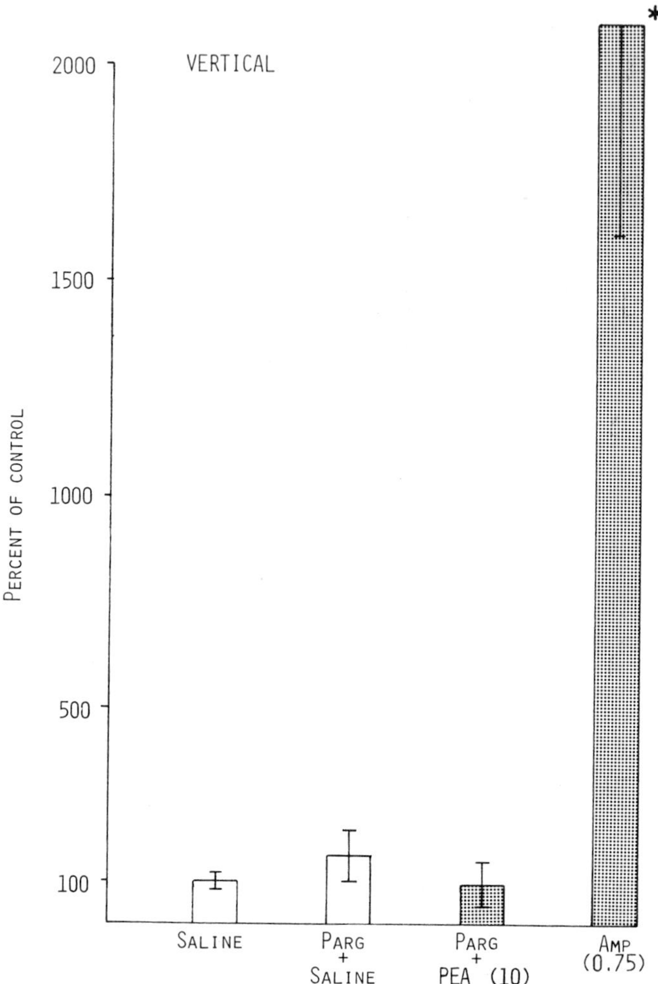

FIG. 2. Increase of vertical movement over 60 min induced by d-amphetamine sulfate (Amp, 0.75 mg/kg, i.v.) and the lack of effect of PEA hydrochloride (PEA, 10 mg/kg, i.v.) in rats pretreated 24 hr with pargyline hydrochloride (Parg, 100 mg/kg, i.p.). The vertical lines represent the range of percent from five animals. *$P < 0.01$ different from the saline control.

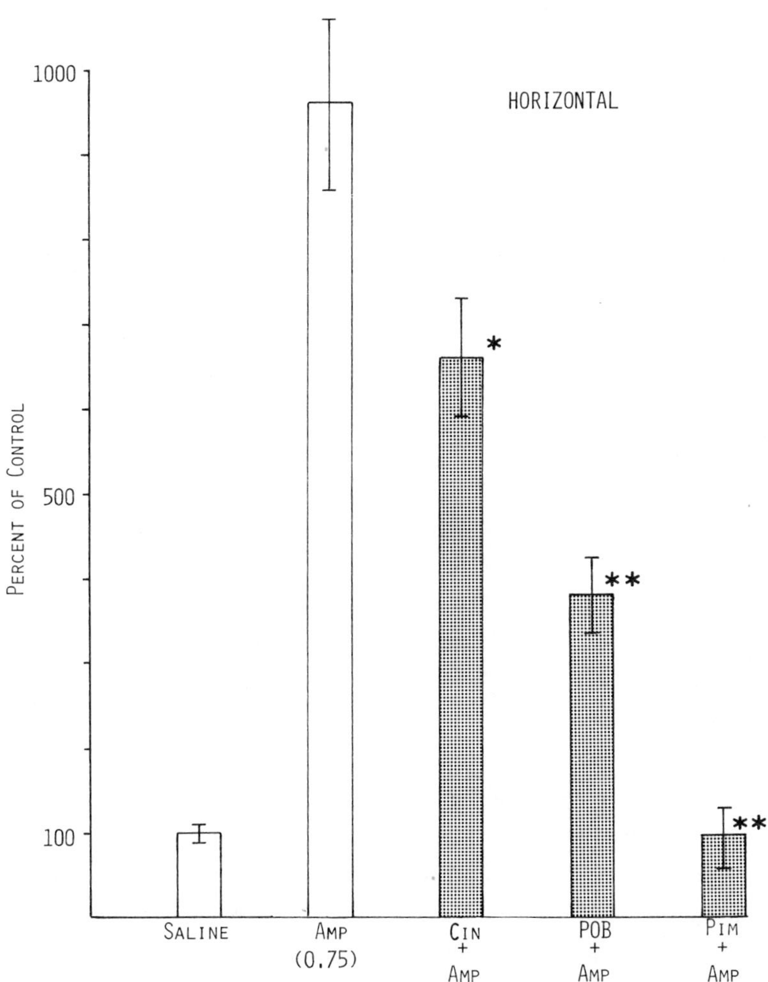

FIG. 3. Antagonism of d-amphetamine sulfate (Amp, 0.75 mg/kg, i.v.) induced increase of horizontal movement in rats over 60 min by pretreatment i.p. with cinanserin hydrochloride (Cin, 15 mg/kg, 30 min before PEA), phenoxybenzamine hydrochloride (POB, 10 mg/kg, 2 hr), and pimozide (Pim, 1 mg/kg, 4 hr). The vertical lines represent the range of percent from five animals. *$P < 0.05$; **$P < 0.001$ different from amphetamine.

Cinanserin, a 5-HT antagonist, significantly reduces the behavioral stimulation of d-amphetamine in horizontal but not the vertical movement (Figs. 3 and 4). This antagonism by cinanserin of d-amphetamine-induced increase in locomotor activity has been reported previously in rats receiving i.p. injection of d-amphetamine (21). Phenoxybenzamine, an α-adrenergic blocker, and pimozide, a dopaminergic blocker, are capable of antagonizing the d-amphetamine-induced increase of both the horizontal and vertical movement of animals (Figs. 3 and 4), while the α-adrenergic blocker, propranolol, and cholinergic blocker, atropine, have no effect on either movement.

In the PEA animals, although the increase of horizontal movement is antagonized by phenoxybenzamine and pimozide (Fig. 5), cinanserin does not have the effect it has with d-amphetamine. Treatment with phenoxybenzamine and pimozide before PEA does not change the vertical movement of animals beyond the saline control level, whereas methysergide, the serotonergic blocker, produces about 10 times the percent increase of the vertical movement in PEA animals (Fig. 6) but not in animals injected with d-amphetamine.

The cholinergic blocker, atropine, exhibits a lack of effect on the amphetamine-induced stimulation in both horizontal and vertical movement. It also does not affect the increase in the horizontal movement induced by PEA. An increase, but not statistically significant, in the vertical movement was observed, however, when the animal was pretreated 30 min with atropine sulfate (2 mg/kg, i.p.) before administration of PEA. Potentiation of PEA behavioral activity in mice by cholinergic agents such as benzhexol, benztropine, atropine, physostigmine, and neostigmine (28), and that of amphetamine in mice and rats by atropine and hyoscine (5,6,37) have been reported. Jackson (28) postulated that the effect of cholinergic agents on PEA behavior is due to the action of these agents on dopaminergic neurons (2,7,8,13). In our studies, animals injected with atropine and other monoaminergic blockers, with the exception of pimozide, did not impair the activity in vertical and horizontal movement when compared to the saline controls.

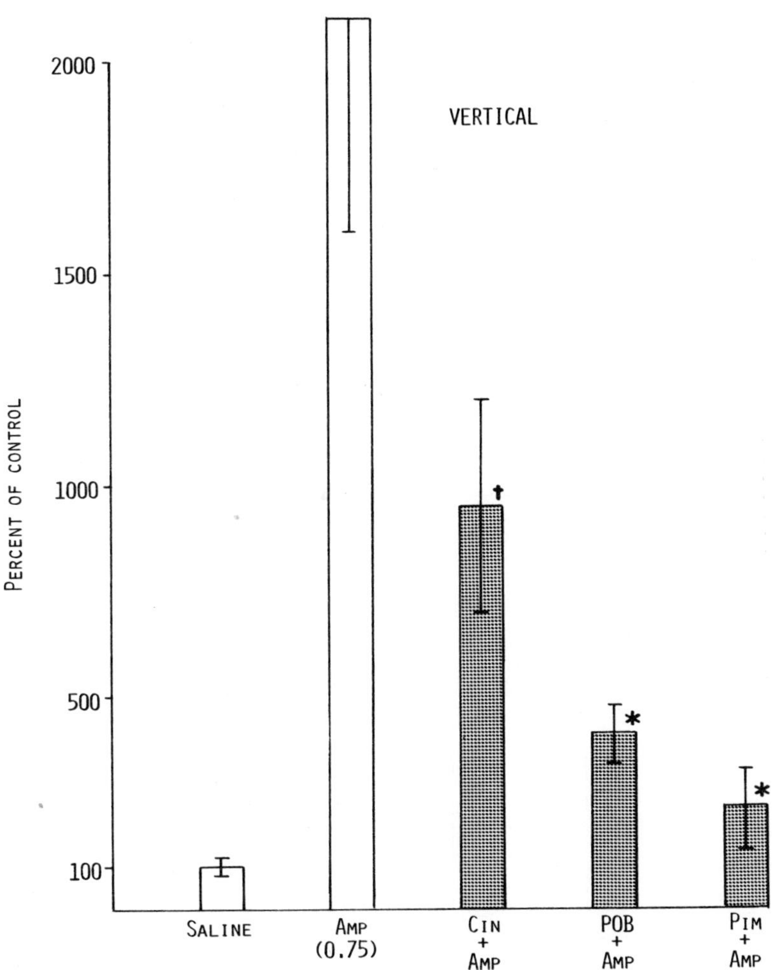

FIG. 4. Antagonism of d-amphetamine sulfate (Amp, 0.75 mg/kg, i.v.) induced increase of vertical movement in rats over 60 min by pretreatment i.p. with phenoxybenzamine hydrochloride (POB, 10 mg/kg, 2 hr before PEA) and pimozide (Pim, 1 mg/kg, 4 hr). The vertical lines represent the range of percent from five animals. †Not significant; *$P < 0.05$ different from amphetamine.

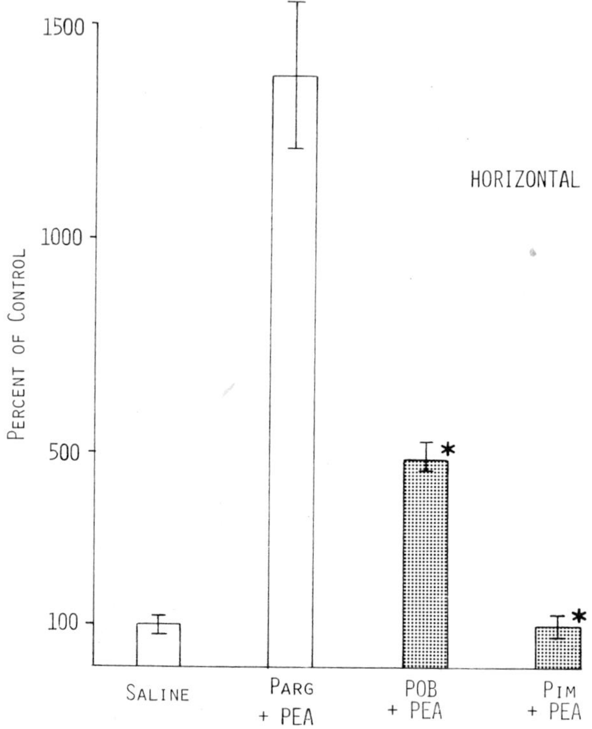

FIG. 5. Antagonistic effect of phenoxybenzamine hydrochloride (POB, 10 mg/kg, 2 hr pretreatment) and pimozide (Pim, 1 mg/kg, 4 hr pretreatment) on PEA hydrochloride (PEA, 10 mg/kg, i.v.) induced increase of horizontal movement in rats pretreated 24 hr with pargyline hydrochloride (Parg, 100 mg/kg, i.p.). The vertical lines represent the range of percent from five animals. *P < 0.001 different from pargyline plus PEA.

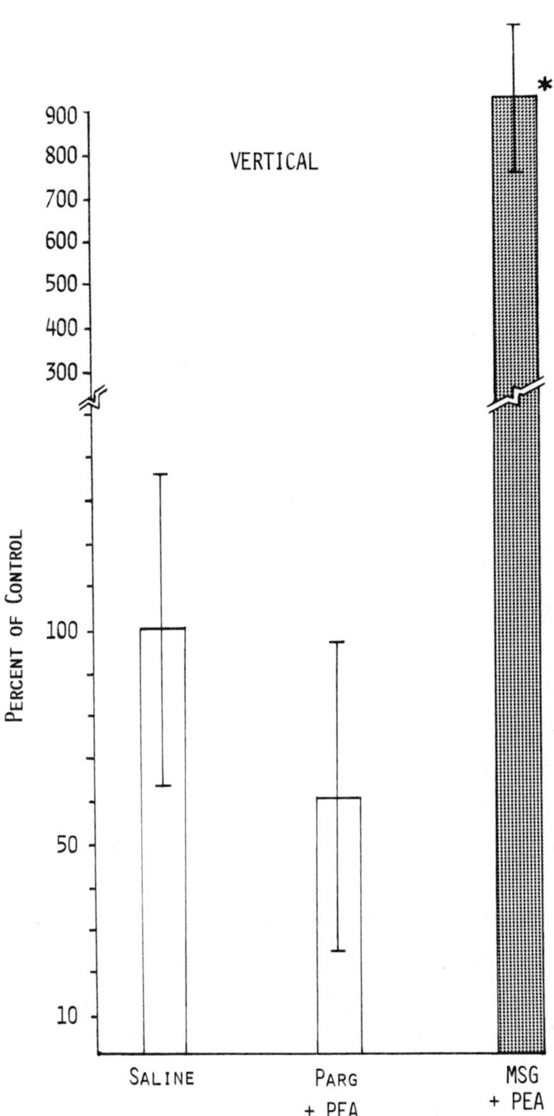

FIG. 6. Enhancement by methysergide maleate (MSG, 5 mg/kg, i.p., 10 min pretreatment) of PEA hydrochloride (PEA, 10 mg/kg, i.v.) induced increase of vertical movement over 60 min in rats pretreated 24 hr with pargyline hydrochloride (Parg, 100 mg/kg, i.p.). The vertical lines represent the range of percent from five animals. *$P < 0.001$ different from pargyline + PEA.

The spontaneous locomotor activity of rodents under the effect of amphetamine is generally measured by the change in the horizontal movement. The involvement of CA in the behavioral effect of d-amphetamine is well known (38,46). Our results (Figs. 3 and 4) agree with this finding, since the amphetamine-induced stimulation of both horizontal and vertical movement is reduced by phenoxybenzamine and pimozide. Jackson (29,30) has also reported the blockade of PEA-induced locomotor stimulation by phenoxybenazmine and pimozide. The data in Fig. 5 shows the similar relationship between the two catecholaminergic blockers and PEA behavior in the horizontal movement of the animals. However, it is not yet clear why cinanserin affects the horizontal movement in amphetamine animals but not in PEA animals. Moreover, methysergide causes a surprisingly large increase in the vertical movement only in animals receiving PEA, while it does not potentiate d-amphetamine action on the vertical movement. Data on the involvement of monoamines in the vertical movement of rodents is scarce. The potentiation of d-amphetamine on this stand-up (rearing) response of rats has been reported (9,20), and it has been suggested that the rearing response of rats is related to the catecholaminergic-cholinergic balance (4). While further work is necessary to define the possible role of 5-HT in the motor activity of both d-amphetamine and PEA, our data has obviously demonstrated some differences between the behavioral effects of the two compounds. Thus, it is difficult to explain the behavioral action of d-amphetamine entirely on PEA. There is reason to believe that some action of d-amphetamine is due in part to PEA, whereas other, for example, that which affects the vertical movement, could relate better with catecholamines.

IV. SUMMARY

Behavioral stimulation, as measured by locomotor activity, can only be induced by i.v. administration of PEA to rats pretreated with an inhibitor of MAO. In contrast to d-amphetamine, PEA produces an increase only in the horizontal movement but not the vertical movement of animals. Other differences in the behavioral action of d-amphetamine and PEA are demonstrated by their responses to cinanserin and methysergide. It would appear that some of the behavioral effects of d-amphetamine are not related to PEA.

ACKNOWLEDGMENTS

The author thanks Mrs. Mary B. O'Brien for her excellent technical assistance. The generous gifts of methysergide maleate from Sandoz Pharmaceuticals, East Hanover, N.J.; pargyline hydrochloride from Abbott Laboratories, North Chicago, Ill.; phenoxybenzamine hydrochloride from Smith Kline & French Laboratories, Philadelphia, Pa.; and pimozide from Professor Paul Janssen of Janssen Pharmaceutical Research Laboratoria, Belgium, are gratefully acknowledged.

REFERENCES

1. A. M. Asatoor and C. E. Dalgliesh (1959). Amines in blood and urine. *Biochem. J.*, 73:26P.

2. G. Bartholini and A. Pletscher (1971). Atropine-induced changes of cerebral dopamine turnover. *Experientia*, 27:1302-1303.

3. A. A. Boulton and L. Milward (1971). Separation, detection and quantitative analysis of urinary β-phenylethylamine. *J. Chromatogr.*, 57:287-296.

4. K. S. Bryan and G. Ellison (1975). Cholinergic modulation of an opposed effect of *d*-amphetamine and methylphenidate on the rearing response. *Psychopharmacologia*, 43:169-173.

5. P. L. Carlton (1961). Augmentation of the behavioral effects of amphetamine by scopolamine. *Psychopharmacologia*, 2:377-380.

6. P. L. Carlton and P. Didamo (1961). Augmentation of the behavioral effects of amphetamine by atropine. *J. Pharmacol. Exp. Ther.*, 132:91-96.

7. H. Corrodi, K. Fuxe, W. Hammer, F. Sjöqvist, and U. Ungerstedt (1967). Oxotremorine and central monoamine neurons. *Life Sci.*, 6:2557-2566.

8. J. T. Coyle and S. H. Snyder (1969). Antiparkinsonian drugs: Inhibition of dopamine uptake in the corpus striatum as a possible mechanism of action. *Science*, 166:899-901.

9. P. C. Dandiya, B. D. Gupta, and M. C. Gupta (1970). Influence of CNS stimulants and hallucinogens on rats in special circumstances. *Pharmakopsychiatr. Neuro-Psychopharmakol.*, 3:349-354.

10. W. G. Dewhurst (1968). New theory of cerebral amine function and its clinical application. *Nature*, 218:1130-1133.

11. D. A. Durden, S. R. Philips, and A. A. Boulton (1973). Identification and distribution of β-phenylethylamine in the rat. *Can. J. Biochem.*, 51:995-1002.

12. D. J. Edwards and K. Blau (1972). Analysis of phenylethylamines in biological tissues by gas-liquid chromatography with electron-capture detection. *Anal. Biochem.*, 45:387-402.

13. L.-O., Farnebo, K. Fuxe, B. Hamberger, and H. Ljungdahl (1970). Effect of some antiparkinsonian drugs on catecholamine neurons. *J. Pharm. Pharmacol.*, 22:733-737.

14. E. Fischer, R. I. Ludmer, and H. C. Sabelli (1967). The antagonism of phenylethylamine to catecholamines on mouse motor activity. *Acta. Physiol. Lat. Am.*, 17:15-21.

15. E. Fischer, B. Heller, and A. N. Miro (1968). β-phenylethylamine in human urine. *Arzneim.-Forsch.*, 18:1486.

16. E. Fischer, H. Spatz, B. Heller, and H. Reggiani (1972). Phenylethylamine content of human urine and rat brain, its alteration in pathological conditions and after drug administration. *Experientia*, 28:307-308.

17. E. Fischer, H. Spatz, J. M. Saavedra, H. Reggiani, A. H. Miro, and B. Heller (1972). Urinary elimination of phenylethylamine. *Biol. Psychiat.*, 5:139-147.

18. K. Fuxe, H. Grobecker, and J. Jonsson (1967). The effect of β-phenylethylamine on central and peripheral monoamine-containing neurons. *Eur. J. Pharmacol.*, 2:202-207.

19. W. J. Giardina, W. A. Pedemonte, and H. C. Sabelli (1973). Iontophoretic study of the effects of norepinephrine and 2-phenylethylamine on single cortical neurons. *Life Sci.*, 12:153-161.

20. B. D. Gupta and K. Gregory (1967). The effects of three drugs and their combinations on the rearing response in two strains of rats. *Psychopharmacologia*, 11:365-371.

21. J.-T. Huang and B. T. Ho (1973). Differentiation of pressor and behavioral effects of *d*-amphetamine and 2,5-dimethoxy-4-methylamphetamine (STP) by cinanserin. *Can. J. Physiol. Pharmacol.*, 51:976-980.

22. J.-T. Huang and B. T. Ho (1974). The effect of pretreatment with iproniazid on the behavioral activities of β-phenylethylamine in rats. *Psychopharmacologia*, 35:71-81.

23. E. E. Inwang, A. D. Mosnaim, and H. C. Sabelli (1973). Isolation and characterization of phenylethylamine and phenylethanolamine from human brain. *J. Neurochem.*, 20:1469-1473.

24. E. E. Inwang, J. H. Sugerman, W. J. DeMartini, A. D. Mosnaim, and H. C. Sabelli (1972). Ultraviolet spectrophotometric determination of β-phenylethylamine-like substances in biological samples and its possible correlation with depression. *Proc. Ann. Meet. Soc. Biol. Psychiat.*, Dallas, April 28-30, p. 13.

25. D. M. Jackson (1970). The occurrence and distribution in blood of β-phenylethylamine in association with cardiac disease. *Comp. Gen. Pharmacol.*, 1:263-272.

26. D. M. Jackson (1971). The effect of β-phenylethylamine on noradrenaline concentrations in guinea-pig brain. *J. Pharm. Pharmacol.*, 23:623-624.

27. D. M. Jackson (1972). The effect of β-phenethylamine upon spontaneous motor activity in mice: A dual effect on locomotor activity. *J. Pharm. Pharmacol.*, 24:383-389.

28. D. M. Jackson (1974). The interaction between β-phenylethylamine and agents which affect the cholinergic nervous system on locomotor activity and toxicity in mice. *Arzneim.-Forsch.*, 24:24-27.

29. D. M. Jackson (1975). β-Phenylethylamine and locomotor activity in mice. *Arzneim.-Forsch.*, 25:622-626.

30. D. M. Jackson (1975). Some further observations of the effect of β-phenylethylamine on locomotor activity in mice. *J. Pharm. Pharmacol.*, 27:278-280.

31. D. M. Jackson and D. B. Smythe (1973). The distribution of β-phenylethylamine in discrete regions of the rat brain and its effect on brain noradrenaline, dopamine and 5-hydroxytryptamine. *Neuropharmacology*, 12:663-668.

32. D. M. Jackson and D. M. Temple (1970). β-Phenylethylamine as a cardiotonic constituent of tissue extracts. *Comp. Gen. Pharmacol.*, 1:155-159.

33. J. B. Jepson, W. Lovenberg, and P. Zaltzman (1960). Amine metabolism studied in normal and phenylketonuric humans by monoamine oxidase inhibition. *J. Biochem.*, 74:5P.

34. J. Jonsson, H. Grobecker, and P. Holtz (1966). Effect of β-phenylethylamine on content and subcellular distribution of norepinephrine in rat heart and brain. *Life Sci.*, 5:2235-2246.

35. R. J. Levine, P. Z. Nirenberg, S. Udenfriend, and A. Sjöerdsma (1964). Urinary excretion of phenethylamine and tyramine in normal subjects and heterozygous carriers of phenylketonuria. *Life Sic.*, 3:651-656.

36. P. Mantegazza and M. Riva (1963). Amphetamine-like activity of β-phenylethylamine. *J. Pharm. Pharmacol.*, 15:472-478.

37. J. H. Mennear (1965). Interaction between central cholinergic agents and amphetamine in mice. *Psychopharmacologia*, 7:107-114.

38. K. E. Moore, L. A. Carr, and J. A. Dominic (1970). Functional significance of amphetamine-induced release of brain catecholamines. In: *Amphetamines and Related Compounds,* Proceedings of the Mario Negri Institute for Pharmacological Research, Milan, Italy, edited by E. Costa and S. Garattini. Raven Press, New York, pp. 371-384.

39. A. D. Mosnaim, E. E. Inwang, J. H. Sugerman, and H. C. Sabelli (1973). Identification of 2-phenylethylamine in human urine by infrared and mass spectroscopy and its quantification in normal subjects and cardiovascular patients. *Clin. Chim. Acta,* 46: 407-413.

40. A. D. Mosnaim and H. C. Sabelli (1971). Quantitative determination of the brain levels of a β-phenylethylamine-like substance in control and drug-treated mice. *Pharmacologist,* 13:283.

41. T. Nakajima, Y. Kakimoto, and I. Sano (1964). Formation of phenylethylamine in mammalian tissues and its effect on motor activity in the mouse. *J. Pharmacol. Exp. Ther.,* 143:319-325.

42. J. A. Oates, P. Z. Nirenberg, J. P. Jepsen, A. Sjverdrma, and S. Udenfriend (1963). Conversion of phenylalanine to phenylethylamine in patients with phenylketonuria. *Proc. Soc. Exp. Biol. Med.,* 112:1078-1801.

43. W. H. Oldendorf (1971). Brain uptake of radiolabeled amino acids, amines, and hexoses after arterial injection. *Am. J. Physiol.,* 221:1629-1639.

44. S. R. Philips, D. A. Durden, and A. A. Boulton (1974). Identification and distribution of *p*-tyramine in the rat. *Can. J. Biochem.,* 52:366-373.

45. S. R. Philips, D. A. Durden, and A. A. Boulton (1974). Identification and distribution of tyrptamine in the rat. *Can. J. Biochem.,* 52:447-451.

46. R. H. Rech and J. M. Stolk (1970). Amphetamine-drug interactions that relate brain catecholamines to behavior. In: *Amphetamines and Related Compounds,* Proceedings of the Mario Negri Institute for Pharmacological Research, Milan, Italy, edited by E. Costa and S. Garattini. Raven Press, New York, pp. 385-413.

47. J. M. Saavedra (1974). Enzymatic isotopic assay for and presence of β-phenylethylamine in brain. *J. Neurochem.,* 22:211-216.

48. J. M. Saavedra and E. Fischer (1970). Antagonism of betaphenylethylamine derivatives and serotonin blocking drugs upon serotonin tryptamine and reserpine behavioral depression in mice. *Arzneim.-Forsch.,* 20:952-957.

49. H. C. Sabelli and W. Giardina (1973). Amine modulation of affective behavior. In: *Chemical Modulation of Brain Function*, edited by H. C. Sabelli. Raven Press, New York, pp. 225-259.

50. H. C. Sabelli, W. J. Giardina, and A. D. Mosnaim (1971). *Proceedings of Central and Peripheral Adrenergic Systems*. Satellite Symposium XXV Int. Congr. Physiol. Sci., Warsaw.

51. H. C. Sabelli, W. J. Giardina, A. D. Mosnaim, and E. E. Inwang (1972). Evidence for a role of β-phenylethylamine as an adrenergic modulator. *Abstr. Commun. V Int. Congr. Pharmac.*, San Francisco, 198 (Paper No. 1184).

52. H. C. Sabelli, A. D. Mosnaim, and A. J. Vazquez (1974). Phenylethylamine: Possible role in depression and antidepressive drug action. In: *Advances in Behavioral Biology, Vol. 10—Neurohormonal Coding of Brain Function*, edited by R. D. Myers and R. R. Drucker-Colin. Plenum Press, New York, pp. 331-357.

53. H. C. Sabelli and A. D. Mosnaim (1974). Phenylethylamine hypothesis of affective behavior. *Am. J. Psychiat.*, 131: 695-699.

54. L. Stein (1964). Self-stimulation of the brain and the central stimulant action of amphetamine. *Fed. Proc.*, 23:836-850.

55. J. Willner, H. F. LeFevre, and E. Costa (1974). Assay by multiple ion detection of phenylethylamine and phenylethanolamine in rat brain. *J. Neurochem.*, 23:847-859.

Chapter 13

β-PHENYLETHYLAMINE: STUDIES ON THE MECHANISM OF ITS STIMULANT EFFECTS

David M. Jackson

Department of Pharmacology
University of Sydney
Sydney, New South Wales
Australia

I. INTRODUCTION

A. The Natural Occurrence of β-Phenylethylamine

Interest in the pharmacology, and possible clinical significance, of β-phenylethylamine (PEA) has slowly increased since the first report of its presence in normal human urine (3). Since then its presence in urine has been confirmed both in normal subjects and phenylketonurics (42,53). Phenylketonurics were found to display [after pretreatment with monoamine oxidase inhibitors (MAOI)] urinary PEA levels some 100 times greater than normal (53). It is also worthy of note that heterozygous carriers of phenylketonuria have been reported to excrete more PEA than normal individuals (44). The latter finding has to the author's knowledge not been repeated. Such a finding could contribute to the understanding of PEA's hypothesised role as a neuromodulator. Alternatively or additionally, it could cast some light on whether or not PEA is merely a waste product resulting from amino acid metabolism.

Since these early findings, PEA has been identified in various rabbit tissues by paper chromatography (52), in ox blood by crystallization with subsequent infrared, nuclear magnetic resonance, and mass spectrometric confirmation (41), and in a number of human and animal tissues including brain, liver, spleen, kidneys, and urine by a variety of chemical and physical methods (7,19,29,51,58,69). In addition, one of the potentially important metabolites of PEA, β-hydroxy-phenylethylamine (OHPEA) has been identified in both animal and human tissues (29,59,60).

B. Occurrence of β-Phenylethylamine in Various Disease States

Since the detection of PEA in abnormal concentrations in phenylketonuria (53), abnormal urinary and blood levels have been reported in various cardiac disorders (30,50). Decreases have been found in the urine of Parkinsonian patients (27,64) and in the urine of patients with various depressive illnesses (6,19,23,49). An increase in urinary PEA excretion has in contrast been reported in schizophrenia and mania (23,24). Moreover it has been shown (see, e.g., 23,24) that tricyclic therapy in depressed subjects raises urinary levels of PEA toward normal. This increase has been suggested to be due to the ability of imipramine to inhibit type B MAO, which enzyme is comparatively specific for PEA (72).

C. Endogenous β-Phenylethylamine Levels in Animals and Man

While there is now extensive evidence supporting the presence of PEA in animal, and normal and pathological human tissues, there is an extremely wide divergency in the actual endogenous levels that have been reported. In rat whole brain, Fischer et al. (20) reported 534 ± 19 ng/g wet weight; Durden et al. (15) found 1·8 ± 0·4 ng/g in rat whole brain, and Saavedra (58) reported a similar concentration of 1·5 ± 1·0 ng/g. In other species, Mosnaim and Inwang (47) reported values of 110 ± 10 ng/g and 400 ± 160 ng/g in cat and rabbit whole brain, respectively.

Similar discrepancies have been reported in urinary levels of PEA in man. For example, Boulton and Milward (6) were unable to detect PEA in the urine of depressives, and reported normal values of 47 ± 44 μg/24 hr (of free PEA). Fischer et al. (23) reported a normal excretion rate of 336 ± 45 μg/24 hr with a total of 114 ± 13 μg/24 hr in patients with endogenous depression. In contrast, Schweitzer et al. (67) found an average excretion of only 10·3 ± 18·2 μg/24 hr (median value = 4·4 μg/day) in the urine of normal subjects. While strain, species, and more particularly methodological differences probably account for these widely differing values, it would appear imperative that further refinements of the assay of PEA in endogenous tissues be made with a view to completely excluding other potentially interfering amines.

II. BIOCHEMICAL STUDIES

A. Effect of β-Phenylethylamine on Brain Amines

Biochemical and pharmacological studies are of considerable importance for the understanding of the mechanism of action of PEA. The first biochemical studies (43) found that high doses of 100 mg PEA/kg i.m. produced an amphetaminelike stimulation in rats, and such doses were able to produce up to 65% depletion of whole brain noradrenaline (NA). Repeated injections of PEA failed to produce a significantly greater depletion of NA. This work was confirmed by Fuxe et al. (26) using both histological and biochemical methods.

However, in contrast to the effects on NA, while Jonsson et al. (43) did not find a significant change in whole brain DA, Fuxe et al. (26) found that a single large dose of PEA produced an increase in rat brain DA, with multiple doses of PEA depleting both DA and NA. Using guinea pigs, Jackson (31) was unable to find any change in whole brain DA levels after single high doses of PEA (100 and 200 mg/kg, i.p.), but found that such doses of PEA caused a long lasting depletion of brain NA for 24 hr, with a peak depletion 1 hr after injection. Moreover, repeated injections of PEA led to a maximum NA depletion of 74% (cf. 43). When discrete areas of the rat brain were examined after acute i.p. administration of PEA (39), significant depletions of endogenous DA were found in the caudate nucleus, midbrain, cortex, and cerebellum, while significant depletions of NA occurred in all areas of the brain examined. In general, of the putative neurotransmitter amines, 5-HT levels appeared to be unaffected by PEA (39), except in mice where a small but significant depletion was found in mouse brain after a dose of 100 mg/kg i.p. (Smythe and Jackson, unpublished observations).

However, more meaningful information is obtained by examining transmitter turnover, rather than endogenous levels. Using the α-methyl-p-tyrosine (AMT) method of catecholamine synthesis inhibition, and measuring the effect of PEA on the rate of brain DA and

NA disappearance in rat brain (40), it was found that PEA accelerated NA disappearance without significantly affecting DA disappearance.

B. Distribution of Endogenous and Exogenous β-Phenylethylamine in Brain

A knowledge of the distribution of PEA (and its metabolites) within the CNS is of importance if the concept of its postulated function as a neuromodulator is to be elucidated. Nakajima et al. (52) reported that rabbit caudate nucleus had the highest PEA concentration in the brain after prior MAOI treatment (plus either phenylalanine or PEA), while the cerebellum had the lowest concentration. Durden et al. (15), using more sophisticated and reliable techniques (preparation of dansyl-PEA, thin layer chromatography with subsequent mass spectrometry), reported endogenous levels of PEA in rat brain to be highest in the hypothalamus (25.3 ± 5.0 ng/g) and the caudate nucleus (8.0 ± 0.3 ng/g), with lowest levels in the cerebellum (3.4 ± 0.5 ng/g) and brain stem (2.2 ± 0.9 ng/g). Edwards and Blau (16), pretreating rats with parachlorophenylalanine, pargyline, and phenylalanine, reported that PEA was fairly uniformly distributed in most parts of the brain with a tendency for it to concentrate most in the corpus striatum (4.5 μg/g) and least in the medulla (2.7 μg/g). Using a different approach, Jackson and Smythe (39), injected tritium-labeled PEA (1 mg/kg) into rats pretreated with iproniazid, and reported the distribution of total radiolabel to be in the order: hypothalamus ≳ midbrain > corpus striatum >> cerebellum > medulla oblongata-pons > cortex. However, even using a high, behaviorally active dose of PEA (100 mg/kg), without MAO I pretreatment, a similar rank order was obtained of total label: corpus striatum ≳ hypothalamus mid brain cerebellum > cortex > medulla-oblongata-pons. The findings of Jackson and Smythe (39) did, however, tend to suggest a somewhat uniform distribution of radiolabel (cf. 16). With such behaviorally effective doses of PEA, correlation of the actual percentage label present in each of these areas of the brain with the endogenous DA present in similar areas produced a highly significant positive correlation (39). This finding suggested a relationship between dopamine (DA) concentration in a particular brain region and radiolabel distribution after high doses of PEA. No significant correlation was obtained with endogenous NA or serotonin (5-HT) levels. In contrast to the foregoing findings, which generally agree that endogenous (or exogenously administered) PEA, or endogenously formed (from exogenously administered phenylalanine) PEA, is concentrated mostly in the hypothalamus and caudate nucleus, and least in the cerebellar and stem regions, are the findings of Mosnaim et al. (48) who found highest endogenous PEA levels (rabbit brain) in the cerebellum (1340 ± 230 ng/g) followed by brain stem (700 ± 320 ng/g) and cortex (310 ± 110 ng/g). Such discrepancies in data from different laboratories may rest partly on strain and species differences, but most probably in methodological differences.

III. BEHAVIORAL STUDIES—LOCOMOTOR STIMULATION

A. General Discussion

Schulte et al. (66) first reported that PEA (80 mg/kg) caused motor stimulation in mice with a duration of response of 4 hr. This work was in part confirmed by Nakajima et al. (52) who found that 10 mg PEA/kg i.v. in groups of five mice produced increased locomotor activity, peaking 8 min and terminating about 12 min after injection. Moreover, as expected, since PEA is a good substrate for MAO (5) prior treatment of the mice with a MAOI potentiated the locomotor stimulation produced by PEA (21,22,25,32,46,52).

However, a number of papers have reported that PEA, without pretreatment of the animals with a MAOI, depresses the activity of mice. Several of these studies measured the activity of individual mice after PEA (21,61). Others used grouped animals (e.g., 62) and measured locomotor activity for 10 min beginning 15 min after PEA (80 mg/kg) injection. Examination of this paper suggested in fact that 80 mg PEA/kg, when compared to the distilled water control, may have in fact produced significant stimulation (from the author's data, t = 2•43, n not reported). This is a period of time in which we have found all doses of PEA (with the exception of 75 mg/kg to be virtually inactive (32). Moreover, we have found in our experiments that PEA does not produce locomotor stimulation in isolated mice (Table 1) which have not been pretreated with a MAOI. Other factors also probably play a role such as strain differences, species differences, different equipment (see, e.g., 63) as well as differences in experimental design. For example, the activity of single mice measured by Fischer et al. (21) for the first 10 min after injection can be considered to be exploratory activity under the influence of a stimulant drug. In contrast to this, Saavedra et al. (62) injected reserpinized-mice with saline at time 0 and placed them in the activity cage; 15 min later the animals received PEA, and their locomotor activity was then measured 15 min later for the subsequent 15 min, that is 30 min after the animals first exposure to a novel environment.

B. Demonstration of a Dual Effect of β-Phenylethylamine upon Locomotor Activity

Grouped animals have been reported by Jackson (32) to respond to PEA administration with what is essentially a biphasic response (Table 1 and Fig. 1). Doses of 25 mg PEA/kg or less were essentially inactive. However, with 50 or 75 mg PEA/kg, a short-lived locomotor stimulation was observed (cf. 52), which peaked around 4 to 6 min after injection and which had a duration of about 15 min (32). Doses of 75 mg PEA/kg showed clear signs of a biphasic effect (Fig. 1), but 50 mg/kg produced only an early phase of activity. The locomotor stimulant phase between 0 and 20 min after injection is hereafter called the early phase of stimulant activity. With doses of 100, 125, and 150 mg/kg, no significant stimulation of locomotor activity was seen in the early phase. With these higher doses, sterotyped behavior was marked during the early phase, in contrast to the co-ordinated locomotor stimulation observed with 50 or 75 mg/kg. However, with doses of 75 mg PEA/kg, or higher, a dose-dependent rise in coordinated locomotor activity was observed between 20 and 60 min after injection

TABLE 1

The Effect of β-Phenylethylamine on Locomotor Activity in Mice[a]

Dose of PEA (mg/kg)	Isolated Animals		Grouped Animals[b]	
	Total Counts Between		Total Counts Between	
	0 to 20 min (n)	20 to 60 min	0 to 20 min (n)	20 to 60 min
Water control	493 ± 61 (7)	627 ± 130	1267 ± 67 (10)	830 ± 165
5	506 ± 40 (7)	358 ± 39	1223 ± 84 (7)	536 ± 81
10	533 ± 84 (7)	765 ± 158	1121 ± 48 (8)	824 ± 184
25	405 ± 51 (6)	561 ± 92	1228 ± 128 (5)	1055 ± 217
50	517 ± 100 (7)	348 ± 63	2476 ± 278 (8)*	1324 ± 193
75	579 ± 139 (6)	500 ± 70	2714 ± 182 (5)*	1925 ± 151*
100	388 ± 62 (6)	645 ± 110	1543 ± 130 (8)	2043 ± 167*
125	326 ± 47 (6)	654 ± 93	1621 ± 136 (5)	2631 ± 195*
150	—	—	1718 ± 188 (5)	3005 ± 185*

[a]The animals, either isolated or in groups of five, were injected with PEA or with water and immediately placed in the activity cage for 1 hr. The values represent the mean number of times the light beams were cut (counts) between 0 and 20 min and between 20 and 60 min after injection. The values in brackets represent the number of experiments. Statistical comparisons were made using Students' t test, and an asterisk signifies that the mean (± SEM) is significantly different from the water control. The mice used were QS strain male mice.

[b]Data partly from D. M. Jackson (32).

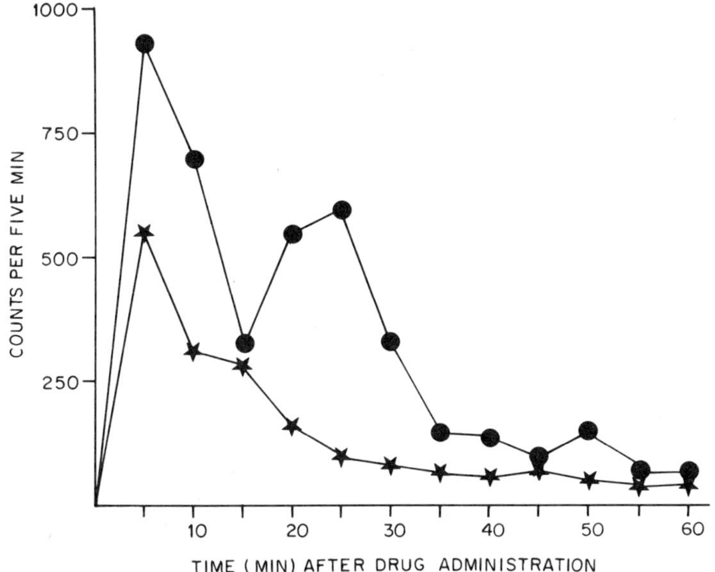

FIG. 1. The effect of β-phenylethylamine, 75 mg/kg i.p., on the locomotor activity of groups of five mice. The data represents the mean counts/5 min for a total of 1 hr, and the SEM have been omitted for clarity. From (32) by permission of the publishers. Closed star, saline; closed circle, β-phenylethylamine.

(hereafter called the late phase), the stereotypies evident between 0 and 20 min after injection having worn off and been replaced by co-ordinated locomotor stimulation. The stimulation during the late phase was pharmacologically quite distinguishable from the early phase stimulation (32,36, Table 2 and below).

C. Pharmacological Analysis of the Early Phase of Locomotor Stimulation

Present evidence would suggest that the PEA-induced early phase of stimulation is due to a release of central catecholamines (32,36, and Table 2). Thus, the early phase was blocked by the DA-β-oxidase inhibitors FLA-63, diethyldithiocarbamate and disulfiram, and by the tyrosine hydroxylase inhibitor AMT. Moreover, the α-adrenergic

TABLE 2

Tabular Summary of the Effect of Various Pharmacological Pretreatments on the Locomotor Stimulation Produced in Groups of Five Mice[a]

Pretreatment	Early phase	Late phase
α-methyl Tyrosine	+	—
Diethyldithiocarbamic acid	+	—
FLA-63	+	—
Disulfiram	+	—
Reserpine	—	—
Phenoxybenzamine	+	—
Phentolamine	+	—
Propranolol	—	—
Chlorpromazine	—	—
Perphenazine	+	—
Pimozide	+	+
Haloperidol	+	+
Protriptyline	—	—
Desipramine	—	—
Amitriptyline	—	—
Clonidine	—	≡
Cyproheptadine	—	—
Parachlorophenylalanine	+ (?)	≡
Atropine	≡	≡
Benzhexol	≡	≡
Physostigmine	—	—
Hyoscine methyl bromide	—	—
Hyosine hydrobromide	≡	≡

[a]Stimulation produced by 50 mg PEA/kg between 0 and 20 min after injection, early phase stimulant activity, and by 100 mg PEA/kg between 20 and 60 min after injection, late phase stimulant activity. Data from Jackson (32-34,36) and unpublished data. +, blocked; -, not blocked; ≡, potentiated.

antagonists phenoxybenzamine and phentolamine (but not the β-adrenergic antagonist, propranolol), and the DA receptor antagonists pimozide and haloperidol, were effective in blocking this early phase of activity. In contrast, reserpine was ineffective (cf. 62) emphasizing the importance of newly synthesised catecholamines from the still functioning tyrosine hydroxylase and DA-β-oxidase. Noradrenaline uptake inhibitors were ineffective in blocking this phase of stimulation, suggesting either that access of PEA to NA neurones or that blockade of re-uptake of released NA was not of critical importance for stimulation. Of relevance here are the findings in rats that NA uptake inhibitors are largely effective in blocking PEA-induced release of NA (26). Consequently, a) with the findings that NA uptake inhibitors are ineffective in blocking early phase locomotor stimulation and b) with data suggesting that such compounds are much more potent in blocking neuronal uptake of NA than of DA (7,28,56), one would conclude that release of DA (rather than NA) is of dominant importance in the manifestation of early phase stimulation. However, a functioning synthetic pathway for NA and available "nonblocked" (nonoccupied) α-receptors also appear of importance. Thus, both dopaminergic and noradrenergic mechanisms appear to be important, but different mechanisms and central sites of action are probably involved in the contribution of each amine. Furthermore, it would not be inconsistent with the data that part of this early phase of locomotor stimulation is due to OHPEA, a metabolite of PEA formed in the neurons by DA-β-oxidase (10,59,60). The blockade of intraneuronal formation of such a metabolite would account for the effectiveness of DA-β-oxidase inhibitors in blocking early phase stimulation (36). Recently, Jackson et al. (38), using the method of direct application of OHPEA to the nucleus accumbens of nialamide pretreated rats, reported that the resultant rise in locomotor activity was completely blocked by prior treatment with AMT, suggesting an indirect action for OHPEA (Fig. 2). The likelihood that OHPEA is contributing significantly to the early phase stimulation is also reduced by the fact that the

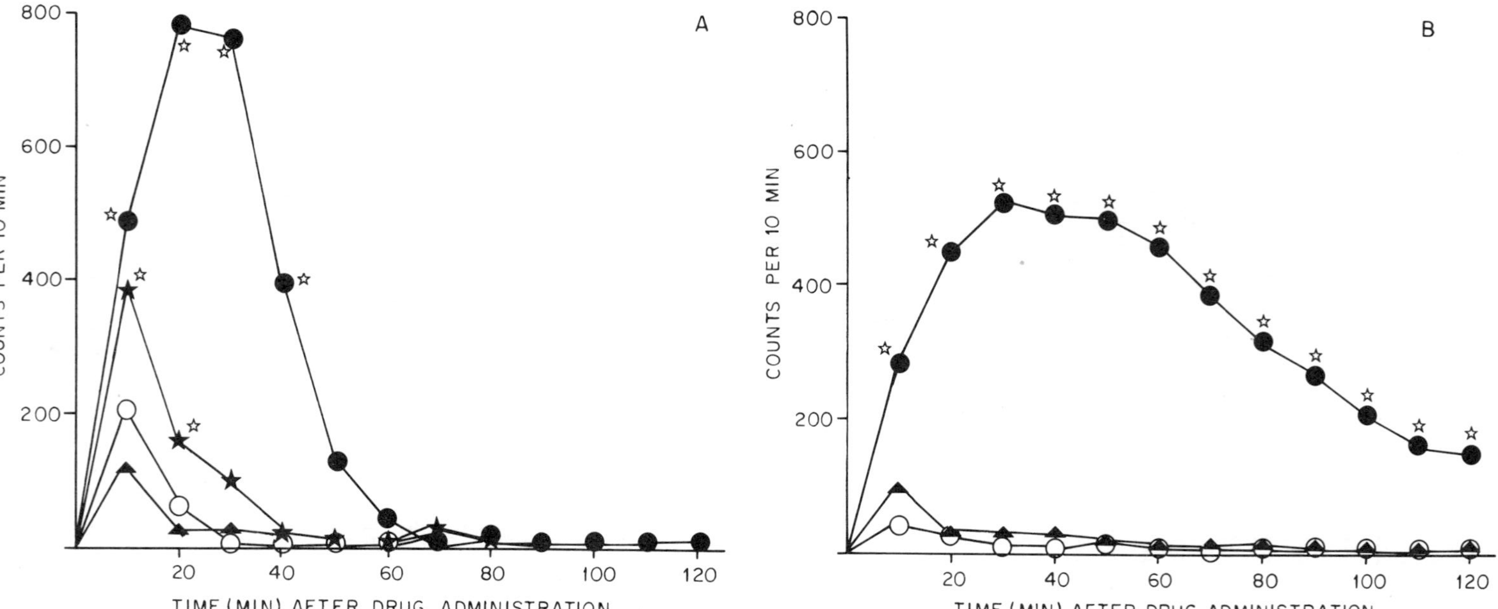

FIG. 2. The effect of phenylethanolamine hydrochloride 57 μg/μl, phenylethylamine hydrochloride, 25 or 50 μg/μl, or saline administered bilaterally (1 μl to each side) to the nucleus accumbens of nialamide (100 mg/kg, i.p., 1 hr premedication) treated rats. Locomotor activity was measured by Animex activity meters each 10 min for 2 hr, and the data are presented as counts/10 min. Standard errors have been omitted for clarity. In some cases, animals were pretreated with AMT, 400 mg/kg (3 hr premedication) prior to the stimulant drug. The number of experiments is given in brackets, and an asterisk on the graph indicates a significance level of $P < 0{\cdot}05$ to the saline groups. Graph A: closed circle, PEA 50 μg bilaterally (4); closed star, PEA 25 μg bilaterally (4); open circle, PEA 50 μg bilaterally plus AMT 400 mg/kg (5); closed triangle, saline, 1 μl bilaterally. Graph B: closed circle, phenylethanolamine 57 μg bilaterally (10); closed triangle, phenylethanolamine 57 μg bilaterally plus AMT 400 mg/kg (4); open circle, saline, 1 μl bilaterally (data from D. M. Jackson and N.-E. Andén, unpublished observations).

main route of metabolism of PEA is by MAO, and only about 10% of PEA appears to be β-hydroxylated (16). That the early phase of stimulation is due to catecholamine release by PEA is supported also by the rapid onset and rapid disappearance of stimulation, which parallels well the rate of appearance and disappearance in blood of exogenously administered PEA (11,25,52, Jackson and Smythe, unpublished observations). That PEA itself exerts its stimulant effect indirectly by releasing catecholamine stores is further supported by the findings of Jackson et al. (38). Bilateral application of DA into the nucleus accumbens of nialamide and reserpine pretreated rats produces an increase in coordinated locomotor activity (37,57). A qualitatively similar response to that elicited by DA can be evoked by the application of PEA (and OHPEA, see previous discussion) into the same area, and the resultant locomotor stimulation is completely blocked by prior treatment of the rat with AMT (Fig. 2). This particular pharmacological model by-passes peripheral mechanisms of metabolism, and probably provides a more reliable and valid view of the action of PEA itself than does systemic administration of PEA. In this model, as in studies utilizing systemic injections of PEA into nialamide-pretreated mice, no evidence of direct receptor stimulation could be demonstrated.

D. Pharmacological Study of the Late Phase of Locomotor Stimulation

The late phase of locomotor stimulation produces by doses of PEA, 75 mg/kg and higher, appeared to involve a different pharmacological mechanism to that hypothesized for the early phase. The stimulation was found to be unaffected by any pharmacological treatment (see Table 2), with the exception of fairly high doses of the DA-receptor antagonists, pimozide and haloperidol (36). These findings confirmed the earlier findings of Jackson (32) that the late phase of stimulation did not require either reserpine sensitive amine stores or newly synthesized catecholamines, and supported the hypothesis that the stimulation was apparently due to a direct stimulation of post synaptic DA receptors. The failure of DA-β-oxidase inhibitors to block the late phase stimulation (32,35,36) suggested that potential

TABLE 3

The Effect on Locomotor Activity in Groups of Three Mice of Metabolites of β-Phenylethylamine, Phenylethanol, and Phenylethanolamine[a]

Pretreatment	Test Drug	Dose of Test Drug (mg/kg)	Locomotor Activity (mean ± SEM) During: 0 to 30 min	30 to 60 min	60 to 90 min	90 to 120 min
—	Phenylethanolamine	Saline (11)	355 ± 29	183 ± 32	85 ± 20	54 ± 8
	Phenylethanolamine	25 (11)	104 ± 14**	285 ± 29*	141 ± 35	59 ± 21
	Phenylethanolamine	50 (11)	32 ± 6**	136 ± 15	237 ± 26**	94 ± 15*
	Phenylethanolamine	100 (11)	18 ± 14**	44 ± 10**	78 ± 14	143 ± 20**
FLA-63	Phenylethanolamine	50 (8)	15 ± 4*	31 ± 9	47 ± 13*	19 ± 4*
	Phenylethanolamine	Saline (8)	36 ± 8	24 ± 6	13 ± 5	6 ± 2
AMT	Phenylethanolamine	50 (8)	51 ± 10*	89 ± 22	84 ± 24*	27 ± 12
	Phenylethanolamine	Saline (8)	144 ± 26	10 ± 6	11 ± 4	12 ± 6
—	Phenylethanol	Saline (11)	278 ± 44	142 ± 40	66 ± 25	35 ± 15
	Phenylethanol	25 (11)	238 ± 37	97 ± 22	37 ± 12	22 ± 11
	Phenylethanol	50 (11)	256 ± 50	124 ± 44	66 ± 35	59 ± 33
	Phenylethanol	100 (11)	253 ± 31	76 ± 19	39 ± 11	24 ± 5

*P < 0•001 compared to the appropriate saline control.

**P < 0•05 compared to the appropriate saline control.

[a]Locomotor activity was measured each 30 min for 120 min after i.p. injection of the drug (or saline, and the data are presented as the total number of counts registered by the Animex activity meter ± SEM. Pretreatments where used were AMT (250 mg/kg, i.p., 4 hr pretreatment) or FLA-63 (25 mg/kg, 2 hr). The number of experiments is in brackets. Data from Dunstan and Jackson (unpublished).

β-hydroxylation was probably not of major importance, and this finding was further supported by studies of Jackson et al. (38). However recent studies in our laboratory (Table 3, Dunstan and Jackson, unpublished data) have shown that OHPEA does in fact possess locomotor stimulant activity per se (i.e., without MAOI pretreatment) and that such stimulation is blocked by neither AMT or FLA-63 pretreatment, suggesting i) a post synaptic mode of action (Table 3) or ii) a possible requirement for reserpine-sensitive stores (not tested). With all doses, the stimulation was preceded by a phase of locomotor depression. A deaminated metabolite of OHPEA, probably formed in the periphery, may be responsible for the rather late onset of locomotor stimulation seen with OHPEA, because Jackson et al. (38) have shown by direct application of OHPEA to the brain that OHPEA itself has only indirect stimulant actions. Although one could suggest that OHPEA formation could contribute to the late phase of PEA-induced direct locomotor stimulant, this is unlikely because DA-β-oxidase inhibitors (see previous discussion) were unable to block this late phase. Providing additional support for the primary functional role of PEA, Sabelli et al. (65), measuring various pharmacological parameters, reported that pretreatment with DA-β-oxidase inhibitors was ineffective in blocking the effects of PEA. Other potentially active candidates are the deaminated metabolites of PEA, such as phenylacetaldehyde, phenylethanol, and phenylacetic acid, or substituted derivatives of these compounds. These metabolites are formed rapidly in the periphery and in the CNS by MAO, and cross the blood-brain barrier if formed peripherally (e.g., for penetration of phenylacetic acid into the CNS, see Pedemonte et al. (54), cited by Ref. 63). These metabolites have been shown to possess some pharmacological activity. Phenylacetaldehyde and phenylacetate depressed locomotor activity in mice (63), but in the case of phenylacetaldehyde, a massive dose of 2 ml/kg was used, and with both drugs, only 5 min of activity was recorded 20 min after injection. Studies in our laboratory have also shown phenylacetic acid (pH 7) and phenylethanol to be essentially inactive when administered i.p. to groups of three mice in doses between 25 and 100 mg/kg and their locomotor activity recorded for 120 min after injection (Table 3, Dunstan and Jackson, unpublished observations). That a deaminated metabolite was probably responsible for the direct stimulant effect of high doses of PEA (Table 1) was supported by further studies in nialamide-pretreated mice. In these studies, stimulation produced by PEA was completely blocked by AMT (35) suggesting that PEA and nondeaminated metabolites were not contributing to the direct receptor stimulation seen only in non-MAOI-pretreated mice. However, it should be noted that Sabelli et al. (64) reported that high doses of AMT (400 mg/kg) were unable to inhibit the stimulant effect of PEA in MAOI-treated mice. Further work is required to clarify these data. Again, in nialamide-treated mice, FLA-63 failed to completely block the locomotor stimulation produced by PEA (35), confirming in these circumstances 1) that OHPEA is probably not of major significance under these conditions and ii) that an intact NA pathway is not of prime importance. However, with nialamide pretreatment, two phases of locomotor stimulation do not occur, and what we are in fact dealing with is probably a greatly exaggerated first phase of locomotor stimulation. Sabelli and Giardini (63) also support the apparent nonimportance of OHPEA since "most of the CNS effects of PEA [were] not prevented by the DA-β-oxidase inhibitor, FLA-63."

Therefore, it would appear from the evidence available that the direct stimulant effect seen in mice is not due to PEA itself, but is due to some metabolite(s) of PEA. β-Hydroxy-phenylethylamine would appear not to be a major contributor, and the active compound would appear to be a deaminated metabolite, the nature of such a moiety at present being unknown. Ring hydroxylation and N-methylation are other metabolic mechanisms likely to produce active and possibly directly acting sympathomimetic amines.

E. Some Pharmacological Studies in Rats

Although locomotor stimulation has been reported by some authors not to occur in rats after PEA administration (26), Fuxe et al. (46) reported that the excitatory behavioral effects of PEA (50-100 mg/kg, i.v.) observed in reserpine-treated rats could be almost completely blocked by AMT, and that administration of small doses of L-DOPA could largely overcome this blockade, supporting the concept that the functional effect in rats is produced by release of newly synthesized DA (and possibly NA).

We have found (Jackson, unpublished observations) that doses of PEA between 25 and 100 mg/kg produce stereotyped behavior in the rat when the animal is restricted to a small wire cage. The intensity and duration of stereotyping was dose-dependent with doses of 75 and 100 mg/kg producing grooming, sniffing, exploratory behavior, up and down front paw movements, some backwards locomotion, and side to side movements of the head. Marked sympathomimetic signs such as salivation and piloerection were also seen. The stereotypies were blocked by prior AMT treatment, but not by prior diethyldithiocarbamate treatment. Haloperidol, perphenazine, and pimozide were all effective in blocking the stereotyped behavior. Reserpine was, however, ineffective (cf. 26). These findings add support to the conclusion that a large component of the action of PEA in the CNS is indirect.

F. Relative Roles of DA and NA for PEA-induced Locomotor Stimulation

There appears to be little doubt that both dopaminergic and noradrenergic mechanisms are involved in the early phase locomotor stimulation elicited by PEA in mice, while dopaminergic mechanisms would seem of dominant importance for the late phase. In rats, at present, only dopaminergic mechanisms have been implicated. Moreover, it is now generally agreed that both DA and NA are of importance in the manifestation of both stimulant-induced and locomotor activity per se (2,14,57,70,71). The involvement and importance of NA receptor activation for the full manifestation of PEA-induced stimulation is seen in the studies of the early phase activity (32,36) and, moreover, in studies where clonidine, an α-adrenergic receptor agonist (1), was combined with PEA. Clonidine by itself is inactive or depressant, but when combined in suitable dosage with apomorphine, a specific

DA agonist, clonidine potentiates the stimulation produced by apomorphine (2,34,45).

Similarly, we have found that clonidine is able to markedly potentiate the locomotor stimulant effect of 100 mg PEA/kg between 20 and 60 min after injection but only under certain pharmacological conditions (34). Potentiation was observed in untreated mice, and in mice which had been pretreated with AMT or with reserpine plus FLA-63. However, no potentiation was observed when the pretreatment was reserpine plus AMT. The latter pretreatment effectively removes granular stores of DA and blocks the synthesis of any new DA (and NA). In contrast, reserpine plus FLA-63 pretreatment, while also depleting granular stores of catecholamines, blocked the synthesis only of NA and was ineffective in influencing the clonidine-induced potentiation of PEA. Thus, in some manner, DA, either in the form of reserpine-sensitive DA or newly synthesized DA, is required for the potentiation of PEA by clonidine. Under the conditions and pretreatments used, clonidine always potentiated the stimulant effect of apomorphine, emphasizing the post-synaptic mode of action of both of these compounds under the experimental conditions. The reason for the failure of clonidine to potentiate existing PEA stimulation in animals with effectively no available DA is unknown, but the data do emphasize the importance of DA for the action of PEA. But, possibly of more interest, the data provide additional evidence for a link between NA and DA, both of which are apparently required for the full manifestation of the stimulant effect of PEA, and highlights the complexity of the link. Moreover, the interaction between clonidine and PEA occurred despite the fact that only DA receptor antagonists are able to block the late phase locomotor stimulation of untreated mice. It is interesting that a single injection of α-methyl DOPA (32), which is converted into α-methyl NA, also potentiated the stimulation produced by PEA (especially, the normally inactive 25 mg/kg dose). α-Methyl DOPA has previously been reported to potentiate the action of amphetamine (68). Since α-methyl NA, although considered a "false transmitter," is, nevertheless, an NA receptor agonist, a mechanism similar to the interaction between clonidine and PEA may underlie its ability to potentiate PEA-induced locomotor activity.

G. Involvement of Cholinergic Mechanisms

Perhaps not surprisingly, antimuscarinic compounds have been shown to potentiate PEA-induced locomotor stimulation (33). Sympathomimetics such as amphetamine and phenylpropanolamine, which produce an increase in locomotor activity, are potentiated by muscarinic antagonists which cross the CNS, but not by agents (such as methylhyoscine), which are only active peripherally (8,9,13). In contrast, drugs which increase the effective concentration of acetylcholine (physostigmine) did not potentiate PEA-induced locomotor activity (33). The mechanism(s) of the interaction is at present unknown but may involve known DA-cholinergic interactions in the corpus striatum (see, e.g., 4) or inhibition of DA uptake into DA neurones by antimuscarinic agents (12,18). However, further definition of the interaction between PEA and antimuscarinic compounds on locomotor activity must await more research.

IV. SUMMARY

β-Phenylethylamine, like amphetamine, produces an increased locomotor activity in mice and this response is potentiated by prior treatment with a MAOI. In animals not pretreated with a MAOI, PEA can be shown to have two distinct phases of action: 1) an early phase which occurs with doses of 50 and 75 mg/kg and which appears to be dependent on an intact synthetic pathway for NA and DA, and also involves both DA and NA receptors; 2) a late phase of increased activity which occurs with doses of 75 mg/kg PEA and higher, and which is apparently produced by a deaminated metabolite of PEA exerting a direct stimulant effect on post-synaptic DA receptors. Moreover, OHPEA appears to be of little importance in the stimulant response of either phase. The importance of, and the complexity of, the NA involvement in PEA-induced locomotor activity under certain carefully defined conditions is discussed. Consideration must also be given to the apparent involvement of a cholinergic mechanism.

ACKNOWLEDGMENTS

I am deeply grateful to Ms. M. Rutherford for excellent technical assistance and to Boehringer Ingelheim; Searle Australia; I.C.I.; Smith, Kline and French Australia; CIBA-Geigy Australia; and to Janssen Pharmaceuticals, Belgium, for generous domations of drugs. The critical assistance of Dr. G. B. Chesher and Dr. F. Caredes is gratefully acknowledged.

REFERENCES

1. N.-E. Andén, H. Corrodi, K. Fuxe, B. Hökfelt, T. Hökfelt, C. Rydin and T. Svensson (1970). Evidence for a central noradrenaline receptor stimulation by clonidine. *Life Sci.*, 9:513-523.

2. N.-E. Andén, U. Strömbom, and T. H. Svensson (1973). Dopamine and noradrenaline receptor stimulation: Reversal of reserpine-induced suppression of motor activity. *Psychopharmacologia*, 29:289-298.

3. A. M. Asatoor and C. E. Dalgliesh (1959). Amines in blood and urine. *Biochem. J.*, 73:26P.

4. G. Bartholini and A. Pletscher (1971). Atropine-induced changes of cerebral dopamine turnover. *Experientia*, 27:1302-1303.

5. H. Blaschko (1952). Amine oxidase and amine metabolism. *Pharmacol. Rev.*, 4:415-458.

6. A. A. Boulton and A. Milward (1971). Separation, detection and quantitative analysis of urinary β-phenylethylamine. *J. Chromatogr.*, 57:287-296.

7. A. K. Carlsson, K. Fuxe, B. Hamberger, and M. Lindquist (1966). Biochemical and histochemical studies on the effects of imipraminelike drugs and amphetamine on central and peripheral catecholamine neurons. *Acta. Physiol. Scand.*, 67:481-497.

8. P. L. Carlton (1961). Augmentation of the behavioural effects of amphetamine by scopolamine. *Psychopharmacologia*, 2:377-380.

9. P. L. Carlton and P. Didamo (1961). Augmentation of the behavioural effects of amphetamine by atropine. *J. Pharmacol. Exp. Ther.*, 132:91-96.

10. C. R. Creveling, J. W. Daly, B. Witkop, and S. Udenfriend (1962). Substrates and inhibitors of dopamine-β-oxidase. *Biochem. Biophys. Acta*, 64:125-134.

11. I. Cohen, J. F. Fischer, and W. H. Vogel (1974). Physiological disposition of β-phenylethylamine, 2, 4, 5-trimethoxyphenylethylamine and β-hydroxymescaline in rat brain, liver and plasma. *Psychopharmacologia*, 36:77-84.

12. J. T. Coyle and S. H. Snyder (1969). Antiparkinsonian drugs: Inhibition of dopamine uptake in the corpus striatum as a possible mechanism of action. *Science*, 166:899-901.

13. W. M. Davis and J. T. Pinkerton III (1972). Synergism by atropine of central stimulant properties of phenyl-propanolamine. *Toxicol. Appl. Pharmacol.*, 22:138-145.

14. P. S. D'Encarnacao, P. D'Encarnacao, and J. T. Tapp (1969). Potentiation of amphetamine induced psychomotor activity by diethyldithiocarbamate. *Arch. Int. Pharmacodyn. Ther.*, 182: 186-189.

15. D. A. Durden, S. R. Phillips, and A. A. Boulton (1973). Identification and distribution of β-phenylethylamine in the rat. *Can. J. Biochem.*, 51:995-1002.

16. D. J. Edwards and K. Blau (1972). Phenylethylamines in brain and liver of rats with experimentally induced phenylketonuria-like characteristics. *Biochem. J.*, 132:95-100.

17. A. M. Ernst (1967). Mode of action of apomorphine and dexamphetamine on gnawing compulsion in rats. *Psychopharmacologia*, 10: 316-323.

18. L.-O. Farnebo, K. Fuxe, B. Hamberger, and H. Ljungdahl (1970). Effect of some antiparkinsonian drugs on catecholamine neurons. *J. Pharm. Pharmacol.*, 22:733-737.

19. E. Fischer, B. Heller, and A. N. Miró (1968): β-Phenylethylamine in human urine. *Arzneim.-Forsch.*, 18:1486.

20. E. Fischer, B. Heller, H. Spatz, and H. Reggiani (1972). Thin-layer chromatographic assay of phenethylamine content of the rat brain and its changes after reserpine and imipramine administration. *Arzneim.-Forsch.*, 22:1560.

21. E. Fischer, R. J. Ludmer, and H. C. Sabelli (1967). The antagonism of phenylethylamine to catecholamines on mouse motor activity. *Acta. Physiol. Lat. Am.*, 17:15-21.

22. E. Fischer, J. M. Saavedra, and B. Heller (1968). Effects of catecholamines, adrenergic substances and their blocking agents on the searching behaviour of mice. *Arzneim.-Forsch.*, 7:780-786.

23. E. Fischer, H. Spatz, B. Heller and H. Reggiani (1972). Phenethylamine content of human urine and rat brain, its alterations in pathological conditions and after drug administration. *Experientia*, 28:307-308.

24. E. Fischer, H. Spatz, J. M. Saavedra, H. Reggiani, A. H. Miró, and B. Heller (1972). Urinary elimination of phenethylamine. *Biol. Psychiat.*, 5:139-147.

25. R. W. Fuller, H. D. Snoddy, B. B. Molloy, and K. M. Hauser (1973). CNS stimulant properties and physiologic disposition of β,β-difluorophenylethylamine in mice. *Psychopharmacologia*, 28:205-212.

26. K. Fuxe, H. Grobecker, and J. Jonsson (1967). The effect of β-phenylethylamine on central and peripheral monoamine-containing neurons. *Eur. J. Pharmacol.*, 2:202-207.

27. B. Heller and E. Fischer (1973). Diminution of phenethylamine in the urine of Parkinson patients. *Arzneim.-Forsch.*, 23: 884-886.

28. A. S. Horn, J. T. Coyle, and S. H. Snyder (1971). Catecholamine uptake by synaptosomes from rat brain-structure activity relationships with differential effect on dopamine and norepinephrine neurons. *Mol. Pharmacol.*, 7:66-80.

29. E. E. Inwang, A. D. Mosnaim, and H. C. Sabelli (1973). Isolation and characterization of phenylethylamine and phenylethanolamine from human brain. *J. Neurochem.*, 20:1469-1473.

30. D. M. Jackson (1970). The occurrence and distribution in blood of β-phenylethylamine in association with cardiac disease. *Comp. Gen. Pharmacol.*, 1:263-272.

31. D. M. Jackson (1971). The effect of β-phenethylamine on noradrenaline concentrations in guinea-pig brain. *J. Pharm. Pharmacol.*, 23:623-624.

32. D. M. Jackson (1972). The effect of β-phenethylamine upon spontaneous motor activity in mice: A dual effect on locomotor activity. *J. Pharm. Pharmacol.*, 24:383-389.

33. D. M. Jackson (1974). The interaction between β-phenylethylamine and agents which affect the cholinergic nervous system on locomotor activity and toxicity in mice. *Arzneim.-Forsch.*, 24:24-27.

34. D. M. Jackson (1974). The involvement of noradrenergic systems in the locomotor activity stimulation produced by β-phenylethylamine. *J. Pharm. Pharmacol.*, 26:651-654.

35. D. M. Jackson (1975). Some further observations on the effect of β-phenylethylamine on locomotor activity in mice. *J. Pharm. Pharmacol.*, 27:278-280.

36. D. M. Jackson (1975). β-Phenylethylamine and locomotor activity in mice: Interaction with catecholaminergic neurones and receptors. *Arzneim.-Forsch.*, 25:622-626.

37. D. M. Jackson (1975). Some functional effects produced by the direct application of sympathomimetic amines to various parts of the brain and their alteration by antipsychotic drugs. In: *Antipsychotic Drugs, Pharmacodynamics and Pharmacokinetics*, edited by Y. Zotterman, B. Uvnäs, and G. Sedvall. Pergamon Press, London, pp. 89-95.

38. D. M. Jackson, N.-E. Andén, and A. Dahlström (1975). A functional effect of dopamine in the nucleus accumbens and in some other dopamine-rich parts of the rat brain. *Psychopharmacologia*, 45:139-149.

39. D. M. Jackson and D. B. Smythe (1973). The distribution of β-phenylethylamine in discrete regions of the rat brain and its effect on brain noradrenaline, dopamine and 5-hydroxytryptamine levels. *Neuropharmacology*, 12:663-668.

40. D. M. Jackson and D. E. Smythe (1974). The effect of β-phenethylamine on noradrenaline and dopamine turnover in rat brain. *J. Pharm. Pharmacol.*, 26:456-458.

41. D. M. Jackson and D. M. Temple (1970). β-Phenylethylamine as a cardiotonic constituent of tissue extracts. *Comp. Gen. Pharmacol.*, 1:155-159.

42. J. B. Jepson, W. Lovenberg, P. Zaltzman, J. A. Oates, A. Sjoerdsma, and S. Udenfriend (1960). Amine metabolism, studied in normal and phenylketonuric humans by monoamine oxidase inhibition. *Biochem. J.*, 74:5P.

43. J. Jonsson, H. Grobecker, and P. Holtz (1966). Effect of β-phenylethylamine on content and subcellular distribution of norepinephrine in rat heart and brain. *Life Sci.*, 5:2235-2246.

44. R. J. Levine, P. Z. Nirenberg, S. Udenfriend, and A. Sjoerdsma (1964). Urinary excretion of phenylethylamine and tyramine in normal subjects and heterozygous carriers of phenylketonuria. *Life Sci.*, 3:651-656.

45. J. Maj, H. Sowinska, L. Baran, and Z. Kapturkiewicz (1972). The effect of clonidine on locomotor activity in mice. *Life Sci.*, 11:483-491.

46. P. Mantegazza and M. Riva (1963). Amphetamine-like activity of β-phenylethylamine after a monoamine oxidase inhibitor in vivo. *J. Pharm. Pharmacol.*, 15:472-478.

47. A. D. Mosnaim and E. E. Inwang (1973). A spectrophotometric method for the quantification of a-phenylethylamine in biological specimens. *Anal. Biochem.*, 54:561-577.

48. A. D. Mosnaim, E. E. Inwang, and H. C. Sabelli (1974). The influence of psychotropic drugs on the levels of endogenous a-phenylethylamine in rabbit brain. *Biol. Psychiat.*, 8:227-234.

49. A. D. Mosnaim, E. E. Inwang, J. H. Sugerman, W. J. De Martini, and H. C. Sabelli (1973). Ultraviolet spectrophotometric determination of 2-phenylethylamine in biological samples and its possible correlation with depression. *Biol. Psychiat.*, 6: 235-257.

50. A. D. Mosnaim, E. E. Inwang, J. H. Sugerman, and H. C. Sabelli (1973). Identification of 2-phenylethylamine in human urine by infrared and mass spectroscopy and its quantification in normal subjects and cardiovascular patients. *Clin. Chim. Acta*, 46: 407-413.

51. A. D. Mosnaim and H. C. Sabelli (1971). Quantitative determination of the brain levels of a β-phenylethylamine-like substance in control and drug-treated mice. *Pharmacologist*, 13:283.

52. T. Nakajima, Y. Kakimoto, and I. Sano (1964). Formation of β-phenylethylamine in mammalian tissue and its effect on motor activity in the mouse. *J. Pharmacol. Exp. Ther.*, 143:319-325.

53. J. A. Oates, P. Z. Nirenberg, J. B. Jepson, A. Sjoerdsma, and S. Udenfriend (1963). Conversion of phenylalanine to phenylethylamine in patients with phenylketonuria. *Proc. Soc. Exp. Biol. Med.*, 112:1078-1081.

54. W. Pedemonte, M. Bulat, and A. D. Mosnaim (1975). Cited by Sabelli et al. (64).

55. A. J. J. Pijnenburg and J. M. van Rossum (1973). Stimulation of locomotor activity following injection of dopamine into the nucleus accumbens. *J. Pharm. Pharmacol.*, 25:1003-1005.

56. S. B. Ross and A. L. Renyi (1967). Inhibition of the uptake of tritiated catecholamines by antidepressant and related agents. *Eur. J. Pharmacol.*, 2:181-186.

57. J. M. van Rossum (1970). Mode of action of psychomotor stimulant drugs. *Int. Rev. Neurobiol.*, 12:307-383.

58. J. M. Saavedra (1974). Enzymatic isotopic assay for and presence of β-phenylethylamine in brain. *J. Neurochem.*, 22:211-216.

59. J. M. Saavedra and J. Axelrod (1973). Demonstration and distribution of phenylethanolamine in brain and other tissues. *Proc. Nat. Acad. Sci. USA*, 70:769-772.

60. J. M. Saavedra, J. T. Coyle, and J. Axelrod (1974). Developmental characteristics of phenylethanolamine and octopamine in the rat brain. *J. Neurochem.*, 23:511-515.

61. J. M. Saavedra and E. Fischer (1970). Antagonism of β-phenylethylamine derivatives and serotonin blocking drugs upon serotonin, tryptamine and reserpine behavioural depression in mice. *Arzneim.-Forsch.*, 7:952-956.

62. J. M. Saavedra, B. Heller, and E. Fischer (1970). Antagonistic effects of tryptamine and β-phenylethylamine on the behaviour of rodents. *Nature*, 226:868.

63. H. C. Sabelli and W. J. Giardini (1973). Amine modulation of affective behaviour. In: *Chemical Modulation of Brain Function*, edited by H. C. Sabelli. Raven Press, New York, pp. 225-259.

64. H. C. Sabelli, A. D. Mosnaim, R. L. Borison, and M. E. Wolf (1975). Possible role of 2-phenylethylamine in the modulation of extrapyramidal function. In: *The Proceedings of a Satellite Symposium to the XXVI International Congress of Physiological Sciences, New Delhi, India: Use of Pharmacological Agents in Elucidation of Central Synaptic Transmission*, edited by B. N. Dhawan and P. B. Bradley. MacMillan and Co., London. In press.

65. H. C. Sabelli, A. J. Vazquez, and D. Flavin (1975). Behavioural and electrophysiological effects of phenylethanolamine and 2-phenylethylamine. *Psychopharmacologia*, 42:117-125.

66. J. W. Schulte, E. C. Reif, J. A. Bacher, Jr., W. S. Lawrence, and M. L. Tainter (1941). Further study of central stimulation from sympathomimetic amines. *J. Pharmacol. Exp. Ther.*, 71:62-74.

67. J. W. Schweitzer, A. J. Friedhoff, and R. Schwartz (1975). Phenethylamine in normal urine: Failure to verify high values. *Biol. Psychiat.*, 10:277-285.

68. C. B. Smith (1963). Enhancement by reserpine and α-methyl DOPA of the effects of d-amphetamine upon the locomotor activity of mice. *J. Pharmacol. Exp. Ther.*, 142:343-350.

69. H. Spatz and N. Spatz (1972). Spectrophotofluorometric determination of betaphenylethylamine in blood and urine. *Biochem. Med.*, 6:1-6.

70. T. H. Svensson (1970). The effect of inhibition of catecholamine synthesis on dexamphetamine induced central stimulation. *Eur. J. Pharmacol.*, 12:161-166.

71. T. H. Svensson and B. Waldeck (1970). On the role of brain catecholamines in motor activity: Experiments with inhibitors of synthesis and of monoamine oxidase. *Psychopharmacologia*, 18:357-365.

72. H.-Y. T. Yang and N. H. Neff (1974). The monoamineoxidases of brain: Selective inhibition with drugs and the consequences for the metabolism of the biogenic amines. *J. Pharmacol. Exp. Ther.*, 189:733-740.

Chapter 14

β-PHENYLETHYLAMINE AND ANIMAL BEHAVIOR

Egidio A. Moja,* David M. Stoff, J. Christian Gillin, and Richard Jed Wyatt

Division of Special Mental Health Research
Intramural Research Program
National Institute of Mental Health
St. Elizabeth's Hospital
Washington, D.C.

I. INTRODUCTION

β-Phenylethylamine (PEA) is a sympathomimetic amine whose presence has been demonstrated in animal and human brain (1-4). While functions are unknown, this book itself is testimony to growing interest about PEA. The reasons are manifold and some of the following have been important in stimulating speculations about the possible role of PEA in mental illness. β-Phenylethylamine is structurally similar to amphetamine (differing only in the absence of a methyl group on

*Current affiliation: Istituto di Clinica delle Malattie Nervose e e Mentali, University of Cagliari, Italy.

the alpha carbon for PEA), and amphetamine-induced behaviors are considered by many as the best pharmacological model of paranoid schizophrenia (5-7). β-Phenylethylamine is the preferred metabolite of type B monoamine oxidase [MAO (8)], an enzyme reported to be lower than normal in blood platelets of some chronic schizophrenics (9); this work has led to further speculations about the role of PEA in schizophrenia (10-11). Furthermore, the report that PEA is lower than normal in the urine of depressed patients (12), although disputed (13), has generated a hypothesis about its role in the depressive illnesses (14).

A preliminary approach in the understanding of the behavioral effects of a compound in humans is the study of behavioral effects of the compound in animals. This could shed light on behavioral processes themselves and the mode of action of the compound. Two very general categories of behavior have been studied: first, the effect of the compound on unlearned ("spontaneous," ongoing) behavior, like motor activity, and second, the effects of the compound on learned behavior (with operant procedures, like bar pressing for food, the most widely used).

This chapter reviews reports of the main animal behavioral effects of PEA (both "unlearned" and "learned") and modification of these effects by drug pretreatment. A summary of details of relevant studies besides our own work is presented in Tables 1A and B. Studies are arranged alphabetically and have been assigned reference numbers from 17 to 44. This review does not claim to be exhaustive and concentrates on behavioral effects of PEA in fields in which the authors have direct experience. The general philosophy underlying these experiments is the attempt to clarify some general effect and some of the functions of this compound.

TABLE 1A

The Effects of PEA on Stereotypy and Locomotor Activity[a]

Ref.	Species/ Strain	MAOI	Dose of PEA	PEA Behavioral Effect		Pretreatment	Alteration of PEA Effect
				Stereotypy	Locomotor Activity		
17	Rabbits (White New Zealand)	Pargyline 100 mg/kg	2-25 mg/kg	"Amphetaminelike effects" (data not shown)	—	—	—
	Mice/Swiss	—	20-50 mg/kg	—	Transient hyperactivity (data not shown)	—	—
		Pargyline 100 mg/kg twice	25 mg/kg	—	Sympathomimetic signs, marked hyperactivity, occasional jumping, vocalization, and agressive displays	—	—
		Pargyline 100 mg/kg twice	50 mg/kg	—	Running and jumping with much vocalization...similar to effect of L-amphetamine	—	—
18	Rats/Wistar	—	40 mg/kg	Minimal stereotypy (occasional sniffing and head movements)	—	—	—
		—	80 mg/kg	Marked stereotypy (continuous sniffing at bottom of cage with head movements from side to side, no locomotion or rearing—licking or biting never observed; onset 10 min after injection lasting 40 min)	—	—	—

TABLE 1A (Continued)

Ref.	Species/ Strain	MAOI	Dose of PEA	PEA Behavioral Effect: Stereotypy	PEA Behavioral Effect: Locomotor Activity	Pretreatment	Alteration of PEA Effect
	Rats/ Wistar	Deprenyl 8 mg/kg	160 mg/kg	Marked stereotypy (same degree of stereotypy as 80 mg/kg PEA without deprenyl)	—	—	—
		Deprenyl 2-8 mg/kg	40 mg/kg	Potentiation and prolongation of stereotypy	—	—	—
		Deprenyl 8 mg/kg	40 mg/kg	—	—	AMT 250 mg/kg	Blockade
		Clorgyline 2-8 mg/kg	40 mg/kg	Minimal stereotypy (same degree of stereotypy as without clorgyline)	—	—	—
19	Guinea pigs/ Dunkin-Harley, DHP Lab	Nialamide 75 mg/kg	100 μg/μl (bilaterally intrastriatal)	Dyskinetic head and neck rocking in three of eight animals (did not observe dyskinesias of gnawing, biting, licking, head and neck twisting, whole body rocking)	No effect	—	—
20,21	Newly hatched chicks	—	0.01-0.5 μmol/ 100 g	Alert behavior (increased cheeping, twitching, desynchronized EEG, enhanced EMG)—threshold dose was 0.01 μmol, optimal dose was 0.5 μmol		—	—
		—	0.01-0.5 μmol/ 100 g			Methysergide 0.01 μmol/100 g	Blockade
22	Mice/ Rockland	—	2.5-10 mg/kg	—	No effect	—	—
		—	40 mg/kg	—	Decrease	—	—
		Iproniazid 200 mg/kg	2.5-10 mg/kg	—	Non-dose-related increase (greatest effect for 5 mg/kg)	—	—

		Iproniazid 200 mg/kg	40 mg/kg	—	Decrease	—	—
		Iproniazid 200 mg/kg	2.5-5 mg/kg	—	Epinephrine-induced decrease	Epinephrine 1-2 mg/kg	Blockade by PEA
		Iproniazid 200 mg/kg	2.5-5 mg/kg	—	DOPA-induced decrease	D,L-DOPA	Blockade by PEA
23	Mice/ Rockland	Iproniazid 200 mg/kg	2.5-10 mg/kg	—	Increase (amphetaminelike effects including aggressive behavior, peripheral signs of adrenergic stimulation)	—	—
		Iproniazid 200 mg/kg	2.5-10 mg/kg	—	DOPA-induced decrease	D,L-DOPA 100-200 mg/kg	Blockade by PEA
		Iproniazid 200 mg/kg	2.5-10 mg/kg	—	Epinephrine-induced decrease	Epinephrine 1-2 mg/kg	Blockade by PEA
24,25	Rats/ Sprague-Dawley	Nialamide 100 mg/kg	50-100 mg/kg	Sniffing	Increase	—	—
		Nialamide 100 mg/kg	50-100 mg/kg	—	—	Reserpine 10 mg/kg plus AMPT 500 mg/kg	Blockade
		Nialamide 100 mg/kg	50-100 mg/kg	—	—	Reserpine 10 mg/kg plus AMPT 500 mg/kg plus L-DOPA 25 mg/kg	Restore
		Nialamide 100 mg/kg	50-100 mg/kg	—	—	Reserpine 10 mg/kg	No effect
26	Mice/ GS Strain	—	50-150 mg/kg	Trembling, twitching, and licking for 100-150 mg/kg during early phase	Increase (for 50 mg/kg, peak at 5 min; for 75-150, biphasic peaks at 5 (early phase) and 30 (later phase) min; early and later peak for 75 mg/kg were significant; early peaks for 100-150 mg/kg was not significant)	—	—

TABLE 1A (Continued)

Ref.	Species/ Strain	MAOI	Dose of PEA	PEA Behavioral Effect: Stereotypy	PEA Behavioral Effect: Locomotor Activity	Pretreatment	Alteration of PEA Effect
		Iproniazid 100 mg/kg	10-25 mg/kg		Prolonged increase (iproniazid, however increased activity alone)	—	—
		—	100-125 mg/kg	—	—	Reserpine 2 mg/kg	No effect (non-significant reduction; changed shape of time course)
		—	25-100 mg/kg	—	—	α-methylDOPA 400 mg/kg	Potentiation
		—	25-100 mg/kg	—	—	Protriptyline 10 mg/kg	Blocked early phase, no effect on later phase
		—	25-100 mg/kg	—	—	Desipramine 25-50 mg/kg	
		—	50-100 mg/kg	—	—	Disulfiram 100 mg/kg	
		—	50-100 mg/kg	—	—	AMT 500 mg/kg	
		—	50-100 mg/kg	—	—	Methysergide 5 mg/kg	No effect
27	Mice/ QS Strain	—	50-100 mg/kg	—	Nonsignificant increase at 50 mg/kg, significant increase at 100 mg/kg	—	—
		—	100 mg/kg	—	—	Atropine 1-10 mg/kg	Potentiation
		—	100 mg/kg	—	—	Hyoscine 0.01-1 mg/kg	
		—	100 mg/kg	—	—	Benzhexol 0.5-10 mg/kg	Potentiation
		—	100 mg/kg	—	—	Benztropine 0.5-2 mg/kg	

		—	100 mg/kg	—	—	Physostigmine 4.25-500 μg/kg	
		—	100 mg/kg	—	—	Neostigmine 6.25-184 μg/kg	No effect
		—	100 mg/kg	—	—	Methylhyoscine 0.01-1 mg/kg	
28	Mice	—	100 mg/kg	—	Increase (20-60 min after injection)	—	—
		—	100 mg/kg	—	—	Clonidine 1.5-10 mg/kg	Potentiation of hyperactivity (accompanied by stereotypies—including backward type of motion...licking of forepaws...so intense that the animal would roll over onto its back ...stereotyped sniffing)
		—	100 mg/kg (plus Clonidine 1,5 mg/kg)	—	—	Reserpine 4 mg/kg plus FLA-63 25-37.5 mg/kg	
		—	100 mg/kg (plus Clonidine 1.5 mg/kg)	—	—	Reserpine 4 mg/kg plus AMT 250-500 mg/kg	No effect on potentiation by clonidine
			100 mg/kg (plus Clonidine 1.5-10 mg/kg)	—	—	Reserpine 4 mg/kg	
			100 mg/kg (plus Clonidine 1.5-10 mg/kg)	—	—	AMT 250-500 mg/kg	
			100 mg/kg (plus Clonidine 1.5-10 mg/kg)	—	—	Reserpine 4 mg/kg plus AMT 250-500 mg/kg	Did not protect potentiation by clonidine

TABLE 1A (Continued)

Ref.	Species/ Strain	MAOI	Dose of PEA	PEA Behavioral Effect: Stereotypy	PEA Behavioral Effect: Locomotor Activity	Pretreatment	Alteration of PEA Effect
29	Mice/ NMRI	Nialamide	5 mg/kg	No signs of stereotypy such as gnawing or licking, but peripheral signs of sympathetic stimulation	Increase (5-35 min after injection)	—	—
		Nialamide 110 mg/kg	5 mg/kg	—	—	Pimozide 1 mg/kg	Blockade
		Nialamide 110 mg/kg	5 mg/kg	—	—	AMT 250 mg/kg	
						FLA-63 25 mg/kg Reserpine 10 mg/kg Phenoxybenzamine 10 mg/kg	No effect
30	Mice/ QS Strain	—	50 mg/kg	—	Increase from 0-20 min after injection	—	—
		—	50 mg/kg	—	—	Phenoxybenzamine 0.63-5 mg/kg	Blockade
		—	50 mg/kg	—	—	Phentolamine 5 mg/kg	
		—	50 mg/kg	—	—	AMT 375 mg/kg	
		—	50 mg/kg	—	—	Diethyldithiocarbonate 200 mg/kg	
		—	50 mg/kg	—	—	Perphenazine 2 mg/kg	
		—	50 mg/kg	—	—	Pimozide 100 μg/kg	
		—	50 mg/kg	—	—	Haloperidol 100 μg/kg	
		—	50 mg/kg	—	—	Apomorphine 1 mg/kg	Potentiation (accompanied by marked stereotyped behavior)

—	50 mg/kg	—	—	Propranadol 10 mg/kg Reserpine 1-4 mg/kg Chlorpromazine 5 mg/kg FLA-63 25 mg/kg Protriptyline 25 mg/kg	No effect
—	100 mg/kg	—	Increase from 20-60 min after injection (Decrease from 0-20 min after injection with signs of stereotyped activity such as sniffing, rearing and backward locomotion)	—	—
—	100 mg/kg	—	—	Pimozide 1-5 mg/kg	Blockade
—	100 mg/kg	—	—	Haloperidol 5 mg/kg	Blockade
				Phenoxybenzamine 0.63-5 mg/kg Phentolamine 5 mg/kg Propranadol 10 mg/kg Reserpine 1-4 mg/kg AMT 375 mg/kg Diethyldithiocarbonate 200 mg/kg FLA-63 25 mg/kg Chlorpromazine 5 mg/kg Perphenazine 2 mg/kg Protriptyline 25 mg/kg Apomorphine 1 μg/kg	No effect

TABLE 1A (Continued)

Ref.	Species/ Strain	MAOI	Dose of PEA	PEA Behavioral Effect: Stereotypy	PEA Behavioral Effect: Locomotor Activity	Pretreatment	Alteration of PEA Effect
31	Rats/ Sprague-Dawley	Nialamide 50 μg/kg	50 μg (Bilaterally in nucleus accumbens)	—	Increase (immediate onset, peak at 10-20 min. No effect 60 min later)	—	—
		Nialamide 50 μg/kg	50 μg (Bilaterally in nucleus accumbens)	—	—	AMT 400 mg/kg	Blockade
32	Male/ Swiss	Iproniazid 200 mg/kg	5-20 mg/kg	—	Non-dose-related increase (greatest effect for 10 mg/kg)	—	—
	Rats/ Wistar	Iproniazid 200 mg/kg	1-50 mg/kg	Symptoms of excitement	No effect	—	—
33	Rats/ Wistar	Deprenyl 8 mg/kg	40-160 mg/kg	Sniffing, head and limb movements (no licking or gnawing)	—	—	—
		Deprenyl 8 mg/kg	40-160 mg/kg	—	—	Reserpine 7.5 mg/kg	Potentiation (in terms of shifting type of stereotypy from sniffing to gnawing and licking
		Deprenyl 8 mg/kg	40-160 mg/kg	—	—	Phenoxybenzamine 20 mg/kg plus clonidine 0.5 mg/kg	Potentiation
		Deprenyl 8 mg/kg	40-160 mg/kg	—	—	FLA-63 30 mg/kg	Potentiation
		Deprenyl 8 mg/kg	40-160 mg/kg	—	—	Phenoxybenzamine 20 mg/kg	Potentiation
34	Rats/ Copenhagen	—	60 mg/kg	—	Continuous sniffing, licking and biting in six of six animals (backward locomotion in four of six)	—	—
35	Mice/ Rockland	—	10-80 mg/kg	—	Non-dose-related increase (greatest effect for 20 mg/kg) (5 mg/kg had no effect)	—	—

		Iproniazid 200 mg/kg	2-10 mg/kg	—	—	—	Dose-related potentiation of hyperactivity
		—	40-80 mg/kg	—	Reserpine-induced decrease (sedation syndrome and tremor)	Reserpine 2 mg/kg	Blockade by PEA
		—	40-80 mg/kg	—	Tryptamine-induced decrease (sedation syndrome and tremor)	Tryptamine 40-80 mg/kg	Blockade by PEA
36	Rats/ Hooded	—	50 mg/kg	—	Parkinsonian-like symptoms (catatonia, akinesia, rigidity)	Reserpine 5 mg/kg	Blockade by PEA
		—	50 mg/kg	—	Parkinsonian-like symptoms	Tryptamine 80 mg/kg	Blockade by PEA
		—	80 mg/kg	Increase	—	Reserpine 2 mg/kg	Blockade
		—	80 mg/kg	Increase	—	Tryptamine 80 mg/kg	No effect
37	Rabbits/ New Zealand	Pargyline 100 mg/kg, twice or Nialamide 50-100 mg/kg, three times	2.5-10 mg/kg	EEG activation, behavioral arousal, and hypertension		—	—
		Pargyline 100 mg/kg, twice or Nialamide 50-100 mg/kg, three times	2.5-10 mg/kg	—	—	Reserpine 5-10 mg/kg	Blocked hypertension but not behavioral and EEG alerting effects
		Isocar-boxazid 50 mg/kg	5 mg/kg	Behavioral and EEG arousal		—	—
		Isocar-boxazid 50 mg/kg	5 mg/kg	—	—	Disulfiram 250 mg/kg	Potentiate

TABLE 1A (Continued)

Ref.	Species/ Strain	MAOI	Dose of PEA	PEA Behavioral Effect		Pretreatment	Alteration of PEA Effect
				Stereotypy	Locomotor Activity		
		Pargyline 100 mg/kg twice	10 mg/kg	—	—	FLA-63 50 mg/kg	Potentiate (motor crisis characterized by rigidity, tremors, and spastic hypertension—followed by death in three of four animals)
		Nialamide 50-100 mg/kg, three times or pargyline 100 mg/kg or isocarboxazid 100 mg/kg	10 mg/kg	Reduced slow negative wave of visual evoked response; PEA 10 mg/kg alone had no effect		—	—
		Nialamide 50-100 mg/kg, three times or pargyline 100 mg/kg or isocarboxazid 100 mg/kg	10 mg/kg	—	—	Reserpine 5 mg/kg	Blockade
		Nialamide 50-100 mg/kg, three times or pargyline 100 mg/kg or isocarboxazid 100 mg/kg	10 mg/kg	—	—	FLA-63 5 mg/kg	Potentiate
		Isocarboxazid 100 mg/kg or pargyline 100 mg/kg	50 mg/kg	(Electroshock induced seizures)		—	Blockade by PEA (20 mg/kg PEA had no effect)

	Isocar-boxazid 100 mg/kg or pargy-line 100 mg/kg	50 mg/kg	—	—	Reserpine 10 mg/kg	No effect on anti-convulsant action of PEA
	Isocar-boxazid 100 mg/kg or pargy-line 100 mg/kg	50 mg/kg	—	—	FLA-63 50 mg/kg	
Male/ Swiss	Pargyline 100 mg/kg	2-100 mg/kg	PEA 2-50 mg/kg induced short phase of excitement and signs of sympathetic stimulation; PEA 100 mg/kg induced marked signs of sympathetic stimula-tion accompanied by depression of posture and reduction of activity; same effect without pargyline		—	—
	Pargyline 100 mg/kg	5-10 mg/kg	—	Increase	—	—
		50 mg/kg	Jumping, fighting	—	—	Slight increase
		5-50 mg/kg	—	—	FLA-63 350 mg/kg	Potentiate
	Pargyline 100 mg/kg or nialamide 100 mg/kg	2-10 mg/kg	Marked hyperactivity, occasional jumping, vocalization, agressive displays; PEA 50 mg/kg induced jumping, vocalization, piloerec-tion and some fighting		—	—
	Pargyline 100 mg/kg, twice or nialamide 100 mg/kg three times	2.5-10 mg/kg	EEG activation, behavioral arousal, hypertension		—	—

		Pargyline 100 mg/kg, twice or nialamide 100 mg/kg three times	2.5-10 mg/kg	—	—	Reserpine 5-10 mg/kg	Blockade of hypertension but not behavioral and EEG alerting effects
		Pargyline 100 mg/kg	5-10 mg/kg	Reserpine induced posture (i.e.,ptosis, hunched back)		Reserpine 10 mg/kg	Blockade by PEA
		Nialamide 100 mg/kg two times	6 mg/kg	Stimulant effect		FLA-63 50 mg/kg	Slight increase
38	Rats	—	80-160 mg/kg	—	Increase	—	—
39	Mice	—	50 µmol/kg	—	Increase	—	—
40	Mice/ Albino	Iproniazid 1600 µmol/kg	50 µmol/kg	—	Heightens increase	—	—
		—	3.2-32 mg/kg	—	No effect	—	—
		Pargyline 150 mg/kg	3.2-32 mg/kg	—	Increase	—	—

[a]For each study, the acute PEA effects (with or without MAO I) were first listed. Pretreatment alteration refers back to the PEA effect. In some cases, drugs other than PEA induced effects, and the ability of PEA to alter these effects was studied (22,23,35, 37). Our studies are not included in the tables. Dashes refer to not applicable.

TABLE 1B

Effects of PEA on Learned Behavior[a]

Ref.	Species/Strain	MAOI	Dose of PEA	Learned Behavior
40	Rats/Albino (Holtzman)	—	1-10 mg/kg	No effect on good performance in signalled, one-way jump avoidance of shock
41	Rats/Wistar	—	40 mg/kg	Disruption on good performance of shuttlebox avoidance (30 mg/kg had no effect, but when this dose was injected 24 hr later after the first 30 mg/kg, there was disruption)
42	Rats/Wistar	—	35-50 mg/kg	Dose response disruption on good performance in signalled one-way, pole-climbing avoidance of shock; same degree of disruption after PEA 35 mg/kg daily for 10 days—no tolerance to disruption? (LD_{50} was determined by using a dose range of 200-275 mg/kg; the LD_{50} was 240-245 mg/kg after one or 10 injections; no deaths at 200 mg/kg)
43	Rats/Sprague-Dawley	—	0.35-1 mg/kg	PEA does not act as discriminative cue in learning an operant response
		Iproniazid 100 mg/kg	1 mg/kg	PEA acts as a discriminative cue
44	Rats	—	10 mg/kg	No effect on hypothalamic self-stimulation (variable internal reinforcement)
		Iproniazid 100 mg/kg	2.5 mg/kg	Potentiates self-stimulation
		Etryptamine 2.5 mg/kg	1.0 mg/kg	Potentiates self-stimulation

[a]For each study, the PEA effects (with or without MAOI) were listed. Our studies are not included in the tables. Dashes refer to not applicable.

II. BEHAVIORAL EFFECTS OF PEA

A. Unconditioned Behavior

The most widely reported effects of PEA in animals were stereotyped behavior and increased motor activity. Because PEA crosses the blood-brain barrier relatively easily (1,41), its behavioral effects can be studied after systemic administration. On the other hand, it must be noted that some authors were able to demonstrate these effects only after pretreatment with monoamine oxidase inhibitors (MAOI): As already mentioned, PEA is the preferred substrate of type B MAO and is rapidly destroyed by this enzyme (8).

As an example of PEA-induced behavioral excitation, our data indicate that PEA (64 mg/kg, i.p.) by itself induced stereotypy and increased motor activity in rats (15). Pretreatment with pargyline (type B MAOI) markedly decreased the dose of PEA necessary to induce stereotypy without noticeably altering the characteristics of the stereotypes. The stereotyped behavior was characterized by head bobbing, sniffing, and backwards locomotion; normal activities such as grooming, eating, and drinking were absent. This activity has a quick onset, 5 min after injection, peaks at about 10 min and disappears after about 30 min. In general, the time course of motor activity paralleled the degree of stereotypy. However, a reduction in motor activity occurred at the most severe degrees of stereotypy.

The description of PEA-induced stereotypy in rats appears to be very similar to what has been frequently reported for amphetamine (e.g., see Ref. 16). Also, similarities between the behavioral effects of amphetamine and PEA have been reported for newly hatched chickens, rabbits, guinea pigs, and mice (17-40). A major difference appears to be that amphetamine produces in rodents compulsive gnawing, licking, and biting in addition to sniffing and head bobbing, while PEA does not. A second difference is that if MAOI are not used, the dose range for PEA stereotypy is much higher than that required for amphetamine stereotypy. It has been suggested that the methyl group that differentiates amphetamine from PEA, while not important for the behavioral actions of the compounds, can protect amphetamine from the action of MAO (32).

In juvenile male *Macaca mullatta*, we have observed dose response related effects of PEA on stereotyped behavior and general activity levels (45). After a dose of 75 mg/kg, i.m., the animals exhibited "checking" behavior (a marked increase in eye movements) and a decrease in general overall physical activity, which lasted approximately 30 min. These effects were in general similar to those observed with D-amphetamine (0.5 mg/kg, i.m.) except that the effects of PEA were more short-lived. We also have recent preliminary data indicating that these monkeys, like man, have type B MAO platelet, although the enzyme activity appears to be lower in the monkey than man.

B. Conditioned Behavior

There are a few reports of the effects of PEA on conditioned behavior. In these studies, PEA was behaviorally active at doses that were subthreshold for stereotypy. This is important because if doses of PEA which produce stereotypy were being used, then this could potentially mask any other behavioral effect. Huang and Ho (43) found that PEA after iproniazid acted as a discriminative cue for reinforcement in rats; similarly, discriminative responding was produced after amphetamine without the MAOI. β-Phenylethylamine did not influence performance of a one-way avoidance response in rats, and amphetamine disrupted this behavior at a dose which produced stereotypy (40). The failure of PEA to influence behavior in this study is probably due to low dose range of PEA used and absence of pretreatment with MAOI. Heller and Lumbreras (42) found that PEA, like methamphetamine, blocked performance of a one-way pole climbing avoidance. Furthermore, these authors found that an acute administration of the drug produced the same degree of disruption after PEA 35 mg/kg daily for 10 days, suggesting no tolerance to disruptive effects of PEA. On the other hand, Cohen et al. (41) recently reported that PEA had a disruptive effect on two-way shuttle-box avoidance and a lower dose was disruptive only after two administrations. These authors suggested a mechanism of hypersensitivity to explain this finding. An accumulation of PEA seems unlikely because of the rapid metabolism of the compound.

C. Chronic Effects of PEA

The possibility that PEA produces either no tolerance or hypersensitivity makes it an even more interesting compound from the view point of psychopathology. Since many mental illnesses, including schizophrenia, are chronic conditions, PEA would be less likely to be involved in the pathophysiology of these disorders if tolerance developed to its behavioral effects. Furthermore, if it were demonstrated that a naturally occurring compound such as PEA could produce hypersensitivity, this could have important implications for its role in psychopathology; sensitization of relevant biological systems may aid in the maintenance of the chronic disease state. For these reasons, we studied the effect of chronic administration of PEA in both rats and monkeys.

We pretreated one group of rats with 19 injections of pargyline (2 mg/kg, i.v.) plus 19 injections of PEA (16 mg/kg, i.p.) over 7 days (46). Pargyline is a type B MAOI (8), and MAO is a stable enzyme that requires several days to be synthesized (47). The appropriate control group was, therefore, a group pretreated with pargyline plus distilled water. A second comparison was made between a group of rats pretreated with distilled water plus PEA and a group pretreated with just distilled water. All four groups were tested at the seventh day after pargyline plus PEA (20th pair of injections) for stereotypy and motor activity. Although in all four comparisons the groups pretreated with PEA had minor scores compared with the control groups, no comparison reached the statistical significance. These data suggest that chronic PEA pretreatment, with or without MAOI, produced partial tolerance, if any. Also, we administered PEA 75 mg/kg three times a day to monkeys and did not observe either tolerance or hypersensitivity to its effects on checking and activity over a 15-day period (45). These data seem to suggest lack of tolerance to multiple administration of PEA. At variance with this suggestion, however, we have recently found pharmacological tolerance in rats to the PEA-induced disruption in the shuttle-box avoidance (48).

We trained rats to successfully perform the conditioned avoidance response in the shuttle-box. β-Phenylethylamine (8-64 mg/kg, i.p.) was given daily for 10 days before avoidance testing on each day. We found that the first administration of PEA produced dose-related disruption (in performance of the conditioned avoidance response) from 16 to 32 to 64 mg/kg; complete disruption occurred at 64 mg/kg (the minimal dose that we found producing marked stereotypy without MAOI). Pharmacological tolerance, and not behavioral, to multiple administrations of PEA developed (a control group was given 32 mg/kg daily for 10 days *after* each conditioning session, was given 32 mg/kg before a conditioning session, and their avoidance rate was at baseline). Acquisition of tolerance to PEA disruption was dose related in the sense that it was apparent after 4 days with 16 mg/kg and after 6 days with 32 mg/kg. No clear tolerance was found after 64 mg/kg.

At the present time, it is, therefore, not clear to what extent tolerance or hypersensitivity develop to chronic administration of PEA. Our data seem to suggest no clear tolerance to the unlearned behavioral effect of PEA in rats and monkeys, but a rapid tolerance to its learned effects in the shuttle-box paradigm (at doses that do not produce stereotypy).

As previously mentioned, Heller and Lumbreras (42) did not find tolerance to PEA in another learned test—one-way, pole-climbing avoidance. Differences in the behavioral task could account for the discrepancies. Longer time period of PEA administration need to be studied. In addition, no one has studied a paradigm in which high blood concentrations of PEA were maintained for long periods of time.

III. MECHANISM OF ACTION OF PEA

The demonstration that PEA could release catecholamines from organs in vivo and in vitro (49,50) raised the possibility that the amine could exert its behavioral effects by influencing the activity of dopamine (DA) and norepinephrine (NE) in the CNS.

Fuxe et al. (24-25) were, in fact, able to show with the histochemical fluorescence technique that PEA released DA and NE from neurons of the CNS in rats. These authors were also able to show that pretreatment with reserpine plus α-methyl-p-tyrosine (AMT) blocked "excitatory behavioral effects" (stereotyped sniffing and increased locomotion) of PEA after a MAOI. Reserpine releases stores of biogenic amines, and AMT is an inhibitor of catecholamine biosynthesis. Following these pretreatments, the stimulatory effects of PEA could be restored by a subthreshold dose of L-DOPA. These data by Fuxe et al. (24-25) seem to form a very coherent picture favoring the view that the action of PEA is mediated by catecholaminergic neurons. Reserpine by itself was not able to block the excitatory effects of PEA, suggesting that among the intraneural catecholaminergic pool, the newly synthesized catecholamines were more important than stored catecholamines in mediating the effects of PEA.

The results of the studies by Fuxe et al. (24-25) have been successfully confirmed. The decrease in brain catecholaminergic concentrations after PEA has been replicated with different biochemical procedures (51,52). The blocking effect by AMT of PEA-induced behavior was confirmed in mice (26,29,30) and in rats (18,31), measuring motor activity, and in monkeys, measuring stereotyped behavior (45). The potentiation of the excitatory effects of PEA by L-DOPA has been confirmed in mice (26) and the blockade by AMT was reversed by administration of L-DOPA in monkeys (45).

Catecholaminergic mechanisms have been postulated in the increased motor activity and stereotyped behavior observed after various drugs (e.g., amphetamine); the relative importance of DA and NE is, however, still in question. Randrup and Munkvad postulated a role of striatal dopaminergic mechanisms for the stereotyped behavior induced by several amphetamines, among which is PEA (see review in Ref. 53).

We have reported recently (54) the effects of various neuroleptic agents on the stereotyped behavior and hyperactivity after PEA in rats pretreated with pargyline. In the following order of potency (mg/kg), haloperidol, pimozide, chlorpromazine, and clozapine were able to block, in a dose-response way, the effects of PEA. This rank order of potency follows exactly the order of potency in which these drugs are able to block DA receptors in striatum (55,56). The effect of these drugs could not be ascribed to an aspecific sedation mechanism because diazepam, a benzodiazepine with sedative properties but no known antidopaminergic activities, was not able to significantly reduce both hyperactivity and stereotypy. These data suggest the importance of the DA system in the excitatory effects of PEA.

Further evidence for a PEA interaction with the dopaminergic system is shown by the finding that [^{3}H]PEA (100 mg/kg, i.p.) in rats tends to concentrate in DA-rich areas (52). No such correlation was found between endogenous NE levels and PEA distribution. It is interesting to note that Silkaitis and Mosnaim (57) have demonstrated the possible conversion of PEA to p-tyramine in vivo in rabbit brain. p-Tyramine, in turn, can be hydroxylated to form DA.

Mogilnicka and Braestrup (33) have proposed that the noradrenergic system modulates the type of PEA-induced stereotypies. These authors were able to show that reserpine, phenoxybenzamine, and a dopamine β-hydroxylase inhibitor (FLA-63), can transform the PEA-induced sniffing in rats pretreated with a MAOI into gnawing and licking. This suggests that drugs which inhibit central noradrenergic mechanisms, could shift the PEA-induced stereotypy to a qualitatively similar stereotypy as amphetamine. The same authors were also able to show that the release of NE, measured as the level of noradrenaline metabolite 3-methoxy-4-hydroxy-phenylglycol in the CNS, in rats after PEA was higher than the release of the same metabolite after amphetamine. This data could explain the different patterns of stereotyped behavior after PEA and after amphetamine.

If the reported evidence seems to imply a role of the catecholaminergic system in the mechanism of action of PEA, data exist in the literature that do not fit into this picture easily. Dewhurst and Marley (20,21) have observed that PEA has an opposite effect compared to catecholamines in newly hatched chickens. β-Phenylethylamine induced alert, exploratory behavior with desynchronization of the EEG, but, on the contrary, DA and noradrenaline, induced a state of drowsiness or sleep with synchronization of the EEG. Due to the selective blockade of methysergide, a tryptamine-receptor blocking agent, Dewhurst and Marley (20,21) proposed a direct mechanism of the action of PEA through central tryptamine receptors. Not specifically denying this hypothesis, Jackson (26) failed to find any effect on PEA-induced spontaneous motor activity in mice pretreated with 5 mg/kg methysergide. The finding, however, of the sedative effects of DA and NE in newly hatched chickens has been replicated (58).

Fischer et al. (22,23) also reported a differential effect on searching behavior in mice between catecholamines (that would depress the searching behavior of mice) and PEA (that would block this effect). Due to the similarities of behavioral effects, Fischer et al. (22) proposed that PEA could play a major role in mediating the CNS actions of amphetamine. Data favoring this hypothesis have been recently reviewed (17). In the experiments by Fischer et al. (22,23), NE and DA were injected i.p., and it might be noted that catecholamines, unlike PEA, cross the blood-brain barrier only at high doses. Evidence exists that when catecholamines are injected in a way to circumvent the blood-brain barrier (e.g., intraventricularly or directly in various nuclei of the brain) an increase of locomotor and other activity is produced (59-62). This criticism, however, cannot be raised about Dewhurst and Marley's data insofar as newly hatched chicks lack a well-developed blood-brain barrier (63,64).

Both a direct mechanism of action and a catecholaminergic-mediated one have been proposed by Jackson (26,28-30) to explain the results of the administration of PEA in mice. This author reported a biphasic effect on motor activity in mice after PEA. A first phase of increased activity occurred about 5 min after the injection and a second after about 30 min. The first peak was increased by previous administration of L-DOPA and apomorphine, and decreased by AMT, by the blockers of dopamine β-hydroxylase disulfiram and diethyldithiocarbonate, by the α-adrenergic blockers phentolamine and phenoxybenzamine and by the neuroleptics haloperidol and pimozide. Reserpine had no significant effect. The second peak was not influenced by any of these drugs, but haloperidol and pimozide were effective at doses at least 10 times higher than those required to block the early phase. Jackson concluded from these data that the first peak was an indirect effect requiring newly synthesized catecholamines, mainly NE, while the second peak appeared to be caused by the direct action of a metabolite of PEA. When mice were pretreated with nialamide, a MAOI, to prevent the production of potentially active deaminated metabolite, AMT was able to completely block PEA-induced stimulation, and no direct receptor stimulation could be seen. Also, AMT was able to completely block the effects of the injection of PEA directly into the nucleus accumbens of rats (31), a method that prevents the peripheral metabolism of PEA. These data suggest that the direct receptor stimulatory effects reported by Jackson are possibly produced by a deaminated metabolite. Jackson (27) also reported that the cholinergic blockers, hyoscine, atropine, benztropine, and benzhexol, increased the motor activity induced by PEA in mice. These data are in agreement with the fact that anticholinergic drugs increase the effects of the sympathomimetics (65).

IV. CONCLUSION

In a general way, PEA induces amphetaminelike behavioral effects in animals. The simple administration of appropriate doses of PEA produces hyperactivity and repetitive, non-goal-directed movements (stereotypy). No clear tolerance has been demonstrated after chronic administration of PEA as far as stereotypy and hyperactivity are concerned. Various drugs, in particular AMT and neuroleptics, are able to attentuate PEA-induced effects. Alpha-methyl-p-tyrosine inhibits catecholaminergic synthesis and neuroleptis block PEA-induced excitatory effects in an order of power that parallels their ability to block DA-receptor sites. These data, together with others, suggest the involvement of the catecholaminergic system in the mechanism of action of PEA.

As we have mentioned in the introduction, PEA may play a role in mental illnesses. The data that PEA induces unusual behavioral effects in animals, that no tolerance has been reported to these effects, and that drugs like neuroleptics, that are efficacious in the treatment of mental illnesses, can block these effects, can only increase the interest in this compound.

We must be cautious before postulating a physiological or pathophysiological role for a naturally occurring compound like PEA from the types of the experiments reported here. A physiological role for such a compound can be hypothesized only after a complex synthesis of anatomical, biochemical, neurophysiological, and behavioral data. Theoretically, the action of a compound in the brain can be attributed to a very specific action of minute quantities of the compound upon precise areas of the brain. In the experiments we have described, however, behavioral effects were observed after high doses of PEA. Behaviorally active doses of PEA could produce pharmacological effects throughout the brain secondary to high tissue levels, thereby masking the significant physiological effects, if any, of the compound.

Despite the limitation of the present knowledge, PEA remains an interesting compound for future research. β-Phenylethylamine is one of the few endogenous compounds which produce recognizable behavioral abnormalities immediately after exogenous administration. If PEA is a natural amphetamine, it may play a role in the regulation of some of the behaviors which amphetamine alters, such as levels of arousal, sleep-wake state, and appetite.

Finally, the PEA parent compound, amphetamine, produces a paranoid psychosis in man and stereotypy in both man and animals. The production of stereotypy has often been used in animal models of schizophrenia and neurological human movement disorders. Whether or not this model is appropriate for these disorders is certainly open to question. The PEA-induced stereotypy could be proposed as a model that is at least just as good, or just as bad, as the amphetamine model, the advantage being that PEA is a naturally occurring compound.

REFERENCES

1. T. Nakajima, Y. Kakimoto, and I. Sano (1964). Formation of β-phenylethylamine in mammalian tissue and its effects on motor activity in the mouse. *J. Pharmacol. Exp. Ther.*, 143:319-325.

2. E. E. Inwang, A. D. Mosnaim, and H. C. Sabelli (1973). Isolation and characterization of phenylethylamine and phenylethanolamine from human brain. *J. Neurochem.*, 20:1469-1473.

3. J. M. Saavedra (1974). Enzymatic isotopic assay for and presence of β-phenylethylamine in brain. *J. Neurochem.*, 22:211-216.

4. J. Willner, H. F. LeFevre, and E. Costa (1974). Assay by multiple ion detection of phenylethylamine and phenylethanolamine in rat brain. *J. Neurochem.*, 23:857-859.

5. P. H. Connel (1958). *Amphetamine Psychosis*, Maudsley Monographs No. 5. Oxford University Press, London and New York.

6. E. H. Ellinwood, Jr. (1969). Amphetamine psychosis: A multidimensional process. *Sem. Psychiat.*, 1:208-226.

7. S. H. Snyder (1973). Amphetamine psychosis, a model schizophrenia mediated by catecholamines. *Am. J. Psychiat.*, 130:61-67.

8. H.-Y. T. Yang and N. H. Neff (1973). β-Phenylethylamine: A specific substrate for type B monoamine oxidase of brain. *J. Pharmacol. Exp. Ther.*, 187:365-371.

9. D. L. Murphy and R. J. Wyatt (1972). Reduced MAO activity in blood platelets from schizophrenic patients. *Nature*, 238:225-226.

10. M. Sandler and G. P. Reynolds (1976). Does phenylethylamine cause schizophrenia? *Lancet*, i:70-71.

11. R. J. Wyatt, J. C. Gillin, D. M. Stoff, E. A. Moja, and J. R. Tinklenberg (1977). β-Phenylethylamine (PEA) and the neuropsychiatric disturbances. In: *Neurotransmitters and Hypotheses of Psychiatric Disorders*, edited by E. Usdin, J. Barchas, and D. Hamburg. Oxford University Press. In press.

12. E. Fischer, H. Spatz, J. M. Saavedra, H. Reggiani, A. H. Miro, and B. Heller (1972). Urinary elimination of phenylethylamine. *Biol. Psychiat.*, 5:139-147.

13. J. W. Schweitzer, A. J. Friedhoff, and R. Schwartz (1975). Phenylethylamine in normal urine: Failure to verify high values. *Biol. Psychiat.*, 10:277-285.

14. H. C. Sabelli and A. D. Mosnaim (1974). Phenylethylamine hypothesis of affective behavior. *Am. J. Psychiat.*, 131: 695-699.

15. E. A. Moja, D. M. Stoff, J. C. Gillin, and R. J. Wyatt (1976). Dose-response effects of β-phenylethylamine on stereotyped behavior in pargyline-treated rats. *Biol. Psychiat.*, 11:731-742.

16. A. Randrup and I. Munkvad (1967). Stereotyped activities produced by amphetamine in several animal species and man. *Psychopharmacologia*, 11:300-310.

17. R. L. Borison, A. D. Mosnaim, and H. C. Sabelli (1976). Brain 2-phenylethylamine as a major mediator for the central action of amphetamine and methylphenidate. *Life Sci.*, 17:1331-1334.

18. C. Braestrup, H. Anderson, and A. Randrup (1975). The monoamine oxidase B inhibitor deprenyl potentiates phenylethylamine behaviour in rats without inhibition of catecholamine metabolite formation. *Eur. J. Pharmacol.*, 34:181-187.

19. B. Costall, R. J. Naylor, and R. M. Pinder (1975). Diskinetic phenomena caused by the intrastriatal injection of phenylethylamine, phenylpyperazine, tetrahydroisoquinoline and tetrahydronaphtalene derivatives in the guinea pig. *Eur. J. Pharmacol.*, 31:94-109.

20. W. G. Dewhurst and E. Marley (1965). Action of sympathomimetic and allied amines on the central nervous system of the chicken. *Br. J. Pharmacol.*, 25:705-727.

21. W. G. Dewhurst and E. Marley (1965). The affects of α-methyl derivatives of noradrenaline, phenylethylamine and tryptamine on the central nervous system of the chicken. *Br. J. Pharmacol.*, 25:682-704.

22. E. Fischer, R. I. Ludmer, and H. C. Sabelli (1967). The antagonism of phenylethylamine to catecholamines on mouse motor activity. *Acta Physiol. Lat. Am.*, 17:12-21.

23. E. Fischer, J. M. Saavedra, and B. Heller (1968). Effects of catecholamines, adrenergic substances and their blocking agents on the searching behavior of mice. *Arzneim.-Forsch.*, 18:780-786.

24. K. Fuxe, H. Grobecker, and J. Jonsson (1967). The effect of β-phenylethylamine on central and peripheral monoamine-containing neurons. *Eur. J. Pharmacol.*, 2:202-207.

25. K. Fuxe and U. Ungerstedt (1970). Histochemical, biochemical and functional studies on central monoamine neurons after acute and chronic amphetamine administration. In: *Amphetamines and Related Compounds*, edited by E. Costa and S. Garattini. Raven Press, New York, pp. 257-288.

26. D. M. Jackson (1972). The effect of β-phenylethylamine upon spontaneous motor activity in mice: A dual effect on locomotor activity. *J. Pharm. Pharmacol.*, 24:383-389.

27. D. M. Jackson (1974). The interaction between β-phenylethylamine and agents which affect the cholinergic nervous systems on locomotor activity and toxicity in mice. *Arzneim.-Forsch.*, 24:24-27.

28. D. M. Jackson (1974). The involvement of noradrenergic systems in the locomotor activity stimulation in mice produced by β-phenylethylamine. *J. Pharm. Pharmacol.*, 26:651-654.

29. D. M. Jackson (1975). Some further observations of the effect of β-phenylethylamine on locomotor activity in mice. *J. Pharm. Pharmacol.*, 27:278-280.

30. D. M. Jackson (1975). β-Phenylethylamine and locomotor activity in mice. *Arzneim.-Forsch.*, 25:622-626.

31. D. M. Jackson, N.-E. Anden, and A. Dahlstrom (1975). A functional effect of dopamine in the nucleus accumbens and in some other dopamine-rich parts of the rat brain. *Psychopharmacologia*, 45:139-149.

32. P. Mantegazza and M. Riva (1963). Amphetamine-like activity of β-phenylethylamine after a monoamine oxidase inhibitor in vivo. *J. Pharm. Pharmacol.*, 15:472-478.

33. E. Mogilnicka and C. Braestrup (1976). Noradrenergic influence on the stereotyped behavior induced by amphetamine, phenethylamine and apomorphine. *J. Pharm. Pharmacol.*, 28:253-255.

34. A. Randrup and I. Munkvad (1966). Dopa and other naturally occurring substances as causes of stereotypy and rage in rats. *Acta Psychiat. Scand. (Suppl.)*, 191:193-199.

35. J. M. Saavedra and E. Fischer (1970). Antagonism of β-phenylethylamine derivatives and serotonin blocking drugs upon serotonin, tryptamine and reserpine behavioral depression in mice. *Arzneim.-Forsch.*, 20:952-957.

36. J. M. Saavedra, B. Heller, and E. Fischer (1970). Antagonistic effects of tryptamine and β-phenylethylamine on the behaviour of rodents. *Nature*, 226:868.

37. H. C. Sabelli, A. J. Vazquez, and D. Flavin (1975). Behavioral and electrophysiological effects of phenylethanolamine and 2-phenylethylamine. *Psychopharmacologia*, 42:117-125.

38. W. Schulte, E. C. Reif, J. R. Bacher, W. S. Lawrence, and M. L. Tainter (1941). Further study of central stimulation from sympathomimetic amines. *J. Pharmacol. Exp. Ther.*, 71:62-74.

39. J. B. Van Der Shoot, E. J. Ariens, J. M. Van Rossum, and J. A. M. Hurkmans (1962). Phenylisopropylamine derivatives, structure and action. *Arzneim.-Forsch.*, 12:902-907.

40. M. R. Vasko, M. P. Lutz, and E. F. Domino (1974). Structure activity relations of some indolealkylamines in comparison to phenethylamine on motor activity and acquisition of avoidance behavior. *Psychopharmacologia*, 36:49-58.

41. I. Cohen, J. F. Fischer, and W. H. Vogel (1974). Physiological disposition of β-phenylethylamine, 2,4,5-trimethoxyphenylethylamine and β-hydroxymescaline in rat brain, liver and plasma. *Psychopharmacologia*, 36:77-84.

42. B. Heller and N. Lumbreras (1976). Studies on the role of phenethylamine in methylamphetamine action mechanisms. *Experientia*, 32:210-211.

43. T.-T. Huang and B. T. Ho (1974). The effect of pretreatment with iproniazid on the behavioral activities of β-phenylethylamine in rats. *Psychopharmacologia*, 35:77-81.

44. L. Stein (1964). Self-stimulation of the brain and the central stimulant action of amphetamine. *Fed. Proc.*, 23:837-850.

45. J. R. Tinklenberg, J. C. Gillin, G. M. Murphy, R. Staub, and R. J. Wyatt (1976). The effect of phenylethylamine (PEA) in Rhesus Monkeys. Presented at the 129th Annual Meeting of the American Psychiatric Association, Miami, Florida.

46. D. M. Stoff, E. A. Moja, J. C. Gillin, and R. J. Wyatt (1977). Disruption of conditioned avoidance behavior by N,N-dimethyltryptamine (DMT) and stereotypy by β-phenylethylamine (PEA): Animal models of attentional defects in schizophrenia. *J. Psychiat. Res.* In press.

47. C. Goridis and N. H. Neff (1971). Monoamine oxidase: An approximation of turnover rates. *J. Neurochem.*, 18:1673-1682.

48. D. M. Stoff, E. A. Moja, D. Jeffrey, J. C. Gillin, and R. J. Wyatt (1977). Tolerance to β-phenylethylamine-induced disruption on rat shuttlebox avoidance. Submitted to *Psychopharmacology*.

49. H. W. Haag, A. Phillipu, and H. J. Schumann (1961). Catechinaminen aus der isoliert durchstromten Nebenniere durch Tyramin und B-Phenylathylamin. *Experientia*, 17:187-188.

50. H. J. Schumann (1960). Uber die Freisetzung von Breuzcatechinaminen durch Tyramin. *Arch. Exp. Pathol. Pharmakol.*, 238:41.

51. G. B. Baker, M. Raiteri, A. Bertollini, R. Del Carmine, P. E. Keane, and I. L. Martin (1976). Interaction of β-phenylethylamine with dopamine and noradrenaline in the central nervous system of the rat. *J. Pharm. Pharmacol.*, 28:456-457.

52. D. M. Jackson and D. B. Smythe (1973). The distribution of β-phenylethylamine in discrete regions of the rat brain and its effect on brain noradrenaline, dopamine and 5-hydroxytryptamine levels. *Neuropharmacology*, 12:663-668.

53. A. Randrup and I. Munkvad (1970). Biochemical anatomical and psychological investigations of stereotyped behavior induced by amphetamines. In: *Amphetamines and Related Compounds*, edited by E. Costa and S. Garattini. Raven Press, New York, pp. 695-713.

54. E. A. Moja, D. M. Stoff, J. C. Gillin, and R. J. Wyatt (1975). Modification of β-phenylethylamine-induced stereotyped behavior by neuroleptic agents. *Neurosci. Abst.*, 1:391.

55. I. Creese, D. R. Burt, and S. H. Snyder (1976). Dopamine receptor binding predicts clinical and pharmacological potencies of antischizophrenic drugs. *Science*, 192:481-483.

56. H. Rollema, B. H. C. Westerink, and C. J. Grol (1976). Correlation between neuroleptic-induced suppression of stereotyped behavior and HVA concentrations in rat brain. *J. Pharm. Pharmacol.*, 28:321-323.

57. R. P. Silkaitis and A. D. Mosnaim (1976). Pathways linking l-phenylalanine and 2-phenylethylamine with p-tyramine in rabbit brain. *Brain Res.*, 114:105-115.

58. M. B. Wallach, E. Friedman, and S. Gershon (1972). Behavioral and neurochemical effects of psychotomimetic drugs in neonate chicks. *Eur. J. Pharmacol.*, 17:259-269.

59. O. Benkert and B. Kohler (1972). Intrahypothalamic and intrastriatal dopamine and norepinephrine injections in relation to motor hyperactivity in rats. *Psychopharmacologia*, 24:318-325.

60. D. Malec and Z. Kleinrok (1972). The spontaneous motility of rats after intraventricular injection of dopamine. *Neuropharmacology*, 11:331-336.

61. D. S. Segal and A. J. Mandell (1970). Behavioural activation of rats during intraventricular infusion of norepinephrine. *Proc. Nat. Acad. Sci. USA*, 66:289-293.

62. E. A. Stone and S. Mendlinger (1974). Effect of intraventricular amines on motor activity in hypothermic rats. *Res. Commun. Chem. Pathol. Pharmacol.*, 7:549-556.

63. L. Bakay (1956). *The Blood Brain Barrier*. Charles C Thomas, Springfield, Ill.

64. A. Lajtha (1957). The development of the blood-brain barrier. *J. Neurochem.*, 1:216-227.

65. W. M. Davis and J. T. Pinkerton (1972). Synergism by atropine of central stimulant properties of phenylpropanolamine. *Toxicol. Appl. Pharmacol.*, 22:138-145.

Chapter 15

2-PHENYLETHYLAMINE AS A NEUROMODULATOR OF WAKEFULNESS, AFFECT, AND EXTRAPYRAMIDAL FUNCTION: RECENT ADVANCES

Hector C. Sabelli*

Department of Pharmacology
Chicago Medical School
University of Health Sciences
Chicago, Illinois

Bruce I. Diamond and Henri S. Havdala

Department of Anesthesiology
Mount Sinai Hospital Medical Center
Chicago, Illinois

Richard L. Borison† and Joan May

Department of Pharmacology
Chicago Medical School
University of Health Sciences
and
Department of Anesthesiology
Mount Sinai Hospital Medical Center
Chicago, Illinois

---

*Current affiliation: Department of Psychiatry, Rush-Presbyterian-St. Luke's Medical Center, Chicago, Illinois.

†Current affiliation: Illinois State Psychiatric Institute, Chicago, Illinois.

I. FORMULATION OF THE PEA THEORY OF AFFECTIVE BEHAVIOR

The apparent contradiction between the experimental evidence that central "adrenergic" pathways facilitate arousal, activity, and euphoria (cf. 1)* and the no less impressive series of observations indicating that CNS effects of administered catecholamines (CA) are primarily depressant (cf. 1) led us to the search for noncatecholic phenylethylamines which could serve as transmitters and/or modulators of the central adrenergic ergotropic pathways. Since both the central stimulant potency and the sensitivity to monoamine oxidase (MAO) of phenylethylamines decrease with OH substitutions (cf. 1), 2-phenylethylamine (PEA) was the best candidate as the central adrenergic ergotropic modulator. This amine had been identified in 1964 by Nakajima et al. (cf. 1) in the brain of rabbits treated with a monoamine oxidase inhibitor (MAOI). Whereas the host of amines found in tissues after MAO inhibition had been considered as false neurotransmitters, Fischer, Ludmer, and Sabelli proposed in 1966 that PEA is an endogenous modulator of affective behavior, activity, and alerting, that its deficit may cause some forms of endogenous and drug-induced depression, and that its excess may play a role in mania and in the mood-elevating effects of amphetamines, antidepressive agents, and so on (cf. 1). In support of the PEA hypothesis, Fischer and co-workers in Argentina demonstrated that the urinary excretion of PEA is reduced in endogenous depression [Fischer et al., 1968 (cf. 1), 2,3] and that the administration of the PEA precursor phenylalanine exerts powerful antidepressant effects (4). Our research group in Chicago identified endogenous PEA [Sabelli et al., 1971 (cf. 1), 5-7] and its metabolites phenylethanolamine (5,6), tyramine (see Mosnaim et al., this volume) and phenylacetic acid (8) in brain, and studied its electrophysiological and behavioral effects (1,9-11), its excretion in several disease entities (6,7,12) (Fig. 1), and its possible role on the mechanism of action of

*For the sake of brevity, we shall refer to our previous review on PEA (1) for older publications from this and other laboratories; this will be indicated by (cf. 1).

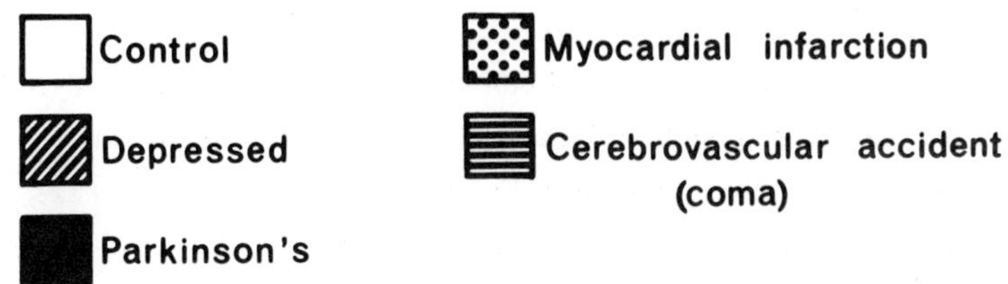

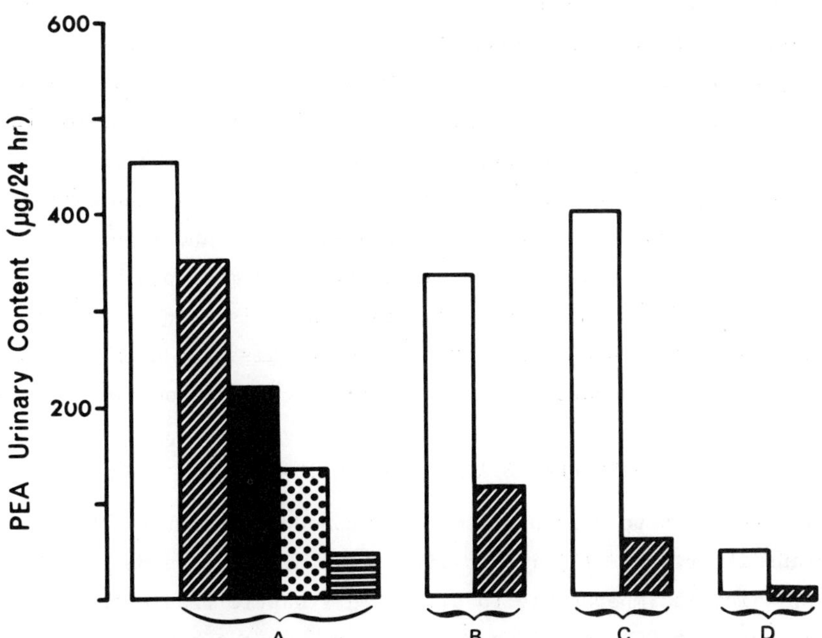

FIG. 1. Urinary PEA excretion in depression and other disease states. (A) Ceric sulfate method; data from Refs. 6,7,12. (B) Dimethylamino-cinnamaldehyde reaction fluorometry method; data from Ref. 2. (C) Gas chromatography method; data from Ref. 3. (D) Alloxan reaction fluorometry method [data from Boulton and Milward, 1971 (cf. 1)].

imipramine [Sabelli et al., 1971 (cf. 1) 9,13], amphetamines (Fischer et al., 1967 (cf. 1), 9,14,15), marijuana (11,16-18), and other psychotropic agents (1,9,12,13,19).

Most of these results have been confirmed and extended by other investigators (as discussed in other chapters of this volume), but two provocative questions have not as yet been answered. First, widely different values for PEA content in brain and urine have been reported (20). The second outstanding problem in this field is the relation between PEA and CA mechanisms to which this paper is addressed.

The methods used in our studies have been published elsewhere and include visually evoked responses in rabbits (cf. 1), gas-liquid chromatography for quantification of PEA as its dinitrophenylsulfonamide derivative (14), intraventricular administration of radiolabeled phenylalanine and PEA for studying PEA synthesis and metabolism in rabbit brain (15), behavioral observations in mice (11), and extracellular recordings of cortical (rabbit brain) (10) and axonal (isolated frog sciatics) (21) unit activity using multiple barrel microelectrodes for simultaneous drug iontophoresis. In all studies involving systemic PEA administration, the animals have been pretreated with a MAOI because otherwise PEA effects are too fleeting.

II. CRITERIA TO DEFINE PEA AS A NEUROMODULATOR

The main arguments to regard PEA as a neuromodulator rather than as a neurotransmitter are: (1) its ability to readily cross the blood-brain barrier, which is impermeable to all known synaptic transmitters. (2) The fact that PEA and CA appear to share enzymatic mechanisms for their synthesis and disposition, binding to storage and active receptors, and so on.

Because the enzymatic and uptake mechanisms for the synthesis, storage, and release of messengers are not entirely specific, a norepinephrine (NE) neuron [containing decarboxylase (L-AAAD), dopamine-β-hydroxylase (DBH), MAO, etc.], by necessity must form the corresponding amino acid precursors, deaminated metabolites, and amine analogs (PEA, phenylethanolamine, octopamine, etc.) (Fig. 2A). Since these compounds share the molecular skeleton of the transmitter, they must also share some of its binding sites, and thus, they should be biologically active (by replacing or competing with the transmitter at receptors or other binding sites). Some modulators may be released by the nerve impulse as co-transmitters [e.g., octopamine in sympathetic junctions, Molinoff et al., 1969 (cf. 1)] or as alternative messengers; they may also act presynaptically, affecting transmitter storage, release, synthesis, or metabolism. Alternatively, because these compounds lack some of the reactive groups of the transmitter, they may have different receptors or may block the transmitter's action, and their inadequate binding to storage sites may result in a continuous diffusion which tonically modifies postsynaptic activity. The rapid turnover of PEA suggests that, in fact, this amine readily diffuses to the extracellular spaces immediately upon its synthesis. The lack of some of the chemical groups of the transmitter also renders these compounds more permeating. Thus, brain accumulates blood-borne PEA (1,14,22), and alteration of the synthesis and disposition of PEA in peripheral tissues drastically alters its brain content. In rabbit brain (PEA content 340.9 ± 45.8 ng/g), the MAO I dimethyl-β-carbolinium iodide (DMCI, 15 mg/kg, 1 hr prior, i.v.) augments (1020.0 ± 63.8 ng/g) (23), and the decarboxylase inhibitor α-methylDOPA hydrazine (AMDH) (200 mg/kg, 4 hr prior, i.p.) decreases (66.9 ± 13.0 ng/g) (14) the brain content of PEA (Fig. 2B) although neither compound crosses the blood-brain barrier. [However, N^1-(DL-seryl)-N^2 (2, 3, 4 trihydroxybenzyl) hydrazine (RO4-4602), which effectively inhibits L-AAAD in peripheral tissues at 25 mg/kg and in the CNS at 400 mg/kg (Table 1), does not reduce brain PEA content unless even higher doses are used (Fig. 3); the interpretation of these and other observations (see the discussion that follows) with RO4-4062 require further experimentation.]

We thus consider that PEA, as well as some of the other precursors, metabolites or by-products of CA metabolism, may be modulators (i.e., endogenous chemicals which modify but are not essential

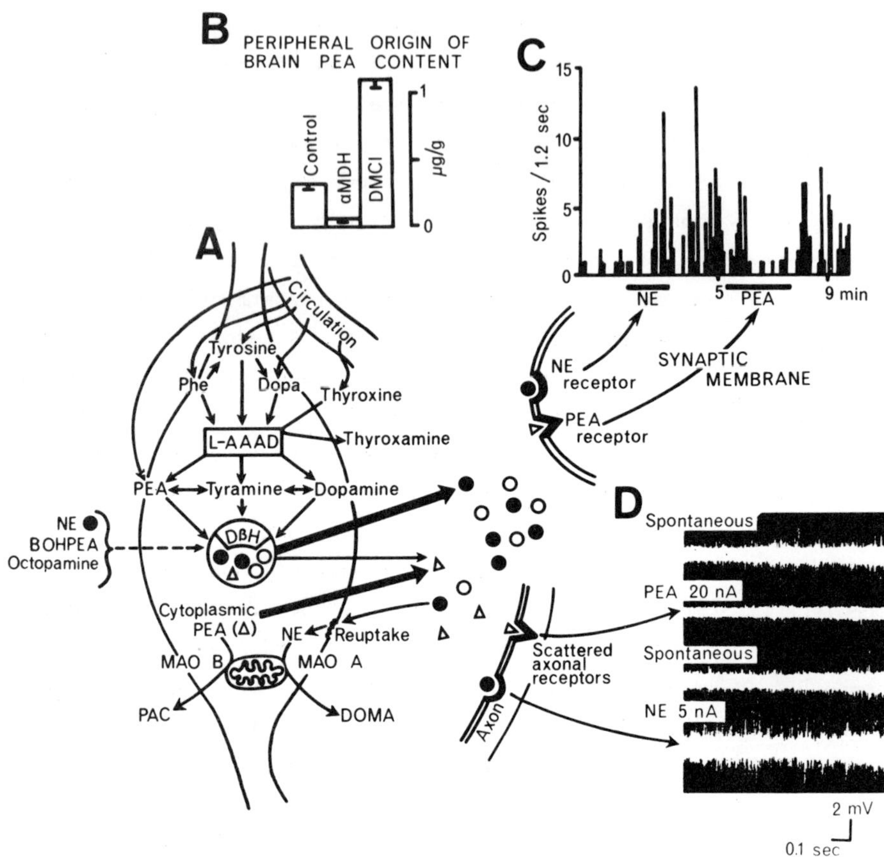

FIG. 2. Hypothetical model for norepinephrine synapses (A) modulated by PEA and other compounds and experimental evidence for this model (B, C, D). Abbreviations: Phe, phenylalanine; NE, norepinephrine; PAC, phenylacetic acid; BOHPEA, phenylethanolamine; DBH, dopamine-β-hydroxylase; DOMA, dihydroxy mandelic acid. B: Peripheral origin of brain PEA (rabbit): depleted by α-methyldopa hydrazine (AMDH), (200 mg/kg, i.p., 4 hr prior) and increased by the MAOI dimethyl-β-carbolinium iodide (DMCI), (15 mg/kg, i.v., 1 hr prior), two agents which do not cross the blood-brain barrier (data from Refs. 14, 38). C: Existence of specific PEA receptors: differential effects of the microiontophoretic administration of PEA and of NE upon a cortical unit (extracellular microelectrode recording in nonanesthetized rabbit, computer output) (data from Ref. 10). D: Presence of extrasynaptic receptors: effects of microiontophoretic NE and PEA upon extracellular microelectrode recordings from frog sciatic axons (isolated) (data from Ref. 21). Excitatory as well as inhibitory effects have been noted with both NE and PEA in (C) and (D); in almost 50% of the experiments, the effects of NE and PEA upon a given unit were different.

TABLE 1

Drug Effects upon the Recovery of Labeled 2-Phenylethylamine from Tissue After the Administration of Labeled L-Phenylalanine

Drug, Route, Dose, and Time before Sacrifice[a]	Species, Organ	Route of L-Phenylalanine Administration; Time Before Sacrifice	Recovery of [^{14}C]PEA (percent of control)	References
Imipramine, i.p. (10 mg/kg daily for 7 days), 2 hr	Mouse brain	i.p.; 10 min	385	Unpublished
	Mouse liver	i.p.; 10 min	49	
	Mouse spleen	i.p., 10 min	59	
Imipramine, i.p. (25 mg/kg) 90 min	Rabbit brain	ivt.; 10 min	337	1
Lithium sulfate, i.p. (50 mg/kg daily for 7 days), 2 hr	Mouse brain	i.p.; 10 min	8	Unpublished
Δ^9-Tetrahydrocannabinol, i.p. (3 mg/kg), 1 hr	Rabbit brain	ivt.; 60 min	323	17
D-Amphetamine, i.p. (10 mg/kg): 30 min	Rabbit brain	ivt.; 10 min	500	15
4 hr	Rabbit brain	ivt.; 10 min	2200	15
24 hr	Rabbit brain	ivt.; 10 min	129	Unpublished
Reserpine, i.p. (4 mg/kg), 4 hr; (0.5 mg/kg, 24 hr	Rabbit brain	ivt.; 10 min	384	Unpublished
α-CH_3 p-Tyrosine, i.p. (100 mg/kg), 3 hr	Rabbit brain	ivt.; 10 min	83	Unpublished
6-Hydroxydopamine, ivt. (400 μg), 24 and 48 hr	Rabbit brain	ivt.; 10 min	38	Unpublished
RO4-4602, i.p., (400 mg/kg), 4 hr	Rabbit brain	ivt.; 10 min	42	Unpublished

[a]ivt. = intraventricular.

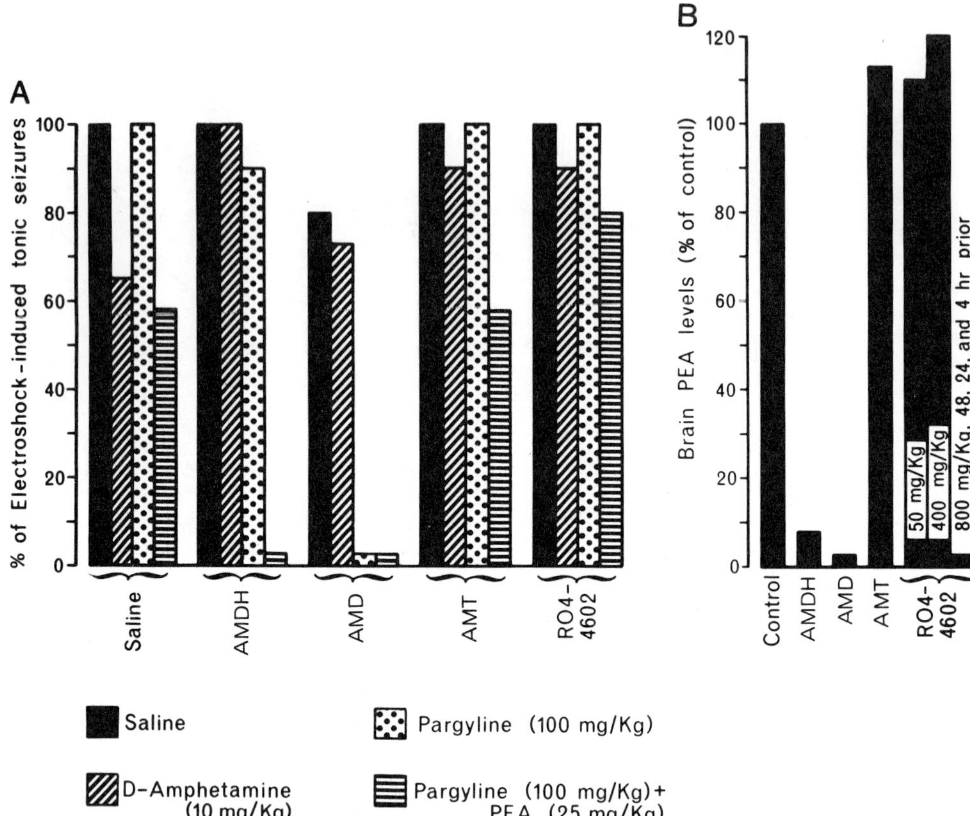

FIG. 3. Influence of enzyme inhibitors upon the anticonvulsant effects of amphetamine and PEA (A) and upon the brain content of PEA (B). Data from Refs. 14 and 15 and unpublished. A: Maximal electroshock seizures: percent of mice displaying maximal tonic convulsion. Pretreatments: α-methyldopa hydrazine (AMDH), 400 mg/kg 4 hr prior; α-methyl-p-tyrosine (AMPT), 250 mg/kg 3 hr prior; N^1-(DL-seryl)-N^2 (2,3,4 trihydroxybenzyl)-hydrazine (RO4-4602), 400 mg/kg 4 hr prior; pargyline 100 mg/kg 24 hr prior; all i.p. Treatments: D-amphetamine (D-AMPH), PEA, or saline, i.p., 30 min prior. B: Brain content of PEA in rabbits measured by gas-liquid chromatography of the dinitrophenylsulfonamide derivative of PEA. Pretreatments: AMDH, 200 mg/kg 4 hr prior; AMD, 200 mg/kg 4 hr prior; AMT, 100 mg/kg 3 hr prior; RO4-4602, 50 or 400 mg/kg 4 hr prior, or 800 mg/kg 48, 24, and 4 hr prior; all i.p.

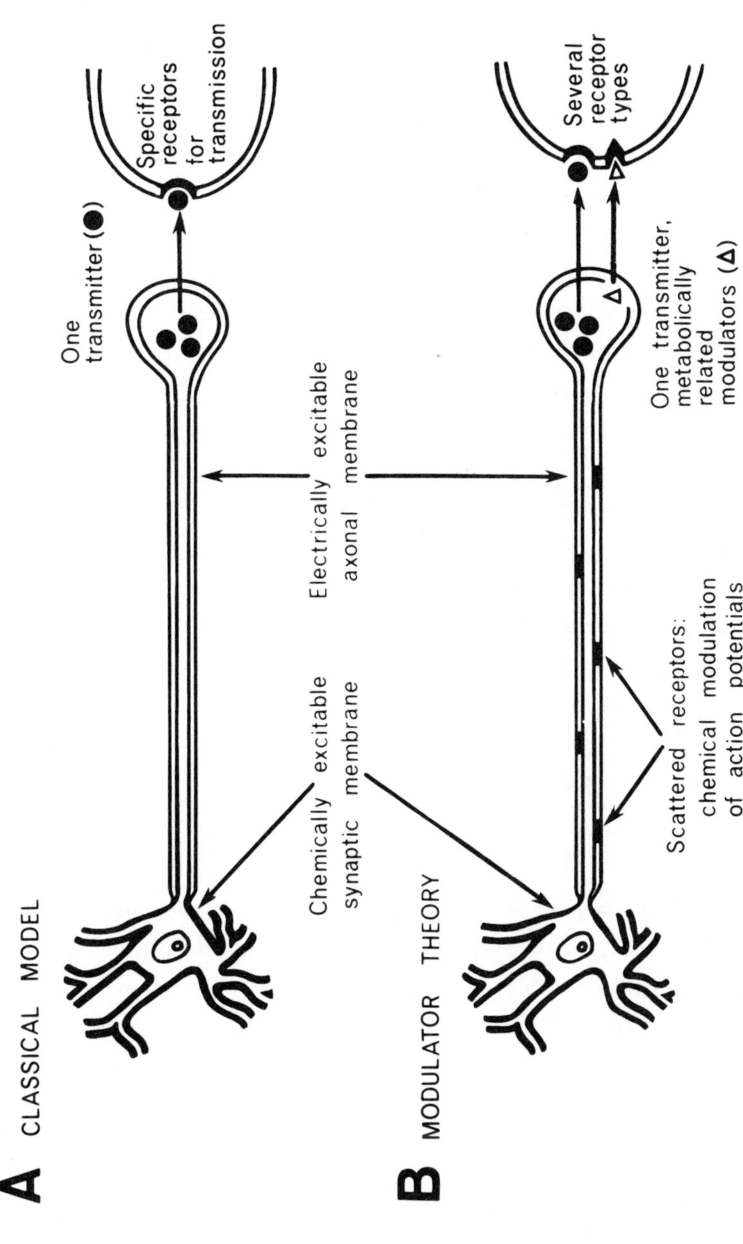

FIG. 4. Chemical modulation of synaptic transmission. A: classical concept; B: modulator theory.

for synaptic transmission) of CA synapses. In the classical model of the synapse (Fig. 4A), each neuron releases one transmitter (Dale's principle), which acts on a specific receptor membrane which is qualitatively different from electrically excitable axonal membranes. Drugs induce selective CNS effects by enhancing or reducing the synaptic action of a specific transmitter. Fig. 4B shows an alternative model (1,19,24) in which each neuron releases a multiplicity of modulators which act on a multiplicity of receptors, some located extrasynaptically (Fig. 2D). Drugs can modify synaptic transmission qualitatively by modifying the relative proportion of modulators released by the synaptic terminal or the relative proportion of free synaptic receptors: They are most selective in their CNS effects when they can modify simultaneously the action of the several messengers involved in the corresponding neural pathway. This view can be embodied into two principles which follow from the metabolic unity of the neuron (Ramon y Cajal). *First*, Dale's principle can be expanded into an "algorithm" to guide the search for new modulators: *each neuron releases from all its endings one transmitter and metabolically related modulators, the relative proportion of which is determined by its physiological state* (1,19). *Second*, the metabolic unity of the neuron probably also requires that *the membrane of each neuron contains the same type(s) of chemosensitive receptors over the entire surface, including the axon* (21) (the greater excitability of synapses probably results from a clustering of receptors).

III. EXTRASYNAPTIC RECEPTORS

If in fact axonal membranes are chemically excitable, diffusable endogenous chemicals and exogenous drugs may directly modulate action potential generation. In support of this view we observed that PEA (7.5-15 mM) induces a local anestheticlike effect in frog sciatic nerves (reduction of spike height, threshold rise, potentiation by partial replacement of NaCl by sucrose). Furthermore, the microiontophoresis of PEA, norepinephrine, and other putative transmitters, triggers or inhibits action potentials in CNS axons (in vivo) and in peripheral nerves (in vitro) (21) (Figs. 2D and 5). Although PEA

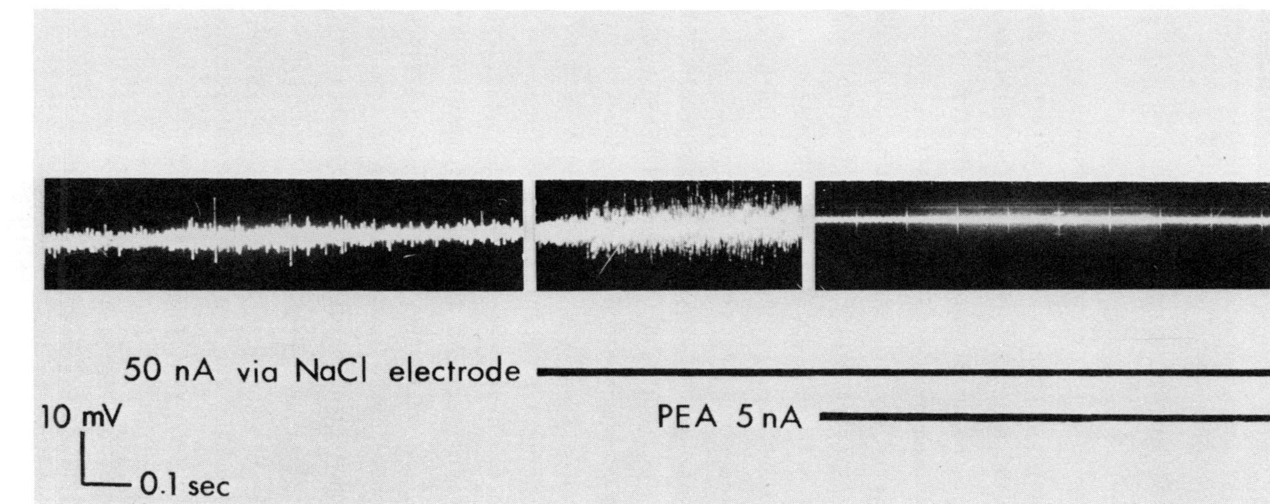

FIG. 5. Effects of microiontophoretic administration of PEA on isolated frog sciatic nerves (5-barrel microiontophoretic system for extracellular recording and iontophoretic drug administration, unpublished data). Current 50 nA elicits spike activity which is inhibited by concomitant administration of PEA.

exerts different effects upon different fibers, these effects are in most instances antagonic to those of CA administered at the same site.

IV. CATECHOLAMINE AND NONCATECHOLAMINIC MECHANISMS IN PEA ACTION

2-Phenylethylamine appears to exert its pharmacological effects by at least three different mechanisms: a) via direct effects on receptors, which may be specific and/or overlap with dopamine receptors; b) via catecholamine release; c) via its metabolites phenylacetaldehyde and phenylethanolamine. A role for the deaminated metabolites is suggested by the observation that in untreated animals, the fleeting stimulant effects of PEA are followed by depression, whereas pretreatment with MAOI prolongs the stimulant phase and abolishes the subsequent depression. Phenylacetaldehyde, and to a lesser extent phenylacetic acid, exert behavioral depressant effects and can markedly modify visual evoked responses (1,19,25). β-Phenylethanolamine (BOHPEA), which we have found in brain and urine (5,6) by mass and ultraviolet spectroscopy [confirmed by radioenzymatic assay by Saavedra and Axelrod (cf. 1)], does not seem to mediate most of the central effects of PEA (11).

Whereas extremely large doses of PEA undoubtedly release CA, such release may be irrelevant to its major behavioral actions (9). Indirect evidence for the existence of specific PEA receptors includes the differences found in the effects of administered PEA versus those of CA upon insulin release (26), spinal reflexes (27), nerve action potentials (see previous discussion), and cortical neuron firing. Thus, using a multiple barrel microelectrode system (10), we have observed that PEA and NE exert opposite effects in 50% of the cortical neurons tested; similarly, dopamine (DA) depressed or did not affect the firing of neurons excited by the microiontophoretic administration of PEA (19).

Phenylethylamine administration induces amphetaminelike effects different and often opposite to the depressant effects of injected CA (even when administered intraventricularly) [Rothballer, 1959 (cf. 1)]. Further, even L-DOPA induces depressant effects (9,30) reducing motor activity at low doses and its stimulant actions (jumping, running, and aggressiveness) do not serve as unambiguous evidence for an ergotropic role of brain CA because they are enhanced (rather than blocked) by decarboxylase inhibition (28). In fact, the ability of DOPA to induce marked aggressivity in animals, and the antihostility effect of the DA blocker haloperidol indicate that CA actually favor a dysphoric mood, in contradiction to their purported role in the modulation of affect. Furthermore, a most characteristic electrophysiological correlate of the arousal evoked by the shift of attention toward a novel stimulus is a reduction in the amplitude of cortical potentials evoked by repetitive stimuli to which the subject has been habituated. Only PEA (or its precursor phenylalanine) but not DOPA (or cholinergic drugs) mimics the effects of alerting stimuli upon visual evoked responses (VER) in rabbits (i.e., a reduction in amplitude of the main slow component of the VER as well as other typical changes in the amplitudes and latencies of the faster components). In contrast, VER are augmented by the administration of DOPA (except at very high doses) (28) or of the CA releaser nicotine [experiments with α-methyl-p-tyrosine (AMT) and α-methylDOPA (AMD) clearly indicate that this action of nicotine is mediated by CA].

Confirming pharmacological evidence comes from studies with CA depleters. The anticonvulsant effect (Fig. 3A) of PEA [which can be demonstrated in the maximal electroshock seizure test with high doses (75-100 mg/kg, 15-30 min prior) of PEA alone or with lower doses (20-50 mg/kg) after MAOI], as well as its behavioral stimulant effect, are not significantly modified by pretreatment with reserpine, AMD, AMT or the selective norepinephrine (NE) depleter [Bis(4-methyl-1-homopiperazinylthiocarbonyl)disulfide] (FLA-63), although AMT prevents some of the neurological effects of PEA, as discussed in this volume by Braestrup and Randrup. Thus, PEA effectively restores to normal the locomotor depression induced by reserpine and by AMD suggesting that the central effects of PEA need not be mediated by brain CA. This is not to preclude that CA release plays a role in some central actions of PEA as illustrated by the ability of DA blockers to prevent them (see Section VII).

Although CA do not fully mediate the CNS actions of PEA, metabolic studies indicate an intimate connection between CA and PEA metabolism. The rate of recovery of labeled PEA after phenylalanine administration (Table 1) depends on the rate of PEA synthesis (phenylalanine uptake plus decarboxylase activity) and on its rate of disposition. The recovery of radiolabeled PEA after its intraventricular administration (Table 2) reflects its rate of disposition (metabolism plus diffusion), which is presumably directly dependent on release and enzymatic destruction and inversely dependent on storage.

We have observed that the intraventricular administration of 6-hydroxydopamine, an agent known to destroy CA neurons, decreased to 40% of control the recovery from brain tissue of radiolabeled PEA after the intraventricular administration of its precursor phenylalanine (Table 1) suggesting that CA neurons significantly contribute to PEA synthesis and/or storage. To further support this view, the two CA depleters, reserpine and AMD (Fig. 3B), also reduce brain PEA content. The ability of AMD to inhibit phenylalanine decarboxylation (Table 1) might contribute to PEA depletion, but the latter is more likely to result from its displacement by the amine metabolites of AMD because these amines mediate the CA-depleting effect of AMD and also because doses of RO4-4602 which inhibit brain decarboxylation fail to deplete brain PEA.

More direct evidence suggesting that PEA is partially stored in CA granules is the depletion induced by reserpine since this agent actually increases the recovery of labeled PEA after intraventricular phenylalanine administration (Table 1). Even at large doses, reserpine fails to fully deplete brain PEA in any of several animal species studied [Sabelli et al., 1971 (cf. 1), 2,13]; thus, a significant proportion of the PEA brain content must represent cytoplasmic stores of amine which diffuse immediately upon synthesis.

TABLE 2

The Effects of Psychotropic Drugs upon the Recovery of Intraventricularly (ivt.) Administered [1-^{14}C]2-Phenylethylamine from Rabbit Brain

Drug, Route, Dose, and Time Before Sacrifice	[1-^{14}C]PEA Administration Time before Sacrifice	Recovery of [1-^{14}C]PEA (percent of control)	References
Imipramine, i.p. (25 mg/kg), 90 min	30 min	37	1
Δ^9-Tetrahydrocannabinol, i.p. (3 mg/kg), 60 min	30 min	361	17
D-Amphetamine, i.p.			
(10 mg/kg): 30 min	10 min	30	15
4 hr	10 min	56	15
(0.5 mg/kg): 30 min	10 min	69	Unpublished
6-Hydroxydopamine, ivt. (400 μg), 24 and 48 hr	10 min	76	Unpublished
α-CH_3 p-Tyrosine, i.p. (100 mg/kg), 3 hr	10 min	188	Unpublished
Reserpine, i.p. (4 mg/kg), 4 hr; (0.5 mg/kg), 24 hr	10 min	76	Unpublished

Catecholamine neurons appear to contribute to the storage and metabolism of PEA since reserpine as well as 6-hydroxydopamine accelerated to the same extent the disposition of injected PEA (Table 2).

Inhibition of CA synthesis by AMT does not alter brain content of PEA (Fig. 3B) and delays PEA disposition (Table 2) while decreasing its synthesis. Such a reduction in PEA turnover suggests that newly synthesized CA (which are preferentially released by physiological stimuli) accelerate PEA synthesis and/or release

V. PEA AND AFFECTIVE BEHAVIOR

We propose the following criteria to demonstrate that a neuroamine plays a positive role in the regulation affect: 1) its (intracerebral) administration must induce behavioral stimulation; 2) a decrease in its turnover must be associated with depression and, conversely, an increase with mania; 3) the administration of its precursor must exert antidepressant effects; 4) its brain levels and/or turnover must be modified in a manner consistent with the hypothesis by drugs which cause or relieve depression, as well as by those with euphoriant or stimulant effects.

Behavioral studies in many species show that PEA is unique among naturally occurring amines in exerting amphetaminelike effects (criterion 1), including EEG alerting, hyperactivity without aggressiveness, and antagonism of the reserpine syndrome (which is considered a model for depression). This is not entirely attributable to its ability to cross the blood-brain barrier because: i) other amines which cross the blood-brain barrier induce depressant effects (tryptamine) or only weak stimulant effects [phenylethanolamine (11)] and ii) increasing brain levels of CA induces a mixture of stimulant and depressant effects (see the previous discussion).

Among monoamines, PEA may satisfy criterion 2 since the urinary excretion of PEA (Fig. 1) and its metabolite phenylacetic acid (3) are reduced in a large proportion of depressed patients.

Phenylethylamine is also the only endogenous amine to fully satisfy criterion 3. In fact, the success of phenylalanine therapy of depression (4) is the most important experimental evidence in support of the PEA theory of affect. L-Phenylalanine exerts stimulant effects in mice (cf. 1) and antagonizes the central depressant effects of DOPA (cf. 1) and of reserpine (9). D-Phenylalanine is more effective than L-phenylalanine in increasing the brain levels of PEA (13), in exerting central stimulant effects (9), and as a therapeutic agent in depressed patients (3,4). In contrast, DOPA is not an effective antidepressant in most patients [Goodwin et al., 1970 (cf. 1)] and in fact induces depressive symptoms in a large proportion of Parkinsonians [Jenkins and Groh, 1970 (cf. 1)].

Biochemical, electrophysiological, and behavioral effects of various psychotropic agents support the view that PEA plays a major role in the modulation of affect (criterion 4). For instance, the brain levels of PEA are decreased by the mood depressants reserpine and AMD (discussed previously) and increased by mood-elevating agents such as ethanol, Δ^9-Tetrahydrocannabinol (THC) (11,16-18), DOPA (12), electroshock [Mosnaim et al., 1972 (cf. 1) , and antidepressive drugs, including classical MAOI (1,2,13) as well as tricyclics (1,2,13).

In his last paper, Fischer (3) described the role of PEA in the maintenance of euthymia as a homeostatic mechanism, postulating the existence of brain deposits of bound PEA and of a feedback mechanism which, whenever the thymic state is reduced to a low state, would trigger the release of PEA in active form which would then restore euthymia. The rapid inactivation of PEA by MAO, as well as the feedback mechanism itself, would prevent pathological euphoria. Depression may thus result from a reduction in brain PEA stores; antidepressive agents would restore these deposits, thereby allowing the normal function of the feedback regulating mechanisms to release PEA when required. In contrast, amphetamines would evoke an unconditional release of PEA thereby causing direct stimulant effects.

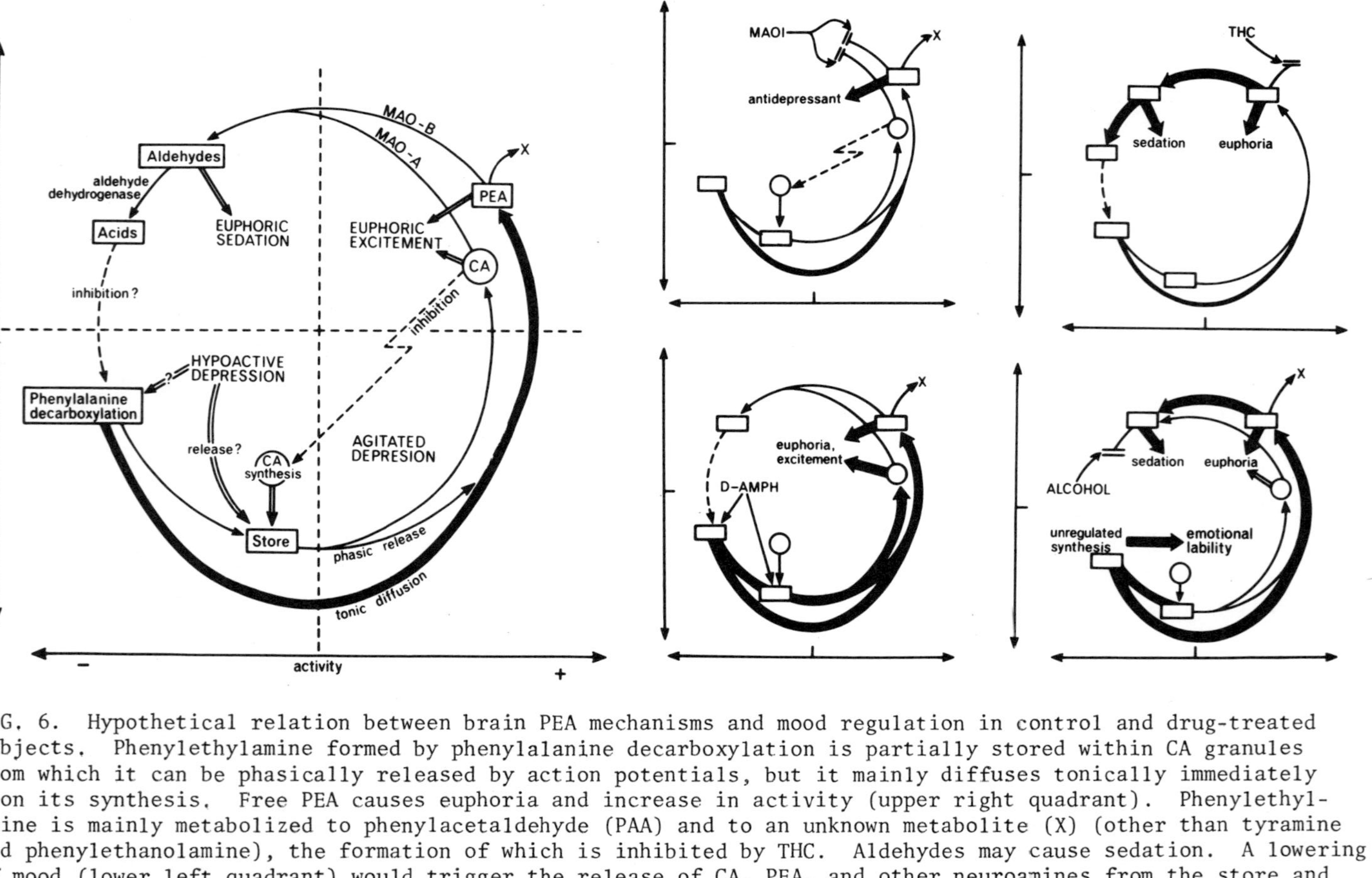

FIG. 6. Hypothetical relation between brain PEA mechanisms and mood regulation in control and drug-treated subjects. Phenylethylamine formed by phenylalanine decarboxylation is partially stored within CA granules from which it can be phasically released by action potentials, but it mainly diffuses tonically immediately upon its synthesis. Free PEA causes euphoria and increase in activity (upper right quadrant). Phenylethylamine is mainly metabolized to phenylacetaldehyde (PAA) and to an unknown metabolite (X) (other than tyramine and phenylethanolamine), the formation of which is inhibited by THC. Aldehydes may cause sedation. A lowering of mood (lower left quadrant) would trigger the release of CA, PEA, and other neuroamines from the store and may also increase PEA synthesis. D-Amphetamine (D-AMPH) appears to induce its stimulant effects also by triggering amine release and simultaneously enhancing PEA synthesis.

Our studies suggest an alternative hypothesis (Fig. 6). First, the importance of stores of bound PEA may be questioned. The fact that PEA has a very fast turnover and cannot be readily stored in granules (29) suggests to us that a significant proportion of newly synthesized PEA continuously diffuses into the extracellular space, thereby exerting a tonic influence on synaptic activity. Thus, a homeostatic regulation of the rate of PEA synthesis may be required for the maintenance of euthymia. The ability of the acid metabolites of several neuroamines to inhibit decarboxylase (30) together with the increased brain levels of PEA induced by alcohol, which inhibits aldehyde dehydrogenase, suggests that the negative feedback regulation of PEA synthesis may be mediated by these neuroacids. A further homeostatic mechanism to prevent excessive euphoria may be the ability of phenylacetaldehyde and other aldehyde metabolites of neuroamines to exert depressant effects. Finally, PEA mechanisms do not appear to function by themselves, but rather operate in conjunction with the CA system. Note that neither the CA depleter AMT nor the PEA depleter AMDH induce depression in humans. In contrast, drugs that can cause depression in man such as AMD and reserpine reduce the brain content of both PEA and CA.

VI. HYPOTHETICAL ROLE OF PEA IN DRUG ACTION

Within this framework, it is possible to relate many of the psychological effects of CNS drugs to their effects upon PEA synthesis and metabolism.

The hypotensive effects of MAOI have been attributed to a relative deficit in CA release by sympathetic neurons, via replacement by false neurotransmitters [Kopin, 1968 (cf. 1)] or decrease synthesis resulting from feedback inhibition of tyrosine hydroxylase [Costa and Neff, 1966 (cf. 1)]. In contradiction, their antidepressant effects have been attributed to the accumulation of brain CA. In our view, MAOI exert their antidepressant effect mainly via inhibition of the disposition of free PEA (a mechanism that per se could account for their ability to elevate mood without causing amphetaminelike stimulation). The fact that PEA exerts clear CNS stimulation suggests that this amine must contribute significantly to the antidepressant action of MAOI. Moreover, MAO appears to play a greater role in the disposition of PEA than in that of other neuroamines. Furthermore, selective inhibitors of type A MAO, which metabolize CA and serotonin (5-HT) (see Braestrup and Randrup, this volume), are devoid of antidepressant effects (e.g., harmine), whereas selective inhibitors of type B MAO [for which PEA is the main naturally occurring substrate (see Braestrup and Randrup, this volume)] are effective antidepressants [e.g., deprenyl (31)].

The antidepressant effect of *imipramine* has been attributed to inhibition of re-uptake in CA neurons. However, the related antidepressant iprindole does not affect re-uptake and the re-uptake inhibitor cocaine is devoid of antidepressant action.

Whereas imipramine does not modify brain content of CA or of serotonin, it markedly increases the brain levels of PEA [Sabelli et al., 1971 (cf. 1), 2,13]. Such augmentation has been attributed to a selective inhibition of type B MAO (32). Such inhibition of type B MAO does not seem significant in vivo because imipramine does not augment the behavioral stimulant effects of PEA (9) to such an extent as typical MAOI do, and moreover, imipramine reduces the recovery from brain of intraventricularly injected PEA (Table 2), suggesting that PEA disposition is actually accelerated. Imipramine augments the recovery from brain tissue of ^{14}C-labeled PEA after the intraventricular injection of labeled phenylalanine (Table 1); following the systemic administration of labeled phenylalanine, the recovery of PEA from brain is augmented while recovery from peripheral tissues is reduced (Table 1). We thus think that imipramine augments the uptake or the synthesis of PEA in brain thereby increasing both the levels and the turnover of brain PEA.

The mood elevating effect of marijuana appears to be mediated by an increase in the levels and turnover of endogenous PEA, whereas its sedative actions may result from an increase in the formation of phenylacetaldehyde and other deaminated metabolites which are behavioral depressants (Fig. 6). This view was suggested to us by the observation that the administration of THC induces euphoria and sedation in untreated subjects, whereas in MAOI pretreated mice, THC induces amphetamine like effects which are enhanced by PEA and by CA depletion. As shown in Fig. 7, the brain levels of PEA in rabbits treated with THC (3 mg/kg, i.v., 1 hr prior to sacrifice) were increased to 2.63 ± 0.06 μg/g in comparison to 0.44 ± 0.05 μg/g in Tween-80 treated controls. We have confirmed by gas-liquid chromatography the results previously obtained (16) using ultraviolet spectroscopy (THC-treated rabbits: 2.30 ± 0. μg/g; vehicle treated animals: 0.58 ± 0.20 μg/g). We have reported that THC enhances threefold the recovery of labeled PEA from brain tissue after the intraventricular administration of labeled phenylalanine (Table 1) or of labeled PEA (Table 2), suggesting that THC inhibits the disposition of PEA without significantly altering its synthesis. Other pharmacological studies support the notion that THC inhibits the disposition of PEA; thus, it enhances the excitatory effects of iontophoretic administration of PEA to cortical neurons and potentiates the behavioral stimulant effects of systemically administered PEA in mice (11,16-18). This potentiation occurs in untreated as well as in MAOI-treated animals. Further, THC administration markedly augments the formation of labeled phenylacetic acid after the administration of labeled PEA (Fig. 7). These studies suggest the existence of an as yet unidentified major metabolite of PEA, the formation of which is inhibited by THC. [This unknown metabolite does not seem to be phenylethanolamine because only 10% of brain PEA is metabolized by DBH [Edwards and Blau, 1972, 1973 (cf. 1)] nor does it seem to be tyramine, which is an even less important metabolite (19, Mosnaim et al., this volume)]. By blocking this pathway, THC augments the brain levels of PEA, thereby causing euphoria, and of phenylacetaldehyde, thereby causing sedation (Fig. 6).

The administration of *ethanol* produces a dramatic rise in PEA brain content (Table 3) and synthesis which may account for the euphoriant effect of alcohol whereas its sedative action may be due to the accumulation of neuroaldehydes (9,33) . Since there is a relation between depressive and alcoholic symptomatology, alcoholism may arise in some patients as an attempt at self-medication.

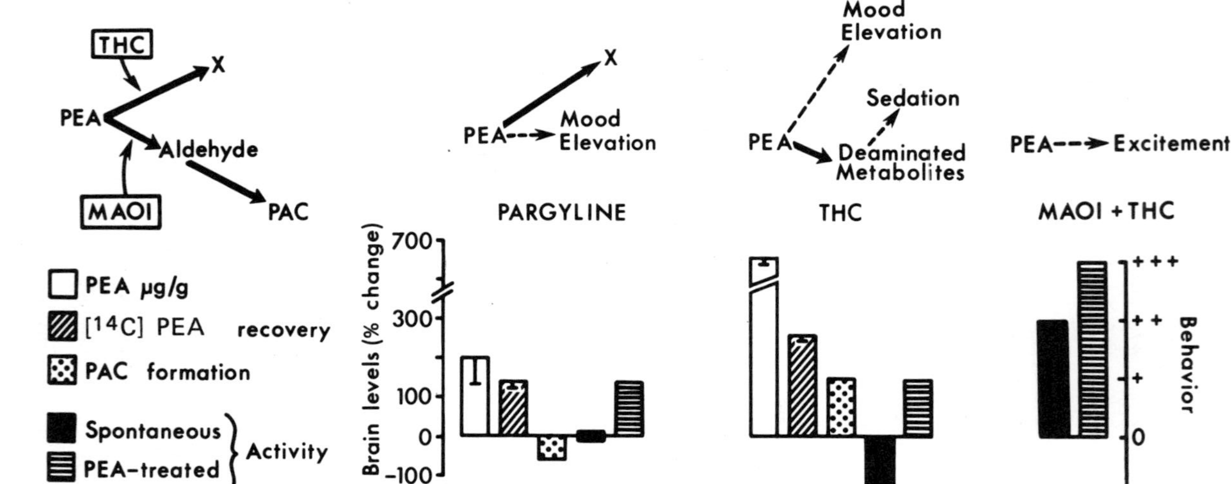

FIG. 7. Schematic representation of the hypothetical role of the deaminated metabolites of PEA in the action of THC and of MAOI with experimental evidence. Biochemical studies in rabbits, behavioral observations in mice. Symbols: (-), sedation; (0), control; (+), hyperactivity; (++), jumping; (+++), jumping and fighting. Data from Refs. 16 and 17, and unpublished.

TABLE 3

The Effects of Ethanol Administration of Endogenous Brain 2-Phenylethylamine in Mice

Treatment (time prior to sacrifice)	Brain PEA (ng/g)[a] $\overline{X} \pm SE$
Saline, 2 hr, i.p.	37.5 ± 7.5
Ethanol, 3 g/kg, 2 hr, i.p.	516.4 ± 42.7

[a]Gas-liquid chromatography [method of Borison et al. (37)].

Phenylethylamine appears to play a major role in the central action of *amphetamines*. Thus, of the naturally occurring amines, only PEA mimics amphetamine in its behavioral effects. Moreover, amphetamine induces only minor changes in the brain content of CA, whereas it markedly affects PEA levels (14,15, Spatz and Spatz, this volume). In rabbits, low doses of D-amphetamine (0.5 mg/kg) initially (30 min) reduce to one-third of its initial value the brain content of PEA (15). Four hours after D-amphetamine administration, the brain PEA content is doubled by low doses (0.5 mg/kg) and enhanced 10-fold by high doses (10 mg/kg). Finally, depletion of brain PEA content by pretreatment with AMDH (which does not deplete other brain monoamines) decreases many of the central actions of amphetamine in mice (15) without reducing the central amphetaminelike actions of PEA. [At doses which partially inhibit phenylalanine decarboxylation in brain without reducing PEA content, RO4-4602 reduces to the same extent the central stimulant actions of amphetamine and of PEA (Fig. 3); the mechanism of this action of RO4-4602 is as yet unexplained.]

D-Amphetamine appears to increase the turnover of PEA in brain since it not only markedly augments its synthesis (Table 1), but it also accelerates its disposition (Table 2). These results together with previous studies [Fischer et al., 1967 (cf. 1), 9], and the observations of Fischer, Spatz and co-workers on the effects of methamphetamine and amphetamine on brain PEA content in rats indicate that PEA release significantly contributes to the central actions of amphetamines.

The hypothesis that the behavioral stimulant action of amphetamine is entirely mediated via CA release is contradicted by the fact that these central actions (as contrasted to the peripheral effects) are neither mimicked by CA or by DOPA nor are they prevented by reserpine. The most important evidence for a role of CA synthesis in the CNS actions of amphetamine is that AMT partially prevents the behavioral stimulant and anti-electroshock effects of both D- and L-amphetamine (while failing to block the central actions of PEA) (Fig. 3), suggesting that a catecholic neuromodulator mediates the stimulant action of amphetamines but not that of PEA. Since AMT slows down PEA turnover (Tables 1 and 2) without altering brain PEA content (Fig. 3), one is tempted to speculate that amphetamine releases newly synthesized CA, which in turn releases PEA, the final mediator for behavioral stimulation (20).

Whereas mood-elevating agents appear to consistently accelerate PEA turnover in brain, *lithium* dramatically reduces PEA synthesis in untreated mice (Table 1) and prevents its increase by amphetamine and by reserpine, an action which may account for its antimanic effect as well as for its prophylactic action in mania and depression (34).

VII. PEA AND EXTRAPYRAMIDAL FUNCTION

Although depression has been attributed to DA deficit, the depressive symptomatology which often accompanies Parkinsonism is not ameliorated and may in fact be worsened by DOPA administration [Jenkins and Groh, 1970 (cf. 1)]. Moreover, Parkinsonism is not a concomitant of depression. On the other hand, there is scanty but suggestive evidence indicating that PEA may modulate extrapyramidal functions (12; see also Heller, this volume) and may thus contribute to the psychological symptomatology of chorea and Parkinsonism.

First, endogenously formed PEA is concentrated in the caudate nucleus [Nakajima et al., 1964; Edwards and Blau, 1972, 1973 (cf.1)] and the cerebellum (13). Second, the urinary excretion of free and conjugated PEA of patients with Parkinson's disease is markedly reduced (12) in comparison to control subjects (Fig. 1). Third, the brain levels of PEA are markedly decreased by drugs which induce Parkinsonism, such as reserpine and AMD, whereas the brain PEA content is increased by L-DOPA.

Fourth, in mice pretreated with a MAOI, the administration of PEA (5-10 mg/kg) relieves the Parkinsonism-induced reserpine (5-10 mg/kg) and markedly potentiates the choreiclike stereotypias of high doses of L-DOPA; moreover, at high doses (50 mg/kg) PEA per se induces choreiclike effects in mice (12) and in guinea pigs (35). Droperidol (0.1 mg/kg), and to a lesser extent chlorpromazine (10 mg/kg), two drugs that can induce Parkinsonism presumably by blocking dopaminergic receptors, prevent the jumping behavior induced by PEA, whereas selective blockers of NE receptors such as phentolamine (alpha) or propranolol (beta) do not (Fig. 8). These results suggest that DA release and/or DA receptors play a major role in the induction of the stereotyped behavior induced by PEA, whereas PEA and CA appear to exert opposite actions in the modulation of exploratory behavior and aggressiveness.

Since amphetamine-induced stereotypy may be considered as an animal model for chorea, and since amphetamine itself is capable of antagonizing the reserpine-induced Parkinsonism in animals and humans, the evidence for a role of PEA in amphetamine action discussed in Section VI may also be considered relevant. Furthermore, the sensitization to the acute extrapyramidal effects of D-amphetamine induced by chronic D-amphetamine treatment (36) in guinea pigs can be maintained by treatment with PEA plus pargyline but not by DOPA (35).

On the basis of these and other studies, we have proposed (12) a model for nigrostriatal synaptic transmission in which dopaminergic transmission is modulated by PEA dihydroxyphenylacetic acid and other compounds metabolically related to dopamine.

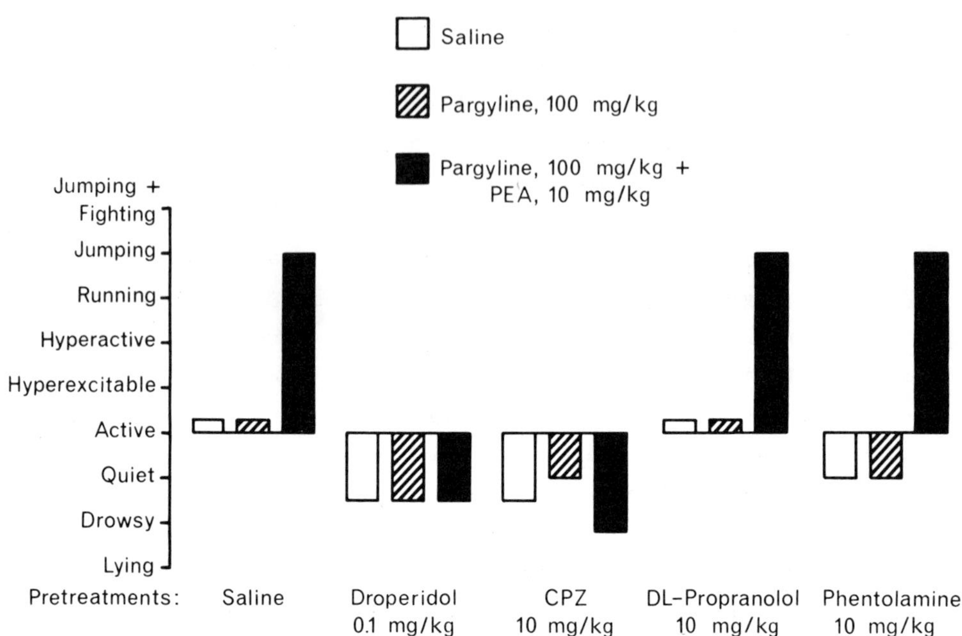

FIG. 8. Influence of catecholamine blockers upon the behavioral effect of PEA. All treatments intraperitoneal.

VIII. PEA AND SYMPATHETIC FUNCTION

Phenylethylamine is not only present in but is also formed by many peripheral tissues including liver (Table 1) and sympathetic neurons because these neurons form and store the PEA metabolite phenylethanolamine. Thus, PEA and its metabolites may modulate the peripheral actions of the CA transmitters, and it may possibly affect the cardiovascular system [Jackson and Temple, 1970 (cf. 1), 7: see also Fig. 1].

Whereas PEA may exert stronger central stimulant effects than CA, it is less effective than the latter in its cardiovascular effects. Thus, the hypotensive effects of pargyline and other MAOI may be accounted for by their ability to augment more markedly the tissue levels of these noncatecholic amines than those of CA [as a result of the fact that PEA is more rapidly destroyed by MAO, that many inhibitors (e.g., pargyline) preferentially affect type B MAO, that

PEA has a much faster turnover than CA, that extracellular CA are metabolized by COMT rather than by MAO, and that CA synthesis is inhibited by MAOI via feedback inhibition of tyrosine hydroxylase]. Thus, endogenous PEA and its metabolites would play the role ascribed to the accumulation of monoamines of exogenous origin in the false neurotransmitter hypothesis.

Phenylethylamines may also play a role in the peripheral action of noncatecholic adrenergic drugs. As in brain, D-amphetamine halves the PEA content of the rabbit heart, whereas L-amphetamine halves the PEA content in both brain and heart (38); hence, the ability of D-amphetamine to release both CA and PEA may account for its strong behavioral effect, whereas the fact that L-amphetamine releases mainly CA may account for its stronger cardiovascular action. Furthermore, imipramine affects differentially not only the central and peripheral actions of amphetamine, but also PEA metabolism, although the relation between these two phenomena is not readily apparent. Thus, imipramine enhances amphetamine CNS stimulation while blocking its sympathomimetic action. As shown in Table 2, following the i.p. administration of radiolabeled phenylalanine, imipramine augments the recovery of radiolabeled PEA from brain tissue while decreasing the recovery from liver and spleen.

IX. CATECHOLAMINES AND PEA AS CO-MODULATORS OF ATTENTION: A REFORMULATION OF THE MONOAMINE THEORY OF AFFECT

Although PEA may be an independent ergotropic messenger which antagonizes the predominantly depressant CA transmitters, recent studies led us (1,38) to view PEA as a modulator which synergizes and sustains (rather than antagonizes) the facilitating action of CA transmitters in pathways subserving ergotropic functions. The observed depressant effects of exogenously administered CA may thus be reconciled with an ergotropic role for central CA neurons if the latter release also PEA and other modulators capable of modifying the action of CA transmitter. One may speculate that altering stimuli and drugs may release brain CA, thereby causing short-lasting arousal; CA release, in turn, may release

PEA which would sustain behavioral excitement. As a modulator of other central as well as peripheral CA synapses, PEA may also play a role in extrapyramidal and cardiovascular functions. Within this framework, we view PEA as a "partial mimetic"—as a dialectic complement to the CA—which can facilitate or antagonize their actions, according to the tissue under consideration and its physiological state, in part by releasing CA, in part by acting via specific PEA receptors, and in part functioning as a partial agonist on CA receptors. In particular, the strong CNS effects of PEA may be essential to sustain the alerting action of CA, whereas its weak peripheral action may actually serve to dampen the cardiovascular effect of CA. The existence of a multiplicity of adrenergic modulators in the CNS can provide a rationale for the multiplicity of behavioral roles attributed to brain CA which include contradictory actions such as maintenance of euthymia, induction of anxiety, triggering of rage, and so on.

Inferring from the effects of PEA and amphetamine in animals and in man, it seems that brain PEA mechanisms modulate simultaneously affect, arousal, and motor activity. The regulation of affect appears to be more closely related to the focusing of attention than to changes in locomotor activity. Thus, amphetamine induces searching in all species, but its effect on activity depends on the pattern for searching in the particular animal (locomotor exploration in mice, searching movements of head and eyes in cats, focusing of attention in humans). Clinical observations are consistent with the view that affect and attention are closely related. Thus, the increased attention of the hypomanic patient (expressed as enthusiasm, attracting attention of others, etc., until hyperactivity itself interferes with attention) contrast with the deficit in attention which appears to occur in depressives (insensitivity to the feelings of others, concentration difficulty, feeling dull, confused and forgetful, lack of interest in anything, failure to notice people or objects even to the point that consciousness may appear fogged, etc.). The work of Magoun and others on reticular activating mechanisms as well as the classical contribution of Pavlov, indicate that attention mechanisms (EEG arousal, the orienting reflex in dogs, exploration in mice, human attention, etc.) ought to be recognized in psychiatry along with sex, fear, anger, and so on, as one of the basic patterns of behavior whose derangement leads to mental illnesses. Human affect may be based on attentive functions (i.e., in the sharing of attention between people) as much as in other basic instincts (sexual, maternal, etc.); a disorder of attention (resulting from a deficit in adrenergic amines) may hence interfere with both love and self-love in depressive disorders.

ACKNOWLEDGMENTS

This investigation was supported by a grant from the State of Illinois Department of Mental Health (510-22-RD). The authors thank Mr. Philip J. Maple for excellent technical assistance and Ms. Anna Maslanka for preparation of the graphics.

REFERENCES

1. H. C. Sabelli, A. D. Mosnaim, and A. J. Vazquez (1974). Phenylethylamine: Possible role in depression and antidepressive drug action. In: *Advances in Behavioral Biology, Vol. 10: Neurohumoral Coding of Brain Function,* edited by R. R. Drucker-Colin and R. D. Myers. Plenum Publishing Co., New York, pp. 331-357.

2. E. Fischer, H. Spatz, B. Heller, and H. Reggiani (1972). Phenylethylamine content of human urine and rat brain, its alteration in pathological conditions and after drug administration. *Experientia,* 28:307-308.

3. E. Fischer (1975). The phenylethylamine hypothesis of thymic homeostasis. *Biol. Psychiat.,* 10:667-673.

4. E. Fischer, B. Heller, M. Nachon, and H. Spatz (1975). Therapy of depression by phenylalanine. *Arzneim.-Forsch.,* 25:132.

5. E. E. Inwang, A. D. Mosnaim, and H. C. Sabelli (1973). Isolation and characterization of phenylethylamine and phenylethanolamine from human brain. *J. Neurochem.,* 20:1469-1973.

6. A. D. Mosnaim, E. E. Inwang, J. H. Sugerman, E. J. DeMartini, and H. C. Sabelli (1973). Ultraviolet spectrophotometric determination of 2-phenylethylamine in biological samples and its possible correlation with depression. *Biol. Psychiat.,* 6(3): 235-257.

7. A. D. Mosnaim, E. E. Inwang, J. H. Sugerman, and H. C. Sabelli (1973). Identification of 2-phenylethylamine in human urine by infrared and mass-spectroscopy and its quantification in normal subjects and cardiovascular patients. *Clin. Chim. Acta*, 46:407-413.

8. U. P. Madubuike, A. D. Mosnaim, and H. C. Sabelli (1974). Brain phenylacetic acid, a major metabolite of 2-phenylethylamine in rabbit brain. *Proc. XXVI Int. Congress of Physiol. Sci.*, New Delhi, India.

9. H. C. Sabelli and W. J. Giardina (1973). Amine modulation of affective behavior. In: *Chemical Modulation of Brain Function*, edited by H. C. Sabelli. Raven Press, New York, pp. 225-259.

10. W. J. Giardina, W. A. Pedemonte, and H. C. Sabelli (1973). Iontophoretic study of the effects of norepinephrine and 2-phenylethylamine on single cortical neurons. *Life Sci.*, 12:153-161.

11. H. C. Sabelli, A. J. Vazquez, and D. F. Flavin (1975). Behavioral and electrophysiological effects of phenylethanolamine, a putative neurotransmitter. *Psychopharmacologia (Berl.)*, 42:117-125.

12. H. C. Sabelli, A. D. Mosnaim, R. L. Borison, and M. E. Wolf (19). Possible role of 2-phenylethylamine in the modulation of extrapyramidal function. In: *Drugs and Central Synaptic Transmission*, edited by B. N. Dhawan and P. B. Bradley. MacMillan & Co., New York. In press.

13. A. D. Mosnaim, E. E. Inwang, and H. C. Sabelli (1974). The influence of psychotropic drugs on the levels of endogenous 2-phenylethylamine in rabbit brain. *Biol. Psychiat.*, 8(2): 227-234.

14. R. L. Borison, A. D. Mosnaim, and H. C. Sabelli (1974). Biosynthesis of brain 2-phenylethylamine: Influence of decarboxylase inhibition and D-amphetamine. *Life Sci.*, 15(10):1837-1848.

15. R. L. Borison, A. D. Mosnaim, and H. C. Sabelli (1975). Brain 2-phenylethylamine as a mediator for the central actions of amphetamine and methylphenidate. *Life Sci.*, 17:1331-1344.

16. H. C. Sabelli, A. J. Vazquez, A. D. Mosnaim, and L. Madrid-Pedemonte (1974). 2-Phenylethylamine as a possible mediator for Δ^9-tetrahydrocannabinol-induced stimulation. *Nature*, 248:144-145.

17. H. C. Sabelli, W. A. Pedemonte, C. Whalley, A. D. Mosnaim, and A. D. Vazquez (1974). Further evidence for a role of 2-phenylethylamine in the mode of action of Δ^9-tetrahydrocannabinol. *Life Sci.*, 14:149-156.

18. P. J. Maple, R. L. Borison, and H. C. Sabelli (1976). Possible mediator of the euphoriant and sedative effect of marihuana by 2-phenylethylamine and by phenylacetic acid. Presented to the *Federation of American Societies for Experimental Biology*, Anaheim, Calif., April 11-16.

19. H. C. Sabelli, A. D. Mosnaim, A. J. Vazquez, W. J. Giardina, R. L. Borison, and W. A. Pedemonte (1976). Biochemical plasticity of synaptic transmission: A critical review of Dale's principle. *Biol. Psychiat.*, 11:481-522.

20. R. L. Borison, N. Narasimhachari, P. J. Maple, H. S. Havdala, and H. C. Sabelli (1976). A reformulation of the phenylethylamine (PEA) theory of affective behavior. Presented at the *Annual Meeting of the Society of Biological Psychiatry*, San Francisco, Calif., June 10-13.

21. H. C. Sabelli and J. May (1975). Chemical excitability of axons: Excitatory and inhibitory effects of putative neurotransmitters and modulators on frog sciatic nerves. *Experientia*, 31:1049-1051.

22. S. R. Snodgrass (1974). Phenylalanine, brain phenylethylamine and motor activity in the rat. *J. Pharm. Pharmacol.*, 26:931-936.

23. R. L. Borison, H. C. Sabelli, and B. Ho (1975). Influence of a peripheral monoamine oxidase inhibitor (MAOI) upon the central nervous system levels and pharmacological effects of 2-phenylethylamine (PEA). *Pharmacologist*, 17:258.

24. H. C. Sabelli (1964). A pharmacological strategy for the study of central modulator linkages. In: *Recent Advances in Biological Psychiatry, Vol. 6*, edited by J. Wortis. Plenum Press, New York, pp. 145-182.

25. W. J. Giardina and H. C. Sabelli (1974). Evidence for the biological activity of the deaminated metabolites of the adrenergic amines. *Biol. Psychiat.*, 9:11-23.

26. J. M. Feldman and H. E. Lebovitz (1971). The nature of the interaction of amines with the pancreatic beta cell to influence insulin secretion. *J. Pharmacol.*, 179:56-65.

27. W. R. Martin and C. G. Eades (1974). Effects of phenethylamine (PEA) on the chronic spinal dog. *Pharmacologist*, 16:205.

28. A. J. Vazquez and H. C. Sabelli (1975). Potentiation of the central nervous system effects of dopa by decarboxylase inhibition: Possible direct role of this neuroamino acid in brain mechanisms. *Exp. Neurol.*, 46:44-56.

29. P. H. Wu and A. A. Boulton (1975). Metabolism, distribution, and disappearance of injected β-phenylethylamine in the rat. *Can. J. Biochem.*, 53:42-50.

30. J. H. Fellman (1956). Inhibition of dopa decarboxylase by aromatic acids associated with phenylpyruvic oligophrenia. *Proc. Soc. Exp. Biol. Med.*, 93:413-414.

31. L. Tringer, G. Haits, and E. Varga (1971). (1-phenyl-isopropyl-methyl-propinylamine HCl) in depression. Societas Pharmacologica Hungarica, *V Conferentia Hungarica Pro Therapia et Investigatione in Pharmacologia*.

32. J. A. Roth and C. N. Gillis (1974). Deamination of β-phenylethylamine by monoamine oxidase: Inhibition by imipramine. *Biochem. Pharmacol.*, 23:2537-2546.

33. A. Feldstein (1971). Effect of ethanol on neurohumoral amine metabolism. In: *The Biology of Alcoholism, Vol. 1: Biochemistry*, edited by B. Kissin and H. Begleiter. Plenum Press, New York, pp. 127-160.

34. R. L. Borison, H. C. Sabelli, B. I. Diamond, P. Maple, and H. S. Havdala (1976). Lithium prevention of amphetamine-induced "manic" excitement and of reserpine-induced "depression" in mice: Possible role of 2-phenylethylamine (PEA). Presented to the Society for Neurosciences, Toronto, Ontario, November 7-11.

35. R. L. Borison and A. Maslanka (1976). 2-Phenylethylamine (PEA) in D-amphetamine induced extrapyramidal disorders. *Fed. Proc.*, 35(3):784.

36. H. L. Klawans, P. Crossett, and N. Dana (1975). Effect of chronic amphetamine exposure on stereotyped behavior: Implications for pathogenesis of L-DOPA-induced dyskinesias. In: *Advances in Neurology, Vol. 9: Dopaminergic Mechanisms*, edited by D. Calne, T. W. Chase, and A. Barbeau. Raven Press, New York, pp. 105-112.

37. E. A. Zeller, A. D. Mosnaim, R. L. Borison, and S. V. Huprikar (1976). Phenylethylamine: Studies on the mechanism of its physiologic action. In: *Advances in Biochemical Psychopharmacology*, edited by E. Giacobini, E. Costa, and R. Paoletti. Raven Press, New York. In press.

38. H. C. Sabelli and R. L. Borison (1976). 2-Phenylethylamine and other adrenergic modulators. In: *Advances in Biochemical Psychopharmacology*, edited by E. Giacobini, E. Costa, and R. Paoletti. Raven Press, New York. In press.

Chapter 16

PHENYLETHYLAMINE AND THE CONDITIONED AVOIDANCE RESPONSE

Wolfgang H. Vogel

Department of Pharmacology
Jefferson Medical College
Thomas Jefferson University
Philadelphia, Pennsylvania

The conditioned avoidance response (CAR) in rats using a shuttle-box has been widely used to detect psychoactive properties of chemicals, to classify such chemicals and to correlate or predict their effects in man (1). Although the CAR, and its alterations by drugs, is not the ideal behavioral test, and not one test is, it is quick, reproducible, easy to employ, and allows a variety of conclusions as to the pharmacological properties of the substance tested.

The interest in phenylethylamine (PEA) as a putative neurotransmitter involved in behavior regulation (2,3,4,5) warrants an examination of its effects on the CAR in rats and a comparison of these results with those obtained with other psychoactive compounds. Furthermore, the availability of brain levels of PEA after administration of the compound allows some speculations as to potency and mode of action.

Using the CAR, rats are usually trained in a conventional two compartment shuttle-box to avoid an electric shock, signalled by a light and buzzer tone, by crossing into the other compartment. In general, a rat gets a 15-sec rest period followed by a 10-sec light and buzzer tone or the conditioned stimulus (CS). If the animal crosses into the other compartment during the CS, the shuttle is called a conditioned avoidance response. If the animal fails to cross, it receives the unconditioned stimulus (US) consisting of a 0.4 mamp shock lasting 30 sec. A shuttle during the shock period is often referred to as the unconditioned response (UR). A shuttle during the rest period is premature response (PR) and the times between onset of CS and CAR and US to UR are called reaction time (RT) and escape time (ET), respectively. This cycle is repeated for 100 trials a day until an animal shows at least 80% CAR during two successive 15-trial periods. If an animal has reached criterion, it is tested for 15 trials at various periods before and after injection of the drug (6).

The concentration of PEA injected into the animal can be determined in the brain, for instance, by a dansylation procedure as described for mescaline (7). In this procedure, PEA is coupled with dansylchloride, and the fluorescence is then measured in a fluorometer. The sensitivity of this method is about 50 ng/g of PEA.

Table 1 shows a comparison between the effects of injected PEA (40 mg/kg) on the CAR and of its concentration in the brain of another group of animals (6). First, it can be seen that PEA interferes with the CAR starting almost immediately after i.p. injection and that interference is very brief. Premature response and RT were only slightly increased, whereas ET was unaffected. Second, PEA reaches the brain quickly but also leaves the CNS rapidly. Third, a good correlation exists between presence of PEA in the brain and interference with the CAR.

These data can be used to speculate on the psychopharmacological properties and mode of action of PEA.

1. PEA might have hallucinogenic effects in man: Although the CAR (shuttle-box) is a crude behavioral test, measurement of various behavioral parameters and their alterations by drugs may give some indication as to the properties of the test material. Similar to the "Bovet-Gatti-Profile" (8) of the Sidman avoidance schedule, drugs with certain psychopharmacological properties can affect behavior of the animal in a CAR situation in a characteristic way.

TABLE 1

Effect of PEA (40 mg/kg, i.p.) on CAR and Its Concentration in the Brain of Rats[a]

Time (min)	Percent CAR	μg/g
-30	82.3 ± 4.8	—
-15	83.9 ± 13.7	—
0*	31.6 ± 24.0**	1.8 ± 0.8
+15	35.0 ± 33.2**	9.4 ± 3.0
+30	38.3 ± 40.9	0.3 ± 0.2
+60	60.0 ± 39.2	n.d.

[a]Mean ± SD from at least five animals.

*5 min for brain levels.

**P = 0.01 (paired t test).

n.d.: not detectable.

(From: Cohen, Fischer, and Vogel (1974). *Psychopharmacology,* 36:77, and reproduced with permission from Springer-Verlag.)

Whereas many drugs such as stimulatory, hypnotic, antipsychotic, or hallucinogenic compounds depress the CAR, they affect the other behavioral parameters differently. For instance, we found for LSD, DMT, mescaline, and related compounds (9), similar to the findings of others (10,11), that these compounds interfere with the CAR but do only slightly increase PR and RT and do not affect ET, a pattern apparently typical for substances with hallucinogenic properties. It has to be emphasized, however, that this is a general rule and that results depend on a large number of variables including the individual animal, species, stress, previous experience and/or dose employed. Since PEA shows a typical hallucinogenic pattern, the speculation can be made that it would act like DMT, LSD, and mescaline and, thus, might be hallucinogenic in man.

2. Phenylethylamine might have a direct mode of action in the CNS: PEA after injection enters the brain very rapidly and also leaves the brain quickly. Presence in the brain seems to parallel behavioral changes closely. This could indicate that PEA causes depression of the CAR by a direct effect on receptors in the CNS since it is conceivable that any indirect effects via a metabolite or changes in brain biogenic amine levels would probably take time and that the behavioral effects would not parallel brain levels that closely. Furthermore, the effect is chemically quite specific; phenylethanolamine or β-hydroxylated PEA is almost devoid of behavioral activity (9). Thus, results would be consistent with the hypothesis that PEA reacts in the brain with a PEA-receptor.
3. Phenylethylamine is not a potent hallucinogenic compound: Since the brain levels of some hallucinogenic and related psychoactive compounds necessary to interfere with the CAR and related responses are known, the potency of PEA as a psychoactive agent can be estimated. From our own data (6,7,12,13) and data from the literature (14-17), minimal brain levels necessary to produce abnormal behavior can be calculated for LSD, lysergic acid, DMT, bufotenin, mescaline, and PEA and are 0.1, 30, 6, 12, 2.4 and 50 nmoles/g, respectively. Based on this comparison, PEA does not appear to be very potent; however, there could be a difference between exogeneously administered (as in our study) and endogenously formed PEA in certain, crucial areas of the brain, with the latter being more potent.

In summary, PEA interacts with the CAR and related responses in the rat in a way in which some typical hallucinogenic compounds do and thus, might be hallucinogenic. Injected PEA does not seem to be a very potent hallucinogenic. PEA seems to have a "direct" mode of action, and data are consistent with the hypothesis that PEA reacts with a specific PEA receptor in the CNS.

REFERENCES

1. R. H. Rech and J. H. Pirch (1971). *Introduction to Psychopharmacology*. Raven Press Books, New York.

2. E. Fischer and B. Heller (1972). *Behav. Psychiat.*, 4:8.

3. A. D. Mosnaim and H. C. Sabelli (1971). *Pharmacologist*, 13:283.

4. H. C. Sabelli and A. D. Mosnaim (1974). *Am. J. Psychiat.*, 131: 695.

5. E. Fischer (1975). *Biol. Psychiat.*, 10:667.

6. I. Cohen, J. F. Fischer, and W. H. Vogel (1974). *Psychopharmacology*, 36:77.

7. J. Cohen and W. H. Vogel (1970). *Experientia*, 26:1231.

8. D. Bovet and G. L. Gatti (1963). *Proc. 2nd Int. Pharm. Meetg., Prague*, 75.

9. W. H. Vogel, unpublished observation.

10. D. M. Stoff, J. J. Mandel, D. A. Gorelick, and W. H. Bridges (1974). *Psychopharmacology*, 36:301.

11. D. A. Gorelick and W. H. Bridges (1975). *Psychopharmacology*, 44:307.

12. I. Cohen and W. H. Vogel (1972). *Biochem. Pharmacol.*, 21:1214.

13. W. H. Vogel, R. A. Carapellotti, B. D. Evans, and A. Der Marderosian (1972). *Psychopharmacology*, 24:238.

14. P. K. Gessner and I. H. Page (1962). *Am. J. Physiol.*, 203:167.

15. J. Axelrod, R. O. Brady, B. Witkop, and E. V. Evarts (1957). *Ann. N.Y. Acad. Sci.*, 66:425.

16. J. R. Smythies, V. S. Johnson, and R. J. Bradley (1969). *J. Psychiat.*, 115:55.

17. E. Sanders and M. T. Bush (1967). *J. Pharmacol. Exp. Ther.*, 158:340.

Part C

CLINICAL STUDIES

Chapter 17

THE SIGNIFICANCE OF PHENYLETHYLAMINE IN PHENYLKETONURIA AND RELATED DISORDERS

Karl Blau

Prenatal Biochemistry Unit
Department of Chemical Pathology
Bernhard Baron Memorial Research Laboratories
Queen Charlotte's Maternity Hospital
London, England

I. INTRODUCTION

Phenylethylamine (PEA), as a metabolite of biological significance to the functioning of the nervous system, first came to our attention in the early studies of Jepson et al. (19) and of Oates et al. (31) on its urinary excretion in patients with phenylketonuria. These were classical studies, as many of the important features of subsequent work, such as monoamine oxidase inhibition (MAOI) and analysis of urinary excretion to assess metabolic turnover, were already described. Our research interests in the role of phenylethylamine and our understanding of its distribution and of its significance in the CNS have developed and spread from this favorable start, and it is, therefore, timely and valuable to take a look at the present status of PEA in phenylketonuria and related states of phenylalaninemia, since the significance of the amine was first recognized in this connection.

II. THE NORMAL METABOLISM OF PHENYLETHYLAMINE

The generally accepted biosynthetic and metabolic pathways are widespread in tissues generally, including the CNS, so that PEA occurs throughout the organism. However, normal tissue concentrations of the amine are relatively low, since the major catabolic route, via MAO-catalyzed oxidation, is an active one, and PEA is rapidly oxidized as soon as it is formed. For this reason, the urinary excretion of PEA in man is also normally relatively low (37), and the major end products of phenylethylamine metabolism are phenylacetic acid (45) and N^{α}-phenacetylglutamine (18). In normal circumstances, almost all of the phenylacetic acid appears to be excreted in the conjugated forms, and besides phenacetylglutamine there may also be a small proportion, of the order of about 5%, in the form of a taurine conjugate (18). Some phenylethylamine is converted to phenylethanolamine in a reaction catalyzed by dopamine-β-hydroxylase (36), and this amine is catabolized in parallel, and its final metabolic end product is mandelic acid (3), which is normally excreted in negligible amounts (35). Minor pathways of the metabolism of PEA and probably also of phenylethanolamine derived from it, include methylation (16) and possibly also acetylation. In addition, a relationship between PEA and p-tyramine, which probably involves microsomal enzymes, has recently been described (4,21,41).

III. THE METABOLISM OF PHENYLETHYLAMINE IN STATES OF PHENYLALANINEMIA

In uncontrolled phenylketonuria and in related states of phenylalaninemia, the metabolic block results in the accumulation of phenylalanine, whose major metabolic route via hydroxylation to tyrosine is completely or partially blocked. Blood and tissue

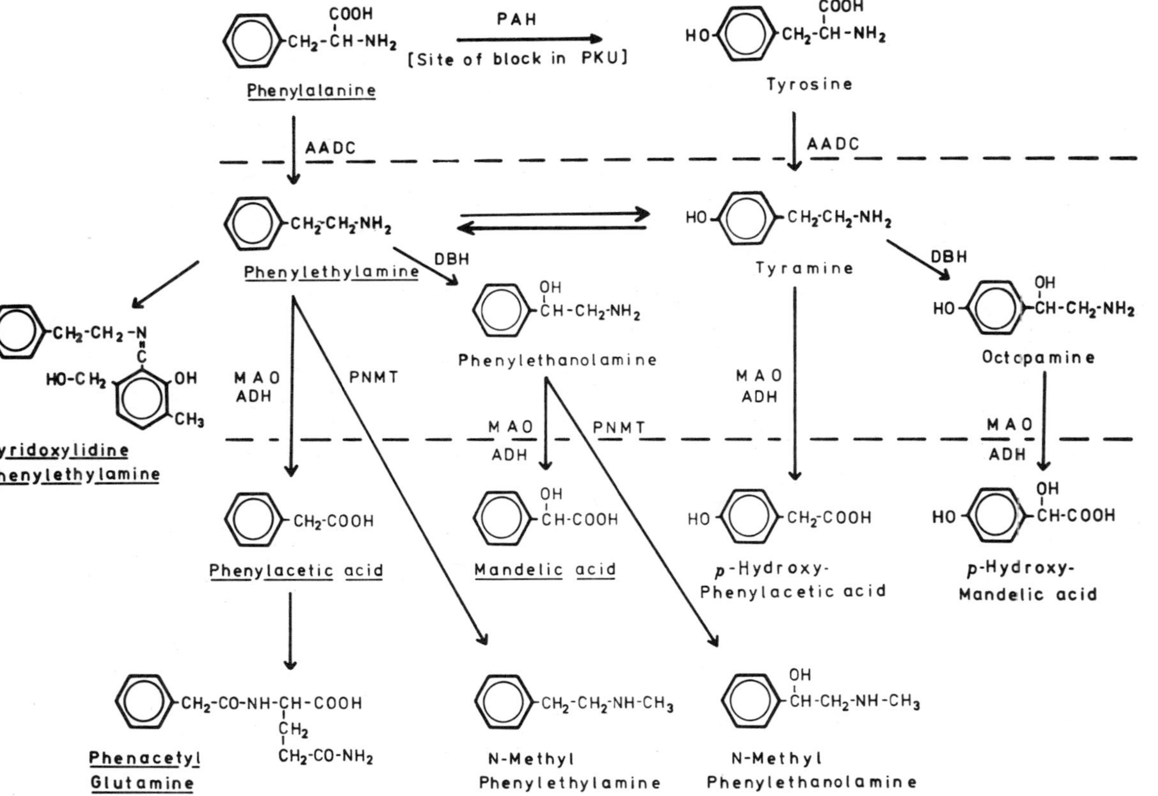

FIG. 1. The biosynthesis and metabolism of PEA. In phenylketonuria and related states of phenylalaninemia there is increased metabolic flux through the pathways of phenylalanine but not of tyrosine metabolism, and increased urinary excretion of those compounds which are underlined has been found. Increased turnover of the compounds between the dashed lines may lead to competition in the biosynthetic pathways of the catecholamines. Abbreviations: AADC, aromatic L-amino acid decarboxylase; ADH, aldehyde dehydrogenase; DBH, dopamine-β-hydroxylase; MAO, monoamine oxidase; PAH, phenylalanine 4-hydroxylase; PNMT, phenylethanolamine N-methyltransferase.

phenylalanine concentrations increase to many times those found normally: In phenylketonuria, the increase may be 20- or even 30-fold. In the various phenylalaninemic variants these increases are much less severe, depending on the extent of the residual phenylalanine hydroxylase activity. In all these clinical conditions, the metabolic flux of phenylalanine into the pathways shown in Fig. 1 is increased in proportion to the increase in its blood and tissue concentrations. Although phenylalanine accumulation elsewhere leads to the production of several new, abnormal metabolites, such as those derived from the transaminated product phenylpyruvic acid (3,11,43), only one such metabolite of PEA has been reported in phenylketonuria (24,25). It has been found, however, that the turnover of octopamine may rise and fall in parallel with that of PEA (13), although the precise metabolic relationship between these two amines has not been established. This applies also to the recently described metabolic relationship between PEA and p-tyramine (4,21,41). The greatly increased turnover of PEA and phenylethanolamine and possibly also of octopamine is accompanied by only small increases in the concentrations of these amines in the tissues and in their urinary excretion, because of the normally high activity of the MAO which catalyze their oxidation. However, their increased turnover is reflected in an increased urinary excretion of phenylacetic acid and mandelic acid which contribute to the characteristic aromatic acid excretion profile found in states of phenylalaninemia (6,12,13). Consequently, the measurement of these aromatic acids gives an approximate indication of the combined central and peripheral turnover of these amines (3,11,43). Although the major source of phenylacetic acid in the urine is the oxidative deamination of PEA, it is generally written in schemes of the phenylalanine metabolism in phenylketonuria that phenylpyruvic acid is metabolized to yield phenylacetic acid. This is probably not the main source of phenylacetic acid, and it has recently been shown that this conversion occurs artifactually in some extraction procedures (42). Care must, therefore, be taken to minimize this artifactual conversion before one can draw valid conclusions about PEA turnover from the excretion of phenylacetic acid in phenylketonuria. This is not a problem in normal metabolism, where no phenylpyruvic acid is being produced, and where phenylacetic acid excretion is very low, reflecting both PEA turnover and dietary sources. The conjugation mechanisms of phenylacetic acid, leading to the formation of phenacetylglutamine and other conjugates, are able to keep pace with its increased production in phenylketonuria (17), so that reliable estimates of PEA turnover based on the total urinary phenylacetic acid output can only be obtained after hydrolysis.

Phenylalaninemic states thus lead to quantitative increases in the turnover of catabolic intermediates and increased output of the end products of PEA metabolism. Qualitatively, abnormal metabolites are not only formed by transamination, but at least one abnormal product has also been reported to be formed from phenylethylamine: an adduct of the Schiff-base type formed from PEA and pyridoxal, identified as pyridoxylidine-β-phenylethylamine (24,25) as shown in Fig. 1. It is uncertain whether this compound, isolated from both brain and urine of rats with artificially-induced phenylketonurialike characteristics and from the urine of uncontrolled patients with phenylketonuria, is formed enzymatically. The mild conditions needed to synthesize the authentic material imply that spontaneous in vivo formation of this "metabonate" is distinctly possible. Nevertheless, regardless of its mode of formation, it has definitely been found in increased amounts in rat CNS, and its presence in the brains of patients with phenylketonuria must, therefore, be regarded as probable.

IV. POSSIBLE TOXIC EFFECTS OF INCREASED PHENYLETHYLAMINE TURNOVER

The brain damage that leads to mental retardation in phenylketonuria occurs during the early post-natal developmental stage, because if affected patients are treated by dietary phenylalanine restriction early enough, preferably within the first 3 weeks of life (22), normal development takes place, and mental retardation is avoided, whereas delay in treatment beyond that time results in progressively more severe impairment. The brain damage is irreversible, and as yet the precise mechanism by which it occurs has not been established. There are several theories, some of which concern structural damage, while others concern functional alterations: The most plausible one seems to be that myelination of axons in the CNS is impaired, because phenylalanine and its metabolites interfere with myelin synthesis, myelin deposition, and myelin stability. This explanation is favored because myelination occurs at the same stage of development as the occurrence of the damage in phenylketonuria, and because it connects the damage with a permanent structural defect, which seems more logical for a disorder with an irreversible course. There is considerable evidence supporting this hypothesis, based on the finding of decreased myelination and immature myelin sheaths in axons of brains from uncontrolled phenylketonurics that were examined post mortem (2,9). The major part of the interference with the synthesis of myelin is most likely ascribable to phenylalanine and phenylpyruvic acid (30,34), because model animal experiments have shown specific effects for these compounds on the synthesis of the basic protein of myelin (1,39) and on lipid biosynthesis (10,38).

However, PEA may contribute to interference with myelin biosynthesis because its increased turnover may deplete vitamin B_6 via obligatory synthesis of the previously mentioned pyridoxylidine-PEA, which possesses anti-vitamin B_6 activity (26). Kurtz et al. (23) have shown that cerebral sphingolipids were reduced by 30 to 60% in rats with neonatal vitamin B_6 deficiency, in addition to the concomitant amino acid imbalance, which probably has a general deleterious effect on protein synthesis. Although the other effects of increased PEA turnover are on the function rather than on structural elements of the CNS, it must be pointed out that the impairment of important functions could, at a susceptible stage of development, lead to permanent damage (stunting) of crucial neurological structures, which has so far escaped notice or which is too subtle to discern by our present methods.

The biochemical effects of PEA include depletion of stored neurotransmitter amines (15,20), interference with synthesis of monoamines by competition for their biosynthetic enzymes (Fig. 1, between the dashed lines), and possible false neurotransmitter action by PEA and octopamine (13,14,36). The pharmacological effects of PEA are described extensively elsewhere in this book and are best characterized as generally amphetaminelike (27). However, there is a significant distinction between amphetamine and PEA in that the latter is a substrate for type B MAO (46), so that even though the turnover of the amine in phenylketonuria may be greatly increased compared with normal (13,19,31), MAO within the organism ensures that actual local concentrations of the amine are only slightly increased (13). It is only after MAO has been inhibited that the expected increased tissue and blood concentrations and increased excretion of PEA occur (13). Another short-coming of attempts to relate the brain damage that occurs in phenylketonuria to the effects of increased PEA turnover is that reversal of the metabolic disturbance by dietary phenylalanine restriction leads to a normalization of PEA turnover apparently without permanent or lasting after-effects. It is more likely that changes in the PEA turnover in states of phenylalaninemia contribute to effects on affective state, learning activity, and general behavior of affected patients, such as hyperactivity, excitability, lability of mood, and short attention span (8), and the increased tendency to epileptiform seizures, possibly via the previously mentioned effect of vitamin B_6 depletion (5).

The main end products of the MAO-catalyzed deamination of PEA are phenylacetic acid and phenacetylglutamine (18,45), and their production and urinary excretion in phenylketonuria is greatly increased in phenylketonuria compared to normal (3,17,35). This forms the basis of the "glutamine-depletion hypothesis" (29,33), which postulates depletion of glutamine via the obligatory conjugation of phenylacetic acid, together with a functional deficit in glutamine-utilization related to the degree of mental impairment. This deficit in utilization of glutamine is an important aspect of the latest version of the hypothesis (33) often overlooked by its critics, but since it postulates a *permanent* metabolic defect, it contributes a further possible explanation for the irreversible brain damage and the resulting mental deficiency. Glutamine is a key metabolite in the defenses of the CNS against the toxic effects of hyperammonemia, and its deficiency at a crucial stage of development might, therefore, lead to damage through the neurotoxic effects of excess of ammonia. There have been conflicting reports (7,28,44), but the subject is technically difficult, and the findings both for and against the hypothesis depend on studies on limited numbers of atypical patients, so that judgment of the hypothesis ought perhaps to be postponed until it has been more extensively evaluated.

Phenylacetic acid has been shown to exert toxic effects in rat cerebellum cultures (40) and in human leucocyte and rabbit lymphocyte cell cultures (6). The concentrations of the acid used in these studies were, however, greater than those found in phenylketonuria or experimental animal hyperphenylalaninemia (12).

V. CONCLUSIONS

If we compare the possible damage caused by the increased turnover of PEA in phenylketonuria and related states of phenylalaninemia with the range of other adverse effects exerted by the accumulation particularly of phenylalanine but also by the products of its transamination (32), it is apparent that PEA makes a relatively minor contribution to the permanent, irreversible structural damage in phenylketonuria. On the other hand, it may well be exerting important although temporary and essentially reversible functional effects on behavior, affective states, and learning ability of affected patients. Whenever we try to gain a better understanding of the properties and function of PEA in the *normal* organism, it may be useful to remember that we should be careful before we ascribe any dramatic actions to the increased turnover of this amine in phenylketonuria, probably because such actions are for the most part prevented through its inactivation via MAO oxidation. We should always consider possible amine mechanisms in conjunction with the activities of the enzymes that inactivate the amines, since the changes in the activities of the inactivating enzymes are at least as likely to lead to alterations in amine concentration and action as changes in the rate of synthesis of the amines.

REFERENCES

1. H. C. Agrawal, A. H. Bone, and A. N. Davison (1970). Effect of phenylalanine on protein synthesis in the developing rat brain. *Biochem. J.*, 117:325-331.
2. E. C. Alvord, L. D. Stevenson, R. S. Vogel, and R. L. Engle (1950). Neuropathological findings in phenylpyruvic oligophrenia. *J. Neuropathol. Exp. Neurol.*, 9:298-310.
3. K. Blau (1970). Aromatic acid excretion in phenylketonuria. Analysis of the unconjugated aromatic acids derived from phenylalanine. *Clin. Chim. Acta*, 27:5-18.
4. A. A. Boulton, L. E. Dyck, and D. A. Durden (1974). Hydroxylation of β-phenylethylamine in the rat. *Life Sci.*, 15:1673-1683.
5. B. D. Bower and P. M. Jeavons (1963). Phenylketonuria presenting as infantile spasms with sudden mental deterioration. *Dev. Med. Child. Neurol.*, 5:577-583.
6. D. M. Bradley and R. F. Mahler (1975). Effect of phenylalanine and its metabolites on the metabolism of leucocytes and lymphocytes. *Clin. Sci. Mol. Med.*, 49:343-351.
7. J. P. Colombo (1971). Plasma glutamine in a phenylketonuric family with normal and mentally defective members. *Arch. Dis. Childhood*, 46:720-721.
8. V. A. Cowie (1971). Neurological and psychiatric aspects of phenylketonuria. In: *Phenylketonuria and Some Other Inborn Errors of Metabolism*, edited by H. Bickel, F. P. Hudson, and L. I. Woolf. Georg Thieme, Stuttgart, pp. 29-39.
9. L. Crome (1971). The morbid anatomy of phenylketonuria. In: *Phenylketonuria and Some Ohter Inborn Errors of Metabolism*, edited by H. Bickel, F. P. Hudson and L. I. Woolf. Georg Thieme, Stuttgart, pp. 126-131.
10. J. N. Cumings, I. K. Grundt, and T. Yanagihara (1968). Lipid changes in the brain in phenylketonuria. *J. Neurol. Neurosurg. Psychiat.*, 31:334-337.
11. H.-Ch. Curtius, J. A. Völlmin, and K. Baerlocher (1972). The use of deuterated phenylalanine in the elucidation of the phenylalanine—tyrosine metabolism. *Clin. Chim. Acta*, 37: 227-285.

12. D. J. Edwards and K. Blau (1972). Aromatic acids derived from phenylalanine in the tissues of rats with experimentally induced phenylketonuria-like characteristics. *Biochem. J.*, 132:495-503.

13. D. J. Edwards and K. Blau (1973). Phenylethylamines in brain and liver of rats with experimentally induced phenylketonuria-like characteristics. *Biochem. J.*, 132:95-100.

14. P. D. Evans, B. R. Talamo, and E. A. Kravitz (1975). Octopamine neurons: Morphology, release of octopamine and possible physiological role. *Brain Res.*, 90:340-347.

15. K. Fuxe, H. Grobecker, and J. Jonsson (1967). The effect of β-phenylethylamine on central and peripheral monoamine-containing neurons. *Eur. J. Pharmacol.*, 2:202-207.

16. L. L. Hsu and A. J. Mandell (1973). Multiple N-methyltransferases for aromatic alkylamines in brain. *Life Sci.*, 13: 847-858.

17. M. O. James and R. L. Smith (1973). The conjugation of phenylacetic acid in phenylketonurics. *Eur. J. Clin. Pharmacol.*, 5:243-246.

18. M. O. James, R. L. Smith, R. T. Williams, and M. Reidenberg (1972). The conjugation of phenylacetic acid in man, sub-human primates and some non-primate species. *Proc. Roy. Soc. London B*, 182:25-35.

19. J. B. Jepson, W. Lovenberg, P. Zaltzman, A. Sjoerdsma, and S. Udenfriend (1960). Amine metabolism, studied in normal and phenylketonuric humans by monoamine oxidase inhibition. *Biochem. J.*, 74:5P.

20. J. Jonsson, H. Grobecker, and P. Holtz (1966). Effect of β-phenylethylamine on content and subcellular distribution of norepinephrine in rat heart and brain. *Life Sci.*, 5: 2235-2246.

21. J. Jonsson, B. Lindeke, and A. K. Cho (1975). Oxidation of phenethylamine yielding tyramine by rat liver microsomes. *Acta Pharmacol. Toxicol. Kbh.*, 37:352-360.

22. E. S. Kang, N. D. Sollee, and P. S. Gerald (1970). Results of treatment and termination of the diet in phenylketonuria (PKU). *Pediatrics*, 46:881-890.

23. D. J. Kurtz, H. Levy, and J. N. Kanfer (1972). Cerebral lipids and amino acids in the vitamin B_6-deficient suckling rat. *J. Nutr.*, 102:291-298.

24. Y. H. Loo (1967). Characterization of a new phenylalanine metabolite in phenylketonuria. *J. Neurochem.*, 14:813-821.

25. Y. H. Loo and P. Ritman (1964). New metabolites of phenylalanine. *Nature*, 203:1237-1239.

26. Y. H. Loo and P. Ritman (1967). Phenylketonuria and vitamin B_6 function. *Nature*, 213:914-916.

27. P. Mantegazza and M. Riva (1963). Amphetamine-like activity of β-phenethylamine after a monoamine oxidase inhibitor in vivo. *J. Pharm. Pharmacol.*, 15:472-478.

28. C. M. McKean and N. A. Peterson (1970). Glutamine in the phenylketonuric central nervous system. *New Eng. J. Med.*, 282:761-766.

29. A. Meister, S. Udenfriend, and S. P. Bessman (1956). Diminished phenylketonuria in phenylpyruvic oligophrenia after administration of L-glutamine, L-glutamate or L-asparagine. *J. Clin. Invest.*, 35:619-626.

30. J. H. Menkes (1968). Cerebral proteolipids in phenylketonuria. *Neurology*, 18:1003-1008.

31. J. A. Oates, P. Z. Nirenberg, J. B. Jepson, A. Sjoerdsma, and S. Udenfriend (1963). Conversion of phenylalanine to phenethylamine in patients with phenylketonuria. *Proc. Soc. Exp. Biol. Med.*, 112:1078-1081.

32. M. S. Patel and I. J. Arinze (1975). Phenylketonuria: Metabolic alterations induced by phenylalanine and phenylpyruvate. *Am. J. Clin. Nutr.*, 28:183-188.

33. T. L. Perry, S. Hansen, B. Tischler, R. Bunting, and S. Diamond (1970). Glutamine depletion in phenylketonuria. A possible cause of the mental defect. *New Eng. J. Med.*, 282:761-766.

34. A. L. Prensky, M. A. Fishman, and B. Daftari (1971). Differential effects of hyperphenylalaninemia on the development of the brain in the rat. *Brain Res.*, 33:181-191.

35. S. Rampini, J. A. Völlmin, H.-R. Bosshard, M. Müller, and H. C. Curtius (1974). Aromatic acids in urine of healthy infants, persistent hyperphenylalaninemia, and phenylketonuria, before and after phenylalanine load. *Pediat. Res.*, 8:704-709.

36. J. M. Saavedra and J. Axelrod (1973). Demonstration and distribution of phenylethanolamine in brain and other tissues. *Proc. Nat. Acad. Sci. USA*, 70:769-772.

37. J. W. Schweitzer, A. J. Friedhoff, and R. Schwartz (1975). Phenethylamine in normal urine: Failure to verify high values. *Biol. Psychiat.*, 10:277-285.

38. S. N. Shah, N. A. Peterson, and C. M. McKean (1972). Impaired myelin formation in experimental hyperphenylalaninemia. *J. Neurochem.*, 19:479-485. Lipid composition of human cerebral white matter and myelin in phenylketonuria. *J. Neurochem.*, 19:2369-2376.

39. F. L. Siegel, K. Aoki, and R. E. Colwell (1971). Polyribosome disaggregation and cell-free protein synthesis in preparations from cerebral cortex of hyperphenylalaninemic rats. *J. Neurochem.*, 18:537-547.

40. D. H. Silberberg (1967). Phenylketonuria metabolites in cerebellum culture morphology. *Arch. Neurol.*, 17:524-529.

41. R. P. Silkaitis and A. D. Mosnaim (1976). Pathways linking L-phenylananine and 2-phenylethylamine with p-tyramine in rabbit brain. *Brain Res.*, 114:105-115.

42. R. M. Thompson, B. G. Belanger, R. S. Wappner, and I. K. Brandt (1975). An artifact in the gas chromatographic analysis of urinary organic acids from phenylketonuric children. Decarboxylation of phenylpyruvic acid during extraction. *Clin. Chim. Acta*, 61:367-374.

43. S. K. Wadman, C. van der Heiden, D. Ketting, and F. J. van Sprang (1971). Abnormal tyrosine and phenylalanine metabolism in patients with tyrosyluria and phenylketonuria; gas-liquid chromatographic analysis of urinary metabolites. *Clin. Chim. Acta*, 34:277-287.

44. P. W. K. Wong, J. L. Berman, M. W. Partington, S. K. Vickery, M. E. O'Flynn, and D. Y. Y. Hsia (1971). Glutamine in PKU. *New Eng. J. Med.*, 285:580.

45. P. H. Wu and A. A. Boulton (1975). Metabolism, distribution, and disappearance of injected β-phenylethylamine in the rat. *Can. J. Biochem.*, 53:42-50.

46. H.-Y. T. Yang and N. H. Neff (1973). β-Phenylethylamine: Specific substrate for type B monoamine oxidase of brain. *J. Pharmacol. Exp. Ther.*, 187:365-371.

Chapter 18

PHARMACOLOGICAL AND CLINICAL EFFECTS OF D-PHENYLALANINE IN DEPRESSION AND PARKINSON'S DISEASE

Bernardo Heller

Laboratory of Psychopharmacology
and Experimental Neuropsychiatry
and
School of Psychiatry
Faculty of Medicine
National University of Buenos Aires
Buenos Aires, Argentina

I. INTRODUCTION

The first attempts to explain the action of the biogenic amines on behavior started with Cannon's (1) research on the role of the sympathoadrenal system in the emergency states. According to his hypothesis, adrenaline is secreted by the suprarenal glands, preparing the organism to cope with these situations. Brodie and Shore (2), based on Hess's (3) hypothesis of the existence of an ergotropic and trophotropic diencephalic system, postulated serotonin (5-HT) and the catecholamines, respectively, as the neurohumoral mediators or modulators of the aforementioned antagonistic system. This theory was also based on the works of Vogt (4), who discovered the presence of noradrenaline in subcortical structures; almost at the same time, Twarog and Page (5) and Amin (6) demonstrated the existence of 5-HT in the same structures, and Weil-Malherbe and Bone (7) found dopamine (DA) in teh basal ganglia. The mechanisms of action of various physiological and pharmacological effects of different drugs were formulated on the basis of this theory. Currently, however, there are many questions regarding the actual role of the catecholamines.

Rothballer (8) in 1954 showed that certain actions of catecholamines in the CNS are excitatory, but others are inhibitory. For example, it has been shown that catecholamines reduce the locomotor activity of mice and at the same time produce an exciting behavior. This duality may be explained by the fact that catecholamines can act both on excitatory and inhibitory receptors. On the other hand, a biogenic amine with ergotropic action, phenylethylamine (PEA), was found by Nakajima, Kakimoto, and Sano (9) in rabbit pretreated with monoamine oxidase inhibitors (MAOI), especially in the basal ganglia. The presence of PEA in the brain of several animals, including man, has been shown more recently in nontreated subjects (10-12). This substance has stimulating effects in animal behavior (13-16), and an antagonism between the depressant effects of catecholamines and the stimulating effects of PEA (apparently of a competitivelike nature) have also been reported by other authors (17-20), using newly hatched chicks which do not have a fully developed blood-brain barrier.

We studied the effects of different drugs on PEA brain levels in rats (21,22), and we found that imipramine and MAOI raise and reserpine lowers its concentration. Our values of PEA in brain of untreated animals are higher (0.5 μg/g) than those found by Saavedra (23) and by Saavedra and Axelrod (24). They reported amounts of brain PEA in the nanogram range. However, with this method, they can measure only free-water soluble-unbound PEA and not the bound fraction to the lipoproteins, as they do not use an efficient deproteinizing procedure. It must be noted that these authors and also Durden et al. (25) observed marked variations in PEA levels after the addition to their systems of MAOI. Durden et al. (25) found in animals pretreated with MAOI brain PEA values 10 to 200 times higher, and Saavedra (23) and Saavedra and Axelrod (24) reported a 40-fold increase of this substance under the same conditions. We found only a threefold increase in the brain PEA levels of animals pretreated with MAOI. We think that low yields can be caused too by losses during the extraction processing and the importance of the extraction methods must be stressed.

Phenylethylamine is present in the urine of control subjects (26,27), and Fischer et al. (28) found a decrease in the daily urinary output of this substance in persons suffering from endogenous depression. These results were confirmed later by other authors (29-32). In view of these findings, Fischer and Heller (33,34) and Fischer (35) postulated that PEA is a stimulating agent with a role in the mechanism of action of different drugs and involved in the etiology of endogenous depression. This hypothesis has been supported by other authors 36,37).

According to our theory, there are pools in the brain of bound PEA and thus when the normal affective state changes to a depressed one, a feedback mechanism triggers the release of PEA in its free active form, restoring the normal conditions by its stimulatory effects. The rapid inactivation of PEA by MAO, as well as the feedback mechanism, avoids an excessive action of free PEA preventing a euphoric state.

II. TREATMENT OF PATIENTS SUFFERING FROM DEPRESSION

Treatment of patients suffering from endogenous depression with tricyclic drugs or MAOI improves their clinical state and raises the daily urinary PEA output. It must be pointed out that Roth and Gillis (38) found that imipramine inhibits the activity of type B monoamine oxidase (MAO) and that PEA is a selective substrate for this enzyme. We have also studied the effects of D- and L-phenylalanine on brain PEA levels in rats. When low doses of D- and L-phenylalanine are injected to rats, both raise the brain PEA content but the effects of D-phenylalanine are significantly more effective than those of L-phenylalanine. Mosnaim (26) postulates that mammals may utilize D-phenylalanine through its transamination and/or oxidative conversion to the corresponding α-ketoacid followed by L-specific reamination.

We believe that D-phenylalanine may follow still another metabolic pathway, which consists of the direct decarboxylation of this amino acid to PEA. It must also be noted that D-phenylalanine has a significantly higher activity than L-phenylalanine in antagonizing reserpine-induced symptoms in animals (39). In view of all of these findings, we decided to investigate the effects of D-phenylalanine in the treatment of depression, using an open field trial and a double-blind study. In the open field trial, patients with diagnoses of depression of different etiologies, some of them having previously been treated unsuccessfully with tricyclics or MAOI, were given D-phenylalanine. After a wash-out period of 7 days in which they did not receive any drug, daily oral dosages that ranged between 100 and 400 mg were administered to them in two doses (8 A.M. and 4 P.M.), for a period of time that varied between 2 and 6 months. Their clinical psychiatric state was evaluated every 5 days by a team of psychiatrists, and also on a weekly basis patients were submitted to physical examinations and for hematological and hepatic tests. Four-hundred-fifty-five patients were treated in this manner. In the cases in which anxiety was a predominant symptom, benzodiazepine drugs were added.

III. DEPRESSION

Patients were qualified as suffering from:

1. Endogenous depression when they fulfilled the following criteria
   a. A family history of unipolar or bipolar affective disorders.
   b. A personal history of repeated episodes of depression, with or without manic episodes which began between the ages of 30 and 55 years of age.
   c. The depressive episodes developed without any apparent psychological or social precipitating problem.
2. Reactive or neurotic depression when
   a. Patients lacked a family history of unipolar or bipolar affective disorders.
   b. They did not have a personal history of repeated episodes of depression with or without manic periods.
   c. The depressive episodes developed with an apparent psychological or social precipitating problem.
3. Involutional depression, when
   a. The disease began after the age of 50 years.
   b. The common antidepressant drugs had relatively little effect on their pathological condition.
   c. Patients deteriorated in their mental state.
4. Other kinds of psychiatric patients suffering from depressive pictures as symptoms of their basic psychiatric disease, as well as patients with depression secondary to prolonged use of amphetamines, were also treated with D-phenylalanine.

Hamilton's, Zung's, and Beck's Rating Scales for Depression were used to assess the clinical state of patients and the results obtained with the treatment with D-phenylalanine were qualified as:

1. Normal Affective State (NAS), when all the depressive symptoms disappeared, and patients recovered their working capacity

2. Improvement (I), when the major part of the depressive symptomatology disappeared, but patients did not recover their working capacity
3. No Improvement (NI), when changes were found in their clinical condition

In the cases of depressive pictures in other mental diseases, only the clinical symptoms corresponding to depression were studied to judge the activity of the drug. The recovery of the working capacity was not a useful parameter in their evaluation. The results obtained are summarized in Table 1.

A. Endogenous Depression

In endogenous depression after 15 days of treatment, 269 patients (73%) out of 370 cases reached NAS; 84 patients (23%) had an I; and in 17 cases (4%) NI was observed. After 60 days of treatment an NAS was seen in 296 patients (80%); 57 patients (15.4%) showed I; and 17 of them (4.6%) had NI in their psychiatric state. After 6 months of treatment, 352 patients (94.6%) had an NAS; 10 patients (2.8%) showed I; and in eight cases (2.6%) NI was observed.

B. Neurotic or Reactive Depression

After 15 days of treatment, 25 of them (53%) showed an NAS; 11 (23%) had an I; and in the other 11 patients (23.5%), NI was observed. After 60 days of treatment, 34 patients (72.3%) showed an NAS; six patients (12.7%) had an I; and in 7 (15%), NI was observed. After 6 months of treatment, 42 patients (89%) showed an NAS and in 5 patients (11%) no appreciable changes in their clinical state were detected.

It is difficult from our theoretical point of view to explain the action of PEA in this disease; it is possible, as Mayer-Gross suggests, that many cases of reactive or neurotic depression are really endogenous depression.

TABLE 1

Open Field Trial—Action of D-Phenylalanine on Different Kinds of Depressive Diseases[a]

Disease	Number of Cases	Daily Oral Dosage of D-Phenylalanine (mg)	State of patients at 15 days of treatment			State of patients at 60 days of treatment			State of patients at 180 days of treatment		
			NAS	I	NI	NAS	I	NI	NAS	I	NI
Endogenous depression	370	100 to 400	296 (73%)	84 (23%)	17 (4%)	296 (80%)	57 (15.4%)	17 (4.6%)	352 (94.6%	10 (2.8%)	8 (2.6%)
Neurotic or reactive depression	47	100 to 400	25 (53%)	11 (23.5)	11 (23.5%)	34 (72.3%)	6 (12.7%)	7 (15%)	42 (89%)		5 (11%)
Involutional depression	15	100 to 400				3 (20%)	6 (40%)	6 (40%)	3 (20%)	6 (40%)	6 (40%)
Post-amphetamine depression	10	100 to 400	10 (100%)			10 (100%)					
Depression in schizophrenia	10	100 to 500	5 (50%)	5 (50%)		8 (80%)	2 (20%				
Arteriosclerotic dementia	4	100 to 500		4 (100%)				4 (100%)			

[a]NAS, normal affective state; I, improvement; NI, no improvement.

C. Involutional Depression

Fifteen patients suffering from this disease were treated with D-phenylalanine. After the administration of this drug, three of them (20%) reached NAS. Six patients (40%) showed an improvement, and the other six patients (40%) had no improvement in their clinical state. Changes were not observed after 6 months of continuous treatment with D-phenylalanine, even after increasing the doses of this drug to 500 and 600 mg daily.

D. Depressive Pictures in Other Mental Diseases

Fourteen patients, 10 suffering from schizophrenia and four of arteriosclerotic dementia, who had depressive symptoms were treated with D-phenylalanine. The schizophrenic patients were receiving neuroleptic drugs, and this medication was not discontinued. D-Phenylalanine was added to the treatment in a dose that ranged between 100 and 400 mg daily. After 15 days of treatment, five schizophrenic patients showed an absence of depressive symptoms and the other five had an improvement in their affective symptomatology. In the four cases of arteriosclerotic dementia, no appreciable changes in their symptomatology were observed. After 60 days of treatment eight schizophrenic patients did not have any depressive symptoms and two of them had an improvement in their affective symptomatology. The depressive state of the patients with arteriosclerotic dementia did not vary.

It is interesting to remark that in preliminary experiments carried out in our laboratory it was observed that treatment with neuroleptic drugs in schizophrenic patients led to a significant decrease in PEA excretion.

E. Post-Amphetamine Depression

Ten patients that developed depressive pictures after prolonged use of D-amphetamine were treated with D-phenylalanine. After 15 days of treatment, a NAS was observed in all of them. However, therapy with this drug had to be maintained for 60 days because if it was discontinued at an earlier time, an impairment in their psychiatric condition was observed. It must be noted that Fischer and Heller (40) found that the prolonged administration of amphetamine to animals leads to a progressive depletion of brain PEA. It must be emphasized that in this open field trial with D-phenylalanine no side-effects or toxic pharmacological actions were observed. To confirm the results obtained in the open field trial, a double-blind trial was carried out with 60 endogenously depressed patients that included 39 males and 21 females whose age ranged between 35 and 55 years and who had an acute untreated depressive episode. In this study, imipramine was given to patients in the control group for ethical and legal reasons. Pills of the same size and color were provided for imipramine (50 mg tablets) and D-phenylalanine (50 mg tablets). Neither the nurses who administered the drugs or the three psychiatrists who evaluated together the clinical state of the patients knew the nature of the medication given. Thirty patients received oral doses of 100 mg of imipramine (50 mg at 8 A.M. and 50 mg at 4 P.M.), and the other 30 patients were administered in the same manner 100 mg of D-phenylalanine. Treatment with these drugs was given for 15 days and then a placebo was administered for 5 days. Subsequently, therapy with D-phenylalanine or imipramine was reinstated. No other drugs were given to these patients during the course of this study.

IV. RESULTS

The results were evaluated in the same form as in the open field trial. Urinary PEA excretion was determined using the method of Spatz et al. (32) in 14 patients (seven treated with imipramine and seven with D-phenylalanine) prior to treatment and subsequent to the 15th and 20th days of the study. Table 2 shows that after 15 days of treatment with imipramine 30% of patients had an NAS; 27% an I; and in 43%, no clinical changes were observed. With D-phenylalanine an NAS was found in 60% of them; 24% had I; and in 16% there were no

TABLE 2

Therapeutic Effects of Imipramine and D-Phenylalanine in Patients with Endogenous Depression

Drug Utilized from Day 1 till Day 15	Clinical State of Patients at Day 15 Number of Cases and Percent with			Drug Utilized from Day 16 till Day 20	Clinical State of Patients at Day 20 Number of Cases and Percent with			Drug Utilized from Day 16 till Day 30	Clinical State of Patients at Day 30 Number of Cases and Percent with		
	NAS	I	NI		NAS	I	NI		NAS	I	NI
Imipramine	9 30%	8 27%	13 43%	Placebo	1 3%	2 6%	27 91%	Imipramine	13 43%	9 30%	8 27%
D-Phenylalanine	18 60%	7 23%	5 17%	Placebo	2 6%	2 6%	26 88%	D-Phenylalanine	20 66%	5 17%	5 17%

[a]NAS, normal affective state; I, improvement; NI, no improvement.

TABLE 3

Phenylethylamine Urinary Elimination in Patients with Endogenous Depression and the Relation with Their Clinical State[a]

Patient	Urinary Elimination of Phenylethylamine (2 g/24 hr)			Clinical State	
	Day 0	Day 15	Day 20	Day 15	Day 20
1	53	528	79	NAS[b]	NI
2	46	53	48	NI	NI
3	21	346	34	I	NI
4	18	27	35	NI	NI
5	58	438	56	NAS	NI
6	39	727	40	NAS	NI
7	43	36	42	NI	NI
X	40 ± 5	308 ± 103	48 ± 6		

[a]Patients were treated with imipramine for 15 days and with placebo for 5 days.

[b]NAS, normal affective state; I, improvement; NI, no improvement.

clinical changes. After discontinuing the administration of both drugs for 5 days, both groups showed an impairment in their clinical state. When drugs were restarted, both groups improved. No side-effects or toxic reactions were observed with treatment with D-phenylalanine; thus, this substance can be considered as a highly effective and safe antidepressant agent. Tables 3 and 4 show the changes in daily urinary PEA output in patients of each group before and after 15 and 20 days of treatment. A strong correlation between the patient's clinical state and the daily urinary PEA output can be seen.

It is very important to remark the correlation between the clinical state of patients, the antidepressant effects of the drugs, and the urinary PEA daily output. These findings, together with those published by us (21,22), Fischer and Heller (40), Spatz and

TABLE 4

Phenylethylamine Urinary Elimination in Patients with Endogenous Depression and the Relation with Their Clinical State[a]

	Urinary Elimination of Phenylethylamine (2 g/24 hr)			Clinical State	
Patient	Day 0	Day 15	Day 20	Day 15	Day 20
1	32	864	47	NAS[b]	NI
2	43	539	51	NAS	NI
3	61	487	55	NAS	NI
4	28	308	42	I	NI
5	36	910	34	NAS	NI
6	50	632	58	NAS	NI
7	64	457	53	NAS	NI
X	45 ± 5	600 ± 82	49 ± 3		

[a]Patients were treated with D-phenylalanine for 15 days and with placebo for 5 days.

[b]NAS, normal affective state; I, improvement; NI, no improvement.

Heller (41) on the effects of MAO inhibitors, imipraminelike drugs, D-phenylalanine, and N-methyl amphetamine on PEA brain levels in the rat encouraged us to propose the use of both animal and clinical tests as a method to study the activity of antidepressant drugs and to differentiate them from euphoriant or stimulant amphetaminelike drugs. These latter do not have antidepressant activity and lower PEA brain in animals as has been shown by Fischer and Heller (40).

V. PHENYLETHYLAMINE AND D-PHENYLALANINE IN PARKINSON'S DISEASE

In 1966, Heller postulated that PEA may be one of the important biogenic amines involved in the control of extrapyramidal activity. Subsequently, rats were treated with agents known to induce Parkinson-like symptoms (42) in animals such as reserpine, tryptamine, and

TABLE 5

Action of Drugs Against Reserpine, Tryptamine, and Thioproperazine Effects[a]

Drugs	Anti-reserpine Effect	Anti-tryptamine Effect	Anti-thioproperazine Effect
Central serotonegic blocking agents Methysergide	+	+	-
Adrenergic blocking agents ( ) propanolol, dihydroergotamine, and phentolamine	-	-	-
Antihistaminic drugs with central action	-	-	+
Muscarinic cholinergic blocking agents atropine and benactyzine	-	-	+
Nicotinic cholinergic blocking agents (anti-Parkinsonian drugs)	+	+	-
L-DOPA	+	+	-
Phenylethylamine and D-amphetamine	+	+	+

[a]+, Blocking action; -, without blocking action

TABLE 6

Action of Neurohumoral Agents in Parkinson's Disease

Anti-Parkinson Biogenic Amines	Pro-Parkinson Biogenic Amines
Dopamine-phenylethylamine	Tryptamine-serotonin Histamine-acetylcholine

thioproperazine and then various drugs were tested for their ability to block such symptoms.

The results obtained are summarized in Tables 5 and 6. It can be seen that anti-serotonergic drugs with central effects are able to antagonize the action of reserpine and tryptamine but not those of thioproperazine. Adrenergic blocking agents (α and β), propanolol, dihydroergotamine, and phentolamine did not have any anti-Parkinsonian action. Muscarinic blocking agents, atropine, and benactyzine antagonized the effects produced by thioperazine, and nicotinic blocking agents blocked the actions of reserpine and tryptamine. Antihistaminic drugs were able to antagonize thioproperazine-induced symptoms. L-DOPA had an antagonistic effect against the symptoms induced by reserpine and tryptamine, and PEA and D-amphetamine were able to block the symptoms produced by the three Parkinson-inducing drugs.

Based on our findings, we advanced the hypothesis of the existence of two neurohumoral systems involved in the control of extrapyramidal activity; the first one is related on one side to 5-HT and tryptamine and on the other to DA and PEA. The second neurohumoral system consists of acetylcholine and histamine on one side and PEA on the other. In the first system any factor that would provoke an increment in the action of the 5-HT-tryptamine group or a diminution in the activity of the DA-PEA group would lead to the production of Parkinsonian symptoms. In the second system, an increment in the activity of the cholinergic-histaminergic group or a diminution of the action of PEA would facilitate such symptoms. The existence of these two neurohumoral systems is supported by the findings of Wada (43) and McGeer (44).

We have also proposed the hypothesis that Parkinsonism and depression are diseases with a certain biochemical similarity and that differences between them may be the result of varying localizations of the metabolic disorder.

Heller and Fischer (43) and Heller, Fischer, and Poch (44) studied also the urinary daily output of PEA in patients suffering from Parkinson's disease, with two different methods. We found that it is diminished in relation to the elimination of this substance by normal persons and these results have been confirmed by Sabelli and Vazquez (45). In view of these findings, patients presenting Parkinson's syndrome of different etiology were treated with D-phenylalanine.

In an open field trial, after a wash-out period of 10 days, 15 patients suffering from Parkinson's disease were treated for a period varying between 6 months and 13 years. The drug was administered for 4 weeks in dosages between 200 and 500 mg daily, distributed in two intakes (8 A.M. and 5 P.M.). In order to assess the actions of D-phenylalanine, five symptoms were evaluated: rigidity, tremor, walking disability, speech difficulties, and depression.

Patients were examined prior to and weekly during treatment. Each symptom was evaluated from 1 to 10. In Table 7, the basis for this form of evaluation is shown. The results obtained were qualified as "Bad" when at the end of the treatment no diminution of the score was obtained and "Regular," "Good," "Very Good," or "Excellent" when the score decreased by 1-40, 41-60, 61-90, or 91-100%, respectively.

The results obtained for each symptom were considered taking into account the total group of patients and they were analyzed by Student's t test.

The general therapeutic effects of D-phenylalanine were evaluated using the test of variance and covariance. The results obtained, as indicated in Table 8, were as follows:

Rigidity. This was an outstanding symptom in all patients. The therapeutic results were "Very Good" in one case; "Good" in three; and "Regular" in nine patients. The rest of them showed no improvement. The Student's t test gave a $P < 0.5$ and < 0.1 (which is not statistically significant).

Walking Disability. The gait was very difficult in 13 patients. The therapeutic effects were "Excellent" in seven patients; "Good" in three; and "Regular" in another three. The Student's t test gave a $P < 0.001$.

TABLE 7

Walking Disability

Rigidity	Tremor	Walking Disability	Speech	Psychological Depression
Score 0	Score 0	Score 0	Score 0	Score 0
No rigidity	No tremor	Normal walking	Normal speech	Normal affective state
Score 2	Score 2	Score 2	Score 2	Score 2
Perceptible only with special maneuvers	Tremor only perceptible when the patient has an object in hand	Slow walking when the patient is tired	Little difficulty in speaking; speech well understandable	Moderate sadness without crying and difficulties in his life in relation to his environment
Score 4	Score 4	Score 4	Score 4	Score 4
Perceptible by inspection; easily modified	Tremor not constant; perceptible at simple inspection	Continuous slow walking	Serious difficulties in speaking; the patient can be understood when close attention is paid to his speech	Sadness more marked; sometimes the patient is crying but he has no difficulties in his life in relation to his environment
Score 6	Score 6	Score 6	Score 6	Score 6
Marked rigidity; only slightly modifiable	Moderate and constant tremor	Typical Parkinsonian march	Very serious difficulties in speaking; only part of the speech can be understood	Sadness accentuated; sometimes the patient is crying and he has difficulties in his life in relation to his environment
Score 8	Score 8	Score 8	Score 8	Score 8
Very marked rigidity; only slightly modifiable	Marked and constant tremor	The patient can walk only with help	Very serious difficulties in speaking; the patient cannot be understood	Very great sadness; the patient cries very much; he renounces his life in relation to his environment
Score 10	Score 10	Score 10	Score 10	Score 10
Very marked rigidity; not modifiable	Tremor forbidding common activities	The patient cannot walk	The patient cannot speak	Very great sadness; the patient cries very much, he renounces his life in rela-
Score 10	Score 10	Score 10	Score 10	Score 10
Very marked rigidity; not modifiable	Tremor forbidding common activities	The patient cannot walk	The patient cannot speak	Very great sadness; the patient cries very much, he renounces his life in relation to his environment; he has suicidal ideas; any of the states with suicidal ideas must be scored 10

TABLE 8

Effects of D-Phenylalanine in Patients with Parkinson's Disease[a]

Patient	Rigidity		Tremor		Walking Disability		Speech		Psychological Depression	
	B	A	B	A	B	A	B	A	B	A
1	6	2	6	4	2	0	2	0	4	0
2	8	4	8	6	6	4	6	2	6	2
3	2	0	4	4	2	0	0	0	2	0
4	6	4	6	4	2	0	2	0	2	0
5	4	2	6	4	4	0	4	2	2	0
6	6	2	6	4	4	2	4	2	6	0
7	10	6	6	4	10	6	8	4	6	2
8	8	6	4	4	8	6	6	4	2	0
9	6	4	6	4	4	2	2	0	4	0
10	4	2	6	4	2	0	0	0	2	0
11	4	2	4	2	2	0	2	2	8	4
12	8	6	6	4	8	6	4	2	4	0
13	2	0	4	2	0	0	0	0	0	0
14	6	2	6	2	0	0	2	0	4	2
15	4	2	4	2	2	0	2	0	2	0

[a]B, Before administration of the D-phenylalanine; A, after administration of the D-phenylalanine.

Depression. This symptom was present in 14 patients. The results obtained were "Excellent" in 10 cases; "Very Good" in two; and "Good" in another two patients. The Student's t test gave a $P < 0.001$.

The test of the variance and covariance applied demonstrated that D-phenylalanine had a highly significant therapeutic activity in Parkinson's disease. No side-effects or toxic action were observed with this drug. A second trial is now being conducted with 25 patients suffering from Parkinson's disease, using a daily dose of 600 mg of D-phenylalanine and comparing its effects in a corssing trial with the pharmacological actions of L-DOPA at a daily dose of 4 g. From these results thus far obtained, no significant differences were observed between the anti-Parkinsonian effects of L-DOPA and D-phenylalanine. The only important variation between these two drugs was the fact that D-phenylalanine showed a considerably stronger antidepressant activity than L-DOPA. We can thus conclude that D-phenylalanine may be considered a very useful drug in the treatment of Parkinson's syndrome.

REFERENCES

1. W. B. Cannon (1915). *Bodily Changes in Pain, Hunger, Fear and Rage.* D. Appleton & Co., New York.
2. B. B. Brodie and P. A. Shore (1957). A concept for a role of serotonin and norephinephrine as chemical mediators in the brain. *Ann. N.Y. Acad. Sci.*, 66:631.
3. W. R. Hess (1949). *Das Swischenhirn.* Schwabe, Bass.
4. M. Vogt (1954). The concentration of sympathin in different parts of the central nervous system under normal conditions and after the administration of drugs. *J. Physiol.*, 123:451.
5. B. M. Twarog, I. H. Page, and H. Bailey (1953). Serotonin content of some mammalian tissues and urine and a method for its determination. *Am. J. Physiol.*, 175:151.
6. A. H. Amin, B. B. Crawford, and J. H. Gaddum (1954). The distribution of substance P and 5-OH-tryptamine in the central nervous system of the dog. *J. Physiol.*, 126:596.
7. H. Weil-Walherbe and A. D. Bone (1957). Intracellular distribution of catecholamines in brain. *Nature,* 180:1050.

8. A. B. Rothballer (1959). The effects of catecholamines on the central nervous system. *Pharmacol. Rev.*, 11:494.

9. T. Nakajima, Y. Kakimoto, and I. Sano (19). Formation of B-phenylethylamine in mammalian tissue and its effects on motor activity in the mouse.

10. A. D. Mosnaim and H. C. Sabelli (1971). Quantitative determination of the brain levels of B-phenylethylamine like substance in control and drug treated mice. *Pharmacologist*, 13:283.

11. E. E. Inwang, A. D. Mosnaim, and H. C. Sabelli (1973). Isolation and characterization of phenethylamine and phenethanolamine from human brain. *J. Neurochem.*, 20:1469.

12. A. A. Boulton and J. R. Majer (1971). Mass-spectrometric identification of the isomers of tyramine; identification of p-tyramine in rat brain. *Can. J. Biochem.*, 49:993.

13. A. D. Mosnaim, E. E. Inwang (1973). A spectrometric method for the quantification of 2-phenylethylamine in biological specimens. *Anal. Biochem.*, 54:561.

14. E. Fischer, R. J. Ludmer, and H. C. Sabelli (1967). The antagonism of phenylethylamine to catecholamine on mouse motor activity. *Acta Physiol. Lat. Am.*, 17:15.

15. E. Fischer and B Heller (1967). Pharmacology of the mechanism of certain effect of reserpine in the rat. *Nature*, 216:1221.

16. J. M. Saavedra, B. Heller, and E. Fischer (1970). Antagonistic effects of tryptamine and P-Phenylethylamine on the behavior of rodents. *Nature*, 226:868.

17. P. Mantegazza and M. Riva (1963). Amphetamine like activity of B-phenylethylamine after a MAO inhibitor in vivo. *J. Pharm. Pharmacol.*, 15:472.

18. B. J. Key and E. Marley (1961). The effect of some sympathomimetic amines on electrocortical activity and behavior of young and adult animals. *J. Physiol.*, 155:39P.

19. B. J. Key and E. Marley (1962). The effect of the sympathomimetic amines on behavior and electrocortical activity of the chicken. *Electroenceph. Clin. Neurophys.*, 14:90.

20. E. Marley (1963). Brain stem receptors for sympathomimetic amines in the chicken. *J. Physiol.*, 165:24P.

21. E. Fischer, B. Heller, H. Spatz, and H. Reggiani (1972). Thin-layer chromatographic assay of phenethylamine content of the rat brain and its changes after reserpine and imipramine administration, *Arzneim.-Forsch.*, 22:1560.

22. E. Fischer, H. Spatz, B. Heller, and H. Reggiani (1972). Phenethylamine content of human urine and rat brain, its alterations in pathological conditions and after drug administration. *Experientia*, 28:307.

23. J. M. Saavedra (1974). Enzymatic isotopic assay for and presence of B-phenylethylamine in brain. *J. Neurochem.*, 22:211.

24. J. M. Saavedra and J. Axelrod (1973). Demonstration and distribution of phenylethanolamine in brain and other tissues. *Proc. Nat. Acad. Sci., USA*, 70:769.

25. D. A. Durden, S. R. Philips, and A. A. Boulton (1973). Identification and distribution of B-phenylethylamine in the rat. *Can. J. Biochem.*, 51:995.

26. A. D. Mosnaim, E. E. Inwang, and H. C. Sabelli (1974). The influence of psychotropic drugs on the levels of endogenous 2-phenylethylamine in rabbit brain. *Biol. Psychiat.*, 8:227.

27. J. B. Jepson, W. Lovenberg, and P. Zaltzman (1960). Amine metabolism studied in normal and phenylketonuric humans by monoamine oxidase inhibition. *Biochem. J.*, 74:5.

28. E. Fischer, B. Heller, and A. H. Miró (1968). B-phenylethylamine in human urine. *Arzneim.-Forsch.*, 18:1486.

29. A. A. Boulton and L. Milward (1971). Separation, determination and quantificative analysis of urinary B-phenylethylamine. *J. Chromatogr.*, 57:287.

30. A. D. Mosnaim, E. E. Inwang, J. H. Sugerman and H. C. Sabelli (1973). Identification of 2-phenylethylamine in human urine by infrared and mass spectroscopy and its quantification in normal subjects and cardiovascular patients. *Clin. Chim. Acta*, 46:40.

31. E. Fischer, H. Spatz, J. M. Saavedra, H. Reggiani, A. H. Miró, and B. Heller (1972). Urinary elimination of phenethylamine. *J. Biol. Psychiat.*, 5:139.

32. H. Spatz, B. Heller, M. Nachón, and E. Fischer (1975). Effects of D-phenylalanine on clinical picture and phenethylaminuria in depression. *J. Biol. Psychiat.*, 10:235.

33. E. Fischer and B. Heller (1972). Phenethylamine as a neurohumoral agent in brain. *Behav. Neuropsychiat.*, 4:8.

34. E. Fischer and B. Heller (1972). Fenetilamina como agente neurohumoral, sus efectos psicofisiológicos, su papel en enfermedades mentales y en el mecanismo de acción de drogas. *Minerva Psiquiátrica Argentina*, 1:18.

35. E. Fischer (1975). The phenethylamine hypothesis of thymic homeostasis. *J. Biol. Psychiatry*, 10:667.

36. H. C. Sabelli and A. D. Mosnaim (1974). Phenylethylamine hypothesis of affective behavior. *Am. J. Psychiat.*, 131:695.

37. H. C. Sabelli, A. D. Mosnaim, and A. J. Vazquez (1973). Phenethylamine: Possible role in Depression and antidepressive drug action. In: *Neurohumoral Coding of Brain Function*, edited by R. R. Drucker-Colin and R. D. Myers. Plenum, New York.

38. J. A. Roth and C. B. Gillis (19). Determination of B-phenylethylamine by monoaminooxidase, inhibition by imipramine. In press.

39. M. M. Fernández Pardal, C. Lopez, J. Fernández Pardal, and E. Fischer (1973). Parkinson experimental y su tratamiento con fenetilamina y fenilalanina. *Annales de Psiquiatría Biológica,* 2:103.

40. E. Fischer and B. Heller (1974). Methylamphetamine and phenethylamine. *Arzneim.-Forsch.,* 24:956.

41. H. Spatz and B. Heller (19). Effects of D- and L-phenylalanine in phenethylamine brain content in the rat. In preparation.

42. B. Heller (1970). Farmacología del Parkinson Experimental, Proceedings of the "1er. 'Simposio Sudamericano sobre el estado actual del tratamiento en la Enfermedad de Parkinson y Parkinsonismo, Buenos Aires.

43. B. Heller and E. Fischer (1973). Diminution of phenethylamine in the urine of Parkinson's patients. *Arzneim.-Forsch.,* 23:884.

44. B. Heller, E. Fischer, and G. Poch (1971). B-Feniletilamina en orina de enfermos parkinsonianos. In: *Parkinsonismos y l-dopa,* edited by Yahr and Poch. Buenos Aires.

45. H. C. Sabelli and J. Vazquez (1974). Comunicacíon en el Congreso Mundial de Psiquiatría Biológica, Buenos Aires.

Chapter 19

POSSIBLE ROLE OF 2-PHENYLETHYLAMINE IN THE PATHOLOGY OF THE CENTRAL NERVOUS SYSTEM

Yen Hoong Loo and George A. Jervis

New York State Institute for Basic Research in Mental Retardation
Staten Island, New York

Marjorie G. Horning

Institute for Lipid Research
Baylor College of Medicine
Houston, Texas

I. INTRODUCTION

Among the numerous genetic disorders of amino acid metabolism, as homocystinuria, maple-syrup-urine disease, hyperprolinemia, cystathionuria, arginosuccinuria, α-ketoadipic aciduria, phenylketonuria has attracted more attention because it very specifically affects CNS function without apparent disturbance of other organs as the liver and kidney. In untreated classical phenylketonuria, the brain usually fails to develop and to reach normal maturation. Because of a deficiency of the enzyme phenylalanine hydroxylase in the liver (EC 1.99.1.2), the major normal pathway of phenylalanine metabolism, the conversion of phenylalanine to tyrosine, is blocked (27). Consequently, tissue levels of phenylalanine are elevated to 20 to 40 times that of normal, bringing alternate routes of metabolism into play, decarboxylation to 2-phenylethylamine (48,52) and transamination to phenylpyruvic acid (28). The ensuing sequence of events culminating in severe mental retardation is not clear. A concensus is being reached that one or more of the abnormal metabolites of phenylalanine directly or indirectly alters brain development and brain function.

In order to determine the pattern of phenylalanine metabolites in the brain and to follow the sequence of biochemical events, we have developed an animal model for our investigations (37). A chemical simulation of the metabolic error is induced by the injection of the phenylalanine hydroxylase inhibitor, p-chlorophenylalanine (22,30), along with L-phenylalanine (32) into the preweanling rat during the developmental phase of the brain. The following biochemical manifestations of clinical phenylketonuria are satisfactorily reproduced in our animal model: a) deficient phenylalanine hydroxylase activity (27), b) high phenylalanine with low or normal tyrosine concentrations in tissues (29), c) depressed serotonin (5-HT) levels (49), d) low brain weight (41), and e) a deficiency of myelin (43). An irreversible mental deficit usually observed in classical phenylketonuria (25,26) has also been demonstrated in the experimental model (2.3).

The greatly diminished stores of 5-HT, a neurotransmitter substance, and the deficit of myelin indicate that defective neuronal development may be the underlying cause of the mental dysfunction in phenylketonuria. In this discussion, an attempt is made to assess the role of 2-phenylethylamine, a major metabolite of phenylalanine, in the pathogenesis of phenylketonuria.

II. PRODUCTION AND DISPOSITION OF 2-PHENYLETHYLAMINE IN PHENYLKETONURIA

The enzyme aromatic L-amino acid decarboxylase (EC 4.1.1.26) occurs in mammalian liver, kidney, brain, heart, and the adrenal glands, with the bulk of the total activity in the liver and kidney. The author (YHL) could not detect any significant amount of this enzyme in blood. The aromatic L-amino acids, DOPA, 5-hydroxytryptophan, tryptophan, tyrosine, and phenylalanine all serve as substrates. The K_M for L-phenylalanine is in the order of 10^{-2} M, and normal plasma concentration of this amino acid is approximately 10^{-4} M. When tissue levels of phenylalanine become elevated as in classical phenylketonuria, decarboxylation of the amino acid to the amine may be the major metabolic pathway. Phenylethylamine is mainly converted to phenylacetic acid (57) by MAO (EC 1.4.3.4) and to a much lesser extent to phenylethanolamine (44,55) by dopamine (DA) hydroxylase (EC 1.14.2.1). Mandelic acid, formed from phenylethanolamine by MAO (56), has been identified in urine of phenylketonuric patients (5). Both 2-phenylethylamine and phenylethanolamine have been unequivocally identified in normal human brain (24) and urine (45). A new metabolic pathway of phenylethylamine (PEA) in brain has been discovered (57), namely, the reversible transformation to p-tyramine, thus linking PEA to the catecholamines. Phenylethylamine is known to be readily and rapidly transported into the brain (17,46) and appears to be the major precursor of phenylacetic acid (16). Thus, the extent of the decarboxylation pathway of phenylalanine metabolism may be approximated by either inhibiting MAO and estimating the amount of PEA which accumulates or by measuring the amount of phenylacetic acid formed.

A. Classical Phenylketonuria

Oates et al. (48) administered monoamine oxidase inhibitors (MAOI) to normal and phenylketonuric subjects and found excessive amounts of PEA in the urine, approximately 50 to 200 times that of normal.

We compared the concentrations of the metabolites derived from phenylalanine in plasma and urine of nonphenylketonuric retarded (control group) and classical phenylketonuric patients residing in the same cottage. Results recorded in Table 1 show that plasma levels of PEA in phenylketonuric patients are significantly elevated above the control. Among the phenylketonuric patients selected to represent varying degrees of mental dysfunction, urinary PEA concentrations are 1.5 to 3.5 times greater than that of the control in the mildly retarded group and 5.5 to 11 times greater in the moderately-severely retarded patients. Rapid metabolism of the amine was reflected in the large amounts of phenylacetic acid excreted in the urine (Table 2).

We found three forms of phenylacetic acid in the urine of our subjects: 1) free acid which could be directly extracted into ether or ethylacetate from the acidified urine at room temperature, 2) a thioesterlike fraction from which the free acid is released at pH 12 to 13 at low temperature, and 3) phenylacetylglutamine (67) which is hydrolyzed by 5 N NaOH at 100°C for 3 hr (65). Wide individual variations were observed in the total amount of phenylacetic acid excreted. However, the less retarded did tend to excrete relatively larger amounts of phenylacetylglutamine. No significant differences in the concentrations of the other aromatic acid metabolites could be discerned among the patients examined.

B. Experimental Phenylketonuria

Metabolites in plasma and urine represent the overall reactions of all the tissues and do not necessarily reflect the course of events in the brain. To gain this information, we turned to our animal model. In the normal rat brain, PEA was found to be compartmentalized

TABLE 1

Tissue Levels of 2-Phenylethylamine[a]

Group	Human Subjects: Concentration of PEA in Plasma (pmol/ml)	Human Subjects: Concentration of PEA in Urine (pmol/mg creatinine)	Weanling Rats: Nerve Endings P_2 Group	Weanling Rats: pmol/g cortex
Controls (5)	3-13		Controls (3)	11-14
PKU (6)	15-18		PKU Model:	
Controls (4)		6-15	24 hr after Ph (3)	38-50
Untreated PKU:			48 hr after Ph (3)	38-50
a. Mild dysfunction (3), IQ 50-70		27-56		Phenylalanine in whole brains µmol/g
			PKU Model:	
b. Moderate dysfunction (4), IQ 30-50		87-172	24 hr after Ph (7)	0.28-0.40
Severe dysfunction (4), IQ 10-30		111-300	Normal controls (8)	0.15-0.20

[a]Abbreviations: PKU, phenylketonuria; Ph, phenylalanine. The number of samples is enclosed in parentheses, and the range of values is recorded. Phenylethylamine concentration in normal urine has been reported to range from 0 to 16 pmol/mg creatinine (58). Edwards and Blau (16) found up to 100 pmol PEA/g tissue in the P_2 fraction isolated from whole brains of rats pretreated with pargyline and p-chlorophenylalanine and killed 1 hr after L-Ph loading.

TABLE 2

Aromatic Acids Derived from Phenylalanine in Urine of Phenylketonuric Patients and Brains of Phenylketonuric Rats[a]

Tissue and Group	Aromatic Acid Metabolites (μmol/mg creatinine)					
	Phenylacetic Unconj.	Phenylacetic Conj.	Mandelic	O-Hydroxy-phenylacetic	Phenyllactic	Phenylpyruvic
Urine from untreated PKU:						
a. Mild dysfunction (3), IQ 50-70	0.13-1.65	3.17-5.72 Av. 4.06	0.02-0.17	0.24-0.76	0.53-4.12	0.98-4.41
b. Moderate dysfunction (4), IQ 30-50	0.08-0.14	1.69-3.85 Av. 2.71	0.03-0.07	0.36-0.64	1.23-2.50	2.80-5.70
severe dysfunction (4), IQ 10-30	0.11-0.38	1.20-3.50 Av. 2.26	0.03-0.07	0.28-0.52	0.63-3.00	2.88-6.98
Non-PKU control (4)	0.005-0.014	0.33-1.35	0.010	0.007-0.012	0.008-0.021	0.11-0.17
	nmol/g wet tissue					
Brains from 12 to 14 rats, 6 hr after L-Ph; range and (percent of animals)	61-114 (42) 22-33 (33) 11-17 (25)	Not detected	5-7 (60) 2-4 (40)	Not detected	10-21 (57) 4-9 (43)	29-41 (29) 11-24 (57) 8-9 (14)

[a]The concentrations of unconjugated (unconj.) phenylacetic and of phenylpyruvic acids in urine of three normal subjects were both reported to be less than 0.03 μmol/mg creatinine (63). The concentration of conjugated (conj.) phenylacetic acid in urine from 10 normal subjects averaged 1.3 μmol/mg creatinine (64). The number of patients examined is indicated in parenthesis.

to an important extent in the nerve ending particles; none was detected in the extraterminal mitochondria (6,16). We, therefore, thought it important to determine whether PEA might be stored and to what extent in the presynaptic terminals of the brain in chronic hyperphenylalaninemia induced during the developmental phase (see Section VI).

The amount of PEA found in nerve endings isolated from the cortices of 21-day-old normal rats ranged from 11 to 14 pmol/g tissue. In contrast to whole brain, Saavedra (54), employing an enzymatic method, found 15 pmol/g, while Willner et al. (66), applying gas chromatography-mass spectrometry reported 18 pmol/g. The phenylketonuric animals stored excessive amounts of PEA in the nerve endings, approximately four times that of the normal (Table 1). It should be emphasized that although PEA is rapidly oxidized by MAO, an excessive amount of the amine remained in the nerve endings of the phenylketonuric animals as long as 24 and 48 hr after L-phenylalanine loading (Table 1). It is also noteworthy that the concentration of PEA in the nerve endings remained elevated although brain phenylalanine levels had dropped from 1.5 to 2.5 μmol/g at 6 hr to 0.3 to 0.4 μmol/g at 24 hr after the injection of L-phenylalanine (saline control values, 0.15-0.20 μmol/g). In view of the finding (14) that p-chlorophenylethylamine was formed in the brains of rats treated with pargyline and p-chlorophenylalanine (p-CPA, total of 450 mg/kg over 24 hr), we injected p-CPA alone into a group of animals, using the same dosage as that of the experimental group which received both p-CPA and L-phenylalanine (see Section VI). No p-chlorophenylethylamine was detected in the nerve endings of either group of animals (see Section VI), and the amount of PEA in the nerve endings of the p-CPA-treated group was similar to that of the saline control rats.

Analyses of brains removed from rats 6 hr after L-phenylalanine loading (see Section VI for details of the procedures employed) showed that phenylacetic acid was the major metabolite derived from phenylalanine in the phenylketonuric brain. Of especial importance is our

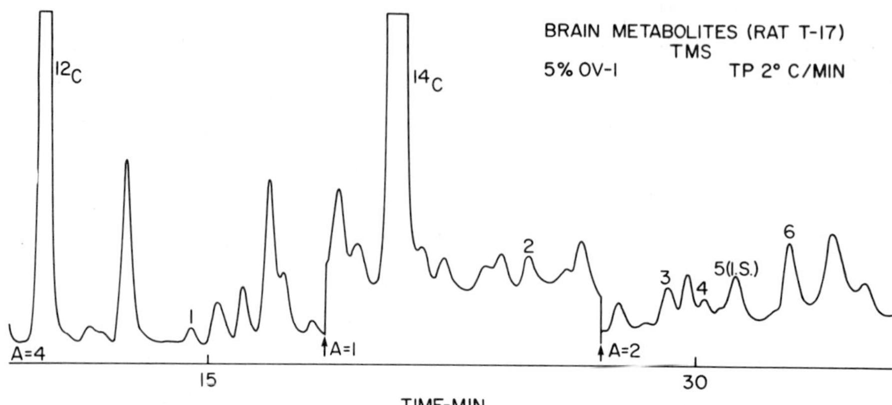

FIG. 1. Gas chromatogram of trimethylsilyl derivatives of aromatic acids derived from phenylalanine prepared from a purified extract of a brain of a phenylketonuric rat. Peaks: (1) phenylacetic, (2) mandelic, (3) O-hydroxyphenylacetic, internal standard, (4) phenyllactic, (5) phenylvaleric, internal standard, (6) phenylpyruvic-oxime. See section on methods for conditions of gas chromatography.

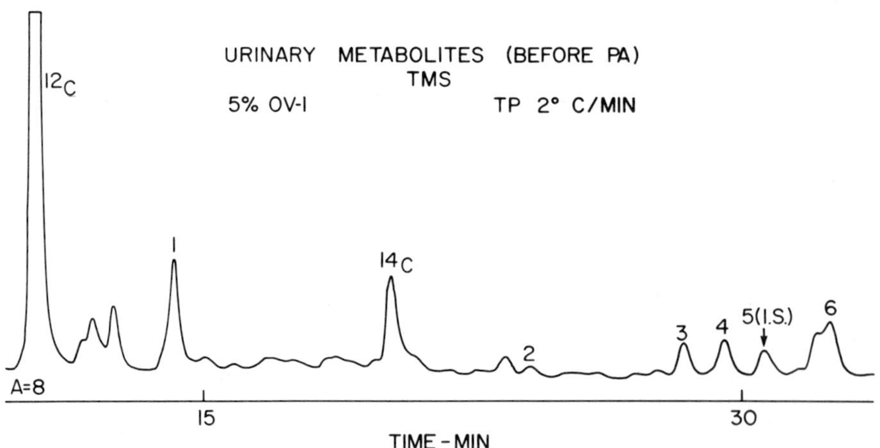

FIG. 2. Gas chromatogram of trimethylsilyl derivatives of aromatic acids derived from phenylalanine prepared from a purified extract of urine of a phenylketonuric patient. Peaks: (1) phenylacetic, (2) mandelic, (3) O-hydroxyphenylacetic, (4) phenyllactic, (5) phenylvaleric, internal standard, (6) phenylpyruvic-oxime.

finding that phenylacetic acid is present in the brain of the phenylketonuric rat in a bound form, stable in acid but labile in alkali. Extraction of the tissue with aqueous acid and organic solvents as employed by many investigators (15) removed approximately 30 to 40% of the total amount of phenylacetic acid present in the brain. Under mild conditions, at pH 12 to 13, 60 to 65°C for 30 min, phenylacetic acid was quantitatively released from the brain.

None of the aromatic acid metabolites derived from phenylalanine could be detected in the brains of normal 15- to 20-day-old rats.

Although wide variations among individual animals were found, the major metabolite, in 80% of the phenylketonuric animals, was phenylacetic acid, while phenylpyruvic acid predominated in the remaining 20%. As much as 100 nmol of phenylacetate/g wet tissue was found in the brains of some of the phenylketonuric animals. In contrast, the highest level of phenylpyruvate was 40 nmol; phenyllactate, 20 nmol; and mandelate, 7 nmol, while o-hydroxyphenylacetate was not detected (limit of detection, 0.5 nmol/g). These results are recorded in Table 2.

As indicated earlier, the bound form of phenylacetate which behaves as a thio ester and the free acid were also found in the urine of phenylketonuric patients. Typical gas chromatograms of a rat brain preparation and of a urinary extract are shown in Figs. 1 and 2 respectively. Analyses were verified by mass spectroscopy.

III. SEROTONIN SYNTHESIS IN PHENYLKETONURIA

A deficit of 5-HT in clinical phenylketonuria has been consistently observed. However, there is no general agreement on the cause of the diminished stores of 5-HT. Pare et al. (50) reported a deficiency of 5-hydroxytryptophan (5-HTP) decarboxylase (EC 4.1.1.28) activity in phenylketonuric patients. Neuhoff et al. (47) identified 5-HTP in the cerebral spinal fluid of phenylketonuric children, the amount being directly related to tissue levels of phenylalanine. Inhibition of the decarboxylase would account for the accumulation of the substrate, 5-HTP, in the brain, and hence for its appearance in the cerebral spinal fluid. In accord with these observations on patients, we found in our animal model, that the greatly diminished synthesis of 5-HT from its precursor, 5-HTP, could largely be accounted for by the decrease in activity of the enzyme, 5-HTP decarboxylase (35).

A. Inhibition of 5-Hydroxytryptophan Decarboxylase by Pyridoxylidene-2-phenylethylamine

Phenylethylamine condenses readily with pyridoxal at physiological pH to form an aldimine Schiff base, pyridoxylidene-2-phenylethylamine (36). Phenylalanine and pyridoxol were found to be precursors for its formation in rat brain (34). The synthetic aldimine was shown to be a very potent inhibitor (K_1, 0.6 μM) of pyridoxal kinase in vitro (38). However, this inhibitory effect was not observed in the phenylketonuric animal, in which no deficiency in the activity of pyridoxal kinase or of the B_6-co-enzymes was found. Instead, a slower turnover of pyridoxal phosphate was demonstrated in the experimental model, suggesting that certain enzymic reactions utilizing pyridoxal phosphate as co-factor might be inhibited (37). The effect of the synthetic Schiff base on the enzyme, 5-HTP decarboxylase was, therefore, investigated.

A partially purified preparation of the enzyme from rat brain (23) with a specific activity of 30 nmol 5-HT formed/hr/mg protein was used. Employing conditions for maximal velocity of the enzyme; 0.13 mM pyridoxal phosphate, 0.13 mM 5-HTP, 0.14 mg enzyme protein, 20 min incubation at 37°C, we observed 50% inhibition of the decarboxylase with pyridoxylidene-2-phenylethylamine at a concentration of 0.4 mM. Under similar conditions, PEA exhibited no effect on the decarboxylase activity. No inhibition of DOPA or glutamic acid decarboxylation was observed even with Schiff base concentrations of 3.5 mM.

The velocity of the enzymic reaction was measured with a fixed concentration of co-factor and different concentrations of the inhibitor and substrate as well as with a constant concentration of substrate and varying concentrations of inhibitor and co-factor. For the preparation of double reciprocal plots of the velocity of the enzyme reaction against substrate (Fig. 3) and co-enzyme concentrations (Fig. 4) the method of Roberts and Simonsen (53) was found very convenient. According to their procedure, velocities are expressed as the ratios of the observed velocity in the presence of the inhibitor (v) to the velocity in the control sample containing concentrations of substrate and co-factor that yield maximal velocity under the conditions of the assay (V). The uninhibited reaction plots of the reciprocals of these ratios (V/v) versus the reciprocals of the molar substrate concentrations (1/S) should give straight lines with an intercept of one. The presence of a competitive inhibitor should only increase the slope. The presence of a noncompetitive inhibitor should give a straight line with the intercept above one. The plots shown in Figs. 3 and 4 indicate a noncompetitive type of inhibition by the Schiff base formed from PEA and pyridoxal.

Following this observation, we decided to determine whether pyridoxylidene-β-PEA may be bound to the enzyme, aromatic amino acid decarboxylase. Our attempts to release intact Schiff base from the enzyme were unsuccessful. Neither could we detect any PEA in the very highly purified enzyme. However, we did find PEA in the 10-fold purified enzyme prepared from whole brains of 15-day-old phenylketonuric rats killed 24 hr after the last injection of phenylalanine. In three different preparations, we found 4.5 to 5.6 pmol PEA in enzyme derived from 1 g of tissue. No PEA was detected in enzyme preparations from control animals. As much as 96 pmol of PEA in enzyme derived from 1 g tissue was found in brains from 21-day-old rats injected s.c. three times with PEA (100 mg/kg) at 20 min intervals and killed 20 min after the last injection.

IV. DEFICIENCY OF MYELIN IN PHENYLKETONURIA

A significant reduction in the amount of myelin in the brain of phenylketonuric patients (59) as well as in hyperphenylalaninemic rats (60) has been reported. We have similarly observed a deficit

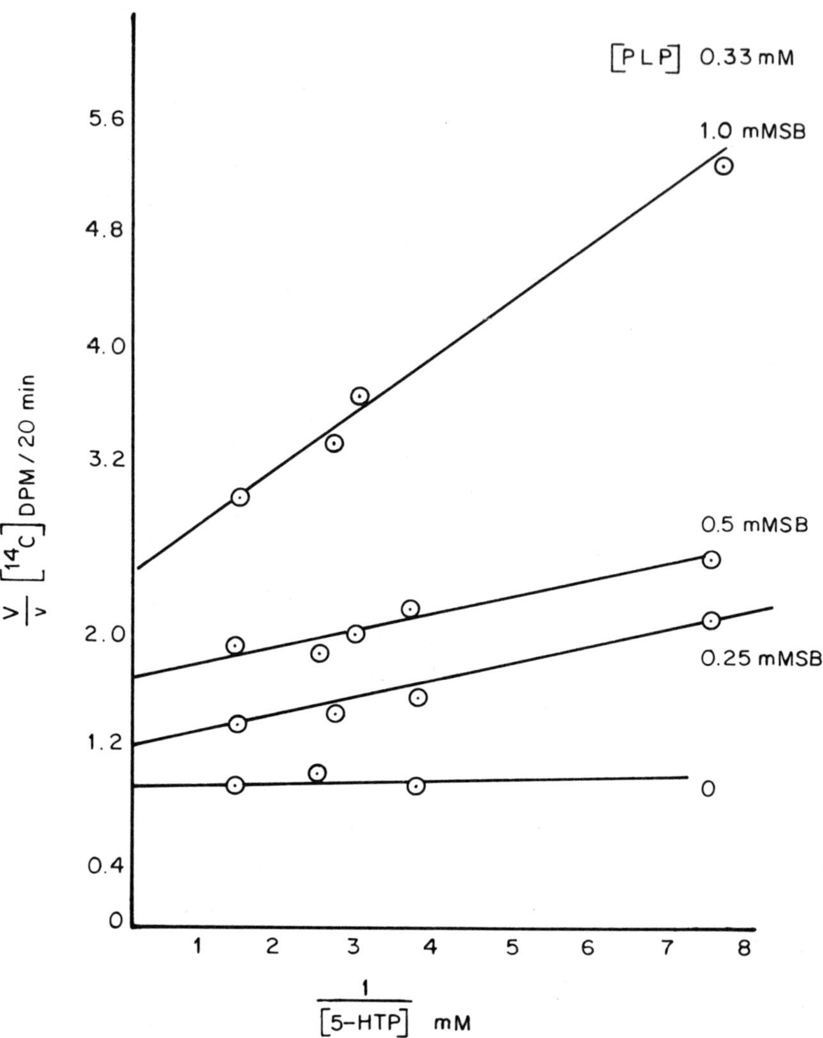

FIG. 3. Inhibition of 5-HTP decarboxylase activity by SB. Reciprocal plots of the velocity of the reaction for three concentrations of the Schiff base at various substrate (5-HTP) concentrations.

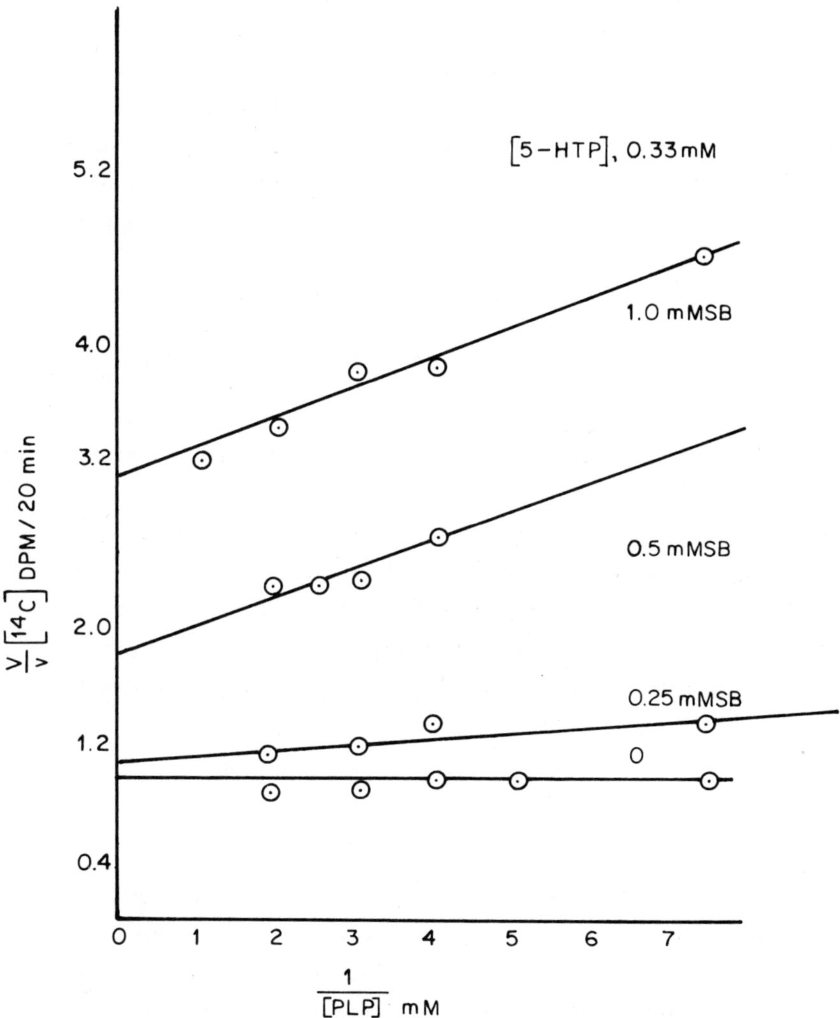

FIG. 4. Reciprocal plots of the velocity of the reaction for three concentrations of the Schiff base at various concentrations of the co-factor PLP.

of myelin in the cerebral hemispheres and the cerebellum of our experimental animals. The biochemical mechanism producing this deficit has not been defined. Based on the in vitro inhibition of enzymes catalyzing reactions concerned with the synthesis of fatty acids and cholesterol and the generation of energy for cellular processes, phenylpyruvic acid has been postulated to be the major metabolite contributing to the deficiency of myelin. Gaull et al. (19) have thoroughly reviewed the literature bearing on the pathogenesis of the brain dysfunction in phenylketonuria.

Present evidence from our experiments indicates that phenylacetic rather than phenylpyruvic acid is more likely to be the toxic metabolite in the brain. The level of phenylacetate in the brains of most of the phenylketonuric rats is much higher than that of phenylpyruvate or phenyllactate (Table 2). Of particular significance is our finding that as much as 70% of it is in a bound form, stable to acid but labile to alkali under mild conditions, behaving as a thio ester. Since phenylacetate does serve as a substrate for the fatty acid activating enzyme (40), it is very probable that it is present in the brain of phenylketonuric animals as the ester of co-enzyme A. In this form it may well interfere with the synthesis of acyl CoA or its utilization for sources of energy and lipid synthesis in the developing brain during myelination.

V. DISCUSSION

The results of our experiments suggest that 2-phenylethylamine and its major metabolic product, phenylacetic acid, contribute to the brain dysfunction characteristic of untreated classical phenylketonuria. The moderately to severely retarded patients which we examined excreted much greater amounts of PEA than the mildly retarded. In the phenylketonuric rat brain, PEA was stored in excess in the cortex nerve endings and was bound to protein fractions with 5-hydroxytryptophan decarboxylase activity, most likely as the Schiff base, pyridoxylidene-2-phenylethylamine.

Among the aromatic acid metabolites of phenylalanine, phenylacetic acid was found in the highest concentration in the phenylketonuric rat brain. In the preweanling rat, phenylacetic acid readily crosses the blood-brain barrier (YHL, unpublished results). The accumulation of this metabolite in the immature brain represents that formed in situ as well as that transported from the peripheral tissues into the brain. In contrast, in the adult cat, phenylacetic acid does not penetrate across the blood-cerebrospinal fluid barrier (51). Because of this difference between the young and the adult animal, excessive production of phenylacetic acid may be noxious to the developing but not to the mature brain. Conversion of this metabolite to phenylacetylglutamine by the liver should reduce its toxicity. Indeed, we did observe that the less retarded patients tend to excrete more conjugated phenylacetic acid (Table 2).

The accumulation of PEA in the brain could well impair the development of a normal neuronal network, and the accumulation of phenylacetic acid could interfere with myelination in the developing brain.

A. Phenylethylamine: A False or Substitute Transmitter?

Mental activity is mediated through an organized system of neuronal transmission, which in turn is dependent on the integrity of synaptic and axonal structures, nerve terminals, and the myelin sheath. For the most part, neuronal development in the rat brain occurs postnatally. At birth, cell bodies of monoaminergic neurons are complete in number, but terminals are very sparsely distributed (33). During the first 10 days after birth, neurons grow rapidly, developing axons and dendrites. Synaptic junctions, still few in number, begin to rise steeply from the 11th to 24th postnatal day (1,9). Terminals continue to proliferate throughout the brain until the fourth to fifth postnatal week, when the adult pattern in all areas of the brain is attained (7).

The increase in levels of monoamines in the rat brain is a consequence of progressive proliferation of nerve terminals, where biosynthesis, storage, release, and metabolism of amines takes place (20). The low levels of biogenic amines in the immature brain may be attributed to fewer storage vesicles (33). The processes for synthesis, storage, and release of a transmitter, however, are not entirely specific. A number of investigators (31) have demonstrated that monoamine precursors, when present in high concentrations in the brain, are capable of entering and being decarboxylated in monoaminergic neurons in which they are apparently normally not detected. Amines, other than the physiological putative transmitter, that accumulate in nerves and displace the normal amine are called false transmitters. For example, the pharmacological effects of α-methylDOPA may be attributed to its metabolic products, α-methyldopamine and α-methylnorepinephrine, formed in tissues and released by nerve stimulation. After L-DOPA loading, DA was found in serotoninergic cells and played the role as a false serotonergic transmitter. The injection of 6-hydroxydopamine, which may act as a false transmitter, into rats during the first postnatal week produced a permanent deficit of norepinephrine (61).

In a similar manner, PEA may act as a false or substitute transmitter preventing the normal development of neuronal pathways. Indeed, phenylketonuric patients have been found deficient in DA, norepinephrine (NE), and 5-HT (10,42). In animals, PEA has been reported to deplete 5-HT (35), DA, and NE stores (13,18). In the animal model nerve terminals remained deficient in 5-HT although this amine was synthesized after 5-HTP loading (35). In the phenylketonuric patient, after phenylalanine loading, 5-HT was found in the cerebralspinal fluid, where it is normally not detected (4). The production of excessive amounts of PEA probably caused the release of 5-HT from storage sites. Decreased monoamine storage capacity appears to be more critical in phenylketonuria than decreased synthesis of the biogenic monoamine.

Because newly synthesized transmitter is selectively released, the availability of the false transmitter precursor may be decisive in determining the extent of interference with normal function (31). Thus, in phenylketonuria, the persistently high phenylalanine concentrations in the tissues would result in the continuous production of PEA, which not only is formed in the brain but is also readily transported from peripheral tissues into the brain where it accumulates in nerve terminals. During the phase of rapid development, excessive stores of PEA could impair neuronal maturation and produce a permanent mental deficit.

B. Phenylacetic Acid, Cause of Myelin Deficit in Phenylketonuria: A Hypothesis

The excretion of abnormally large amounts of phenylpyruvic, phenyllactic, and o-hydroxyphenylacetic acids but normal amounts of phenylacetate by two sisters with normal mental development (64) suggests that phenylpyruvic probably does not significantly contribute to the mental aberration in classical phenylketonuria. Similarly, we could not correlate the amounts of phenylpyruvic, phenyllactic, and o-hydroxyphenylacetic acids excreted to the degree of mental dysfunction exhibited by the patients we examined.

The accumulation of phenylacetic acid in a bound form, very likely as an ester of co-enzyme A, in the brain may very well inhibit myelin synthesis. Any postulated mechanism must explain why lipid synthesis is inhibited in the brain but not in the liver.

The conjugation of phenylacetate with glutamine in the liver would serve to detoxify the acid. In contrast, the brain lacks the enzyme for carrying out this reaction. Another difference in the metabolism of these two tissues was discovered by Edmond (11) who found that acetoacetate and 3-hydroxybutyrate are incorporated preferentially into fatty acids and sterols by the brain, spinal cord, and skin of normal developing rats and only minimally in the liver fatty acids and sterols. The brain is rich in the enzyme catalyzing the synthesis of acetoacetyl CoA, 3-oxoacid : succinyl CoA transferase (EC 2.8.35), which is lacking in the liver (12). Inhibition of the synthesis of acetoacetyl CoA and 3-hydroxybutyryl CoA or their utilization in the developing brain by phenylacetic acid or S-phenylacetyl CoA would seriously impair energy production, lipid synthesis, and consequently myelination, without affecting the liver. Ketone bodies are not utilized by the liver but play a major role in the brain during myelination not only as sources of energy but also of carbon for lipid biosynthesis.

The accumulation of 2-phenylethylamine in nerve terminals during postnatal development of the brain could impair the development of serotonergic, dopaminergic, and nonadrenergic pathways, while its major metabolic product, phenylacetic acid, could interfere with myelination as well as the production of energy for cellular processes, thus culminating in the mental dysfunction, characteristic of phenylketonuria.

VI. METHODS

A. Human Experiments

Mentally retarded adult male and female patients diagnosed as classical phenylketonurics at the Letchworth State Hospital cooperated in this investigation. Age and sex matched non-phenylketonuric patients who were housed in the same hospital unit and consumed the same diet, served as controls. Blood was collected into tubes containing the MAOI pargyline (50 μg/ml). Plasma was prepared and stored frozen for 2 to 3 days before analysis. Random specimens of urine were collected and kept frozen for 2 to 3 days before analysis.

B. Animal Experiments

1. Induction of Phenylketonuria

To obtain an animal model of phenylketonuria, a regimen of subcutaneous injections of a uniform suspension of p-CPA and L-phenylalanine into suckling rats, beginning at 4 to 5 days of age, was adopted (35). An initial injection of p-CPA (175 mg/kg) was followed by a daily injection of a mixture of p-CPA (90 mg/kg) and L-phenylalanine (360 mg/kg). Phenylalanine hydroxylase was effectively blocked and tissue levels of phenylalanine were elevated to those reported for phenylketonuric patients. Tyrosine levels remained normal. Animals were allowed to nurse with normal mothers, maintained on a Purina chow diet.

2. Phenylalanine Loading Experiment

For the determination of metabolites in the brain, animals were injected with p-CPA only (175 mg/kg) on the 14th day. On the following day, a solution of L-phenylalanine was injected intraperitoneally (360-400 mg/kg). Six hours later, animals were killed by decapitation, and whole brains were dissected out, blotted on moist filter paper to remove blood, weighed, and immediately frozen. Frozen tissues were stored at -60°C till analyzed.

3. Preparation of Nerve Ending Particles

For the preparation of subcellular fractions, animals were sacrificed at 21 days of age. Animals were injected with p-CPA (175 mg/kg) on the fifth postnatal day followed by daily doses of a mixture of p-CPA (90 mg/kg) and L-phenylalanine (360 mg/kg) until the 19th day, when they were divided into two groups. On the 19th and 20th days a group was injected with p-CPA only (90 mg/kg) and killed on the 21st day, 48 hr after the last injection of phenylalanine. Another group received p-CPA only (90 mg/kg) on the 20th day and was killed on the 21st day, 24 hr after phenylananine administration. A third group of rats received only p-CPA at the same dosage from postnatal day 5 through 20.

The cerebral hemispheres were removed from the decapitated animals, and myelin was carefully scraped off. Four cortices were pooled for homogenization in 0.32 M sucrose containing pargyline, 50 μg/ml. The crude mitochondrial fraction, P_2, composed of nerve endings and free mitochondria was separated by the standard procedure of Gray and Whittaker (21). This particulate fraction was washed once with 0.4 M sucrose containing pargyline (20 μg/ml) and was sedimented to a firm pellet, which was then stored at -70°C until analyzed.

4. Preparation of 5-Hydroxytryptophan Decarboxylase

Normal 21 day old animals were decapitated. Whole brains were removed and homogenized in 0.2 M Tris-HCl buffer at pH 8.5 containing dithiothreitol (1 μmol/ml). The enzyme was isolated from the high speed supernatant extract and purified 10-fold by repeated fractionation with ammonium sulfate according to the procedure of Ichiyama et al. (23). The precipitate of the enzyme was stored at -70°C and used within 3 to 4 days.

C. Analytical Procedures

1. Estimation of Phenylethylamine

The intensity of the fluorescence emitted by the reaction product of phenylethylamine with fluaram (purchased from Hoffman-La Roche) is used as a quantitative measure of the concentration of the amine (62). The procedure described by Mosnaim et al. (45) was followed for the extraction of the amine from enzyme protein, plasma, urine, and a suspension of P_2 prepared by sonification in saline containing pargyline (20 μg/ml). To the solid residue of PEA-HCl collected from the second acidified hexane extract was added a minimal volume (75-100 μl) of IN NaOH, and the free amine was extracted into 10 volumes of ethylacetate. Excess fluaram dissolved in acetonitrile was added, and after 5 min the ethylacetate solution was concentrated under N_2 to approximately 0.2 to 0.3 ml over a period of approximately 10 min. Triethanolamine dissolved in acetonitrile (10 μl of a 2% solution, by volume) was then added to stabilize the fluorphor. The reaction mixture was further concentrated to a volume convenient for spotting on a plate precoated with silica gel (Applied Science, Adsorbosil 5). The phenylethylamine-fluorescamine derivative (R_f 0.38) was well separated from excess reagent and tissue contaminants but not from the derivative of p-chlorophenylethylamine when the chromatogram was developed in chloroform : methanol : triethanolamine in a v/v ratio of 125 : 25 : 5. To separate these two derivatives, the fluorescent area with R_f 0.38 was scraped off the plate and the silica gel was eluted with acetone. A second chromatograph on another silica gel plate in chloroform : methanol : hexane : glacial acetic acid in a v/v ratio of 60 : 6 : 60 : 5 separated the fluophors of p-chlorophenylethylamine, R_f 0.60, and phenylethylamine, R_f 0.66. Standard solutions of varying concentrations of an authentic sample of phenylethylamine were carried through the same procedure and spotted on the same plate with samples. The overall recovery of the amine ranged from 75 to 85%. The yield of phenylethylamine added to normal plasma was in this same range. The intensity of the fluorescence emitted by the fluorphor of PEA at 475 nm when activated at 390 nm was determined in a Schoeffel spectrodensitometer equipped for measuring fluorescence on thin layer chromatograms. Ten picomoles of the amine could be detected and 25 to 200 pmol could be quantitatively measured. The fluorphor on the thin layer chromatogram was stable for 20 hr at 0°C in the dark.

2. Simultaneous Determination of the Aromatic Acid Metabolites Derived from Phenylalanine

Gas chromatography was the method of choice for the estimation of all of the aromatic acid metabolites derived from phenylalanine in the same tissue. Its successful application to the quantitative measurement of nanomole amounts in brain depended on the complete release of the metabolites from tissue constituents, the stabilization of phenylpyruvate, and adequate purification of the extract. Aqueous acid, commonly used for the extraction of metabolites from tissues, released approximately 30% of the total amount of phenylacetate in the brain. Treatment of the tissue homogenate in the first step with hydroxylamine at pH 12 to 13 under nitrogen at 60 to 65°C for 30 min quantitatively released the metabolites and stabilized phenylpyruvate by converting it to its oxime derivative. Sufficient purification was attained by extraction of the aqueous solution at pH 11 to 12 with ether and ethylacetate to remove neutral and basic contaminants, adsorption of the aromatic acids on deactivated carbon at pH 3.5 to 4.5, and elution of the metabolites with a mixture of benzene : ether : ethylacetate : methanolic 1 N HCl in a v/v ratio of 1 : 1 : 0.1 : 0.1. o-Hydroxyphenylacetic and phenylvaleric acids were added as internal standards to the tissue before homogenization and phenylvaleric acid was added to urine before processing.

The trimethylsilyl (TMS) derivatives of the metabolites were prepared by reaction with bis-(trimethylsilyl)trifluoroacetamide (Regis Chemical Co.) with 2% piperidine as catalyst at 70°C for 50 min (100 μl/tissue). The TMS derivatives were separated by gas chromatography over a column of 5% OV-1 on Chromosorb W-HP-80/100 mesh with temperature programming at 2°C/min from 65° to 145°C. Flow rate of the carrier gas, helium, was 40 ml/min. A flame ionization detector set at 280°C was employed. Temperature at the injector port was 250°C.

The TMS derivatives of the acids were identified by their methylene unit (MU) values and quantitated by measuring their peak areas. The identity and purity of the TMS derivatives of the metabolites separated by gas chromatography were confirmed by mass spectroscopy.

Urine was treated similarly. In addition, after conversion of phenylpyruvic acid to the oxime derivative and removal of excess hydroxylamine by extraction with ethylacetate, an aliquot was removed for the estimation of phenylacetylglutamine by the procedure of Wadman et al. (65).

3. Measurement of 5-Hydroxytryptophan Decarboxylase Activity

The procedure, which was devised from the publications of Christenson et al. (8) and Ichiyama, et al. (23), was described in detail in an earlier publication (35). The amount of $[^{14}C]O_2$ liberated from the substrate, 5-HTP ($[^{14}C]OOH$) served as a measure of the activity of the enzyme.

The Schiff base pyridoxylidene-2-phenylethylamine was synthesized as described earlier (34). The inhibitory action of the Schiff base was observed when it was preincubated with the enzyme at 0°C for 5 min before addition to the solution of substrate and co-factor.

ACKNOWLEDGMENTS

The skillful assistance of Joseph Scotto and Miss Lois Scotto is very much appreciated, and cooperation of Mr. Lawrence Black, librarian, is gratefully acknowledged. This investigation was supported by the New York State Department of Mental Hygiene, a grant from the National Institute of Child Health and Human Development, HD-06843-01, and a grant from Eli Lilly and Co.

REFERENCES

1. G. K. Aghajanian and F. E. Bloom (1967). The formation of synaptic junctions in developing rat brain: A quantitative electron microscopic study. *Brain Res.*, 6:716-727.

2. A. E. Anderson and G. Guroff (1972). Enduring behavioral changes in rats with experimental phenylketonuria. *Proc. Nat. Acad. Sci. USA*, 69:863-867.

3. A. E. Anderson, V. Rowe, and G. Guroff (1974). The enduring behavioral changes in rats with experimental phenylketonuria. *Proc. Nat. Acad. Sci. USA*, 71:21-25.

4. A. W. Behbehani, C.-D. Quentin, F. J. Schulte, and V. Neuhoff (1974). Microanalysis with ^{14}C-dansyl chloride of amino acids and amines in the cerebrospinal fluid of patients with phenylketonuria. II. Determination of 5-hydroxyindole derivatives after loading with L-phenylalanine. *Neuropädiatrie*, 5:258-270.

5. K. Blau (1970). Aromatic acid excretion in phenylketonuria. Analysis of the unconjugated aromatic acids derived from phenylalanine. *Clin. Chim. Acta*, 27:5-18.

6. A. A. Boulton and G. B. Baker (1975). Subcellular distribution of 2-phenylethylamine, p-tyramine, tryptamine in rat brain. *J. Neurochem.*, 25:477-482.

7. D. W. Caley and D. S. Maxwell (1968). An electron microscopic study of neurons during postnatal development of the rat cerebral cortex. *J. Comp. Neurol.*, 133:17-44.

8. J. G. Christenson, W. Dairman, and S. Udenfriend (1970). Preparation and properties of a homogeneous aromatic L-amino acid decarboxylase from hog kidney. *Arch. Biochem. Biophys.*, 141:356-367.

9. C. Cotman, D. Taylor, and G. Lynch (1973). Ultrastructural changes in synapses in the dentate gyrus of the rat during development. *Brain Res.*, 63:205-213.

10. H.-Ch. Curtius, K. Baerlocher, and J. A. Völlmin (1972). Pathogenesis of phenylketonuria: Inhibition of dopa and catecholamine synthesis in patients with phenylketonuria. *Clin. Chim. Acta,* 42:235-239.

11. J. Edmond (1974). Ketone bodies as precursors of sterols and fatty acids in the developing rat. *J. Biol. Chem.,* 249:72-80.

12. J. Edmond and G. Popjak (1974). Transfer of carbon atoms from mevalonate to n-fatty acids. *J. Biol. Chem.,* 249:66-71.

13. D. J. Edwards and S. M. Antelman (1975). Phenylethylamine: Interaction with dopaminergic neurons. In: *The Round Table Discussion. PEA: Its Possible Role as a Neuromodulator, Am. Soc. for Neurochem.* VI Annual Meeting, Mexico City.

14. D. J. Edwards and K. Blau (1972). The in vivo formation of p-chloro-2-phenylethylamine in young rats injected with p-chlorophenylalanine. *J. Neurochem.,* 19:1829-1832.

15. D. J. Edwards and K. Blau (1972). Aromatic acids derived from phenylalanine in the tissues of rats with experimentally induced phenylketonuria-like characteristics. *Biochem. J.,* 130:495-503.

16. D. J. Edwards and K. Blau (1973). Phenethylamines in brain and liver of rats with experimentally induced phenylketonuria-like characteristics. *Biochem. J.,* 132:95-100.

17. E. Fischer and B. Heller (1972). Phenethylamine as a neurohumoral agent in brain. *Behav. Neuropsychiat.,* 4:8-12.

18. K. Fuxe, H. Grobecker, and J. Jonsson (1967). The effect of 2-phenylethylamine on central and peripheral monoamine-containing neurons. *Eur. J. Pharmacol.,* 2:202-207.

19. G. E. Gaull, H. H. Tallan, A. Lajtha, and D. K. Rassin (1975). Pathogenesis of brain dysfunction in inborn errors of amino acid metabolism. In: *Biology of Brain Dysfunction, Vol. 3,* edited by G. E. Gaull. Plenum Press, New York, pp. 47-143.

20. N. K. Gonatas, L. A.-Gambetti, P. Gambetti, and B. Shafer (1971). Morphological and biochemical changes in rat synaptosome fraction during neonatal development. *J. Cell. Biol.,* 51:484-498.

21. E. C. Gray and V. P. Whittaker (1962). The isolation of nerve endings from brain: An electronmicroscopic study of cell fragments derived by homogenization and centrifugation. *J. Anat.,* 96:79-88.

22. G. Guroff (1969). Irreversible in vivo inhibition of rat liver phenylalanine hydroxylase by p-chlorophenylalanine. *Arch. Biochem. Biophys.,* 134:610-611.

23. A. Ichiyama, S. Nakamura, Y. Nishizuka, and O. Hayaishi (1970). Enzymic studies on the biosynthesis of serotonin in mammalian brain. *J. Biol. Chem.,* 245:1699-1709.

24. E. E. Inwang, A. D. Mosnaim, and H. C. Sabelli (1973). Isolation and characterization of phenylethylamine and phenylethanolamine from human brain. *J. Neurochem.*, 20:1469-1473.

25. G. A. Jervis (1937). Phenylpyruvic oligophrenia: Introductory study of fifty cases of mental deficiency associated with excretion of phenylpyruvic acid. *Arch. Neurol. Psychiat.*, 38:944-963.

26. G. A. Jervis (1939). The genetics of phenylpyruvic oligophrenia. *J. Ment. Sci.*, 85:719-762.

27. G. A. Jervis (1947). Studies on phenylpyruvic oligophrenia. The position of the metabolic error. *J. Biol. Chem.*, 169:651-656.

28. G. A. Jervis (1950). Excretion of phenylalanine and derivatives in phenylpyruvic oligophrenia. *Proc. Soc. Exp. Biol. Med.*, 75:83-86.

29. G. A. Jervis, R. J. Block, D. Bolling, and E. Kanze (1940). Chemical and metabolic studies on phenylalanine. II. The phenylalanine content of the blood and spinal fluid in phenylpyruvic oligophrenia. *J. Biol. Chem.*, 134:105-113.

30. B. K. Koe (1967). Inhibiting action of p-chlorophenylalanine and α-methyl-p-chlorophenylalanine on liver phenylalanine and tyrptophan hydroxylase. *Med. Pharmacol. Exp.*, 17:129-138.

31. I. J. Kopin (1972). Aminergic neurons and substitute transmitters in brain. In: *Neurotransmitters*, edited by I. J. Kopin. The Association for Research in Neurons and Mental Disease, William and Wilkins, Baltimore, pp. 207-215.

32. M. A. Lipton, R. Gordon, G. Guroff, and S. Udenfriend (1967). p-Chlorophenylalanine-induced chemical manifestations of phenylketonuria in rats. *Science*, 156:248-250.

33. L. A. Loizou (1972). The postnatal ontogeny of monoamine containing neurons in the central nervous system of the albino rat. *Brain Res.*, 40:395-418.

34. Y. H. Loo (1967). Characterization of a new phenylalanine metabolite in phenylketonuria. *J. Neurochem.*, 14:813-821.

35. Y. H. Loo (1974). Serotonin deficiency in experimental hyperphenylalaninemia. *J. Neurochem.*, 23:139-147.

36. Y. H. Loo and W. M. Cort (1972). Assay of pyridoxal and its derivatives. In: *Methods of Neurochemistry, Vol. 2*, edited by R. Fried. Marcel Dekker, New York, pp. 169-204.

37. Y. H. Loo and K. Mack (1972). Effect of hyperphenylalaninemia on vitamin B_6 metabolism in developing rat brain. *J. Neurochem.*, 19:2377-2383.

38. Y. H. Loo and V. P. Whittaker (1967). Pyridoxal kinase in brain and its inhibition by pyridoxylidene-2-phenylethylamine. *J. Neurochem.*, 14:997-1011.

39. W. Lovenberg, H. Weissbach, and S. Udenfriend (1962). Aromatic L-amino acid decarboxylase. *J. Biol. Chem.*, 237:89-92.

40. H. R. Mahler and S. J. Wakil (1953). Enzymatic activation of fatty acids. *J. Biol. Chem.*, 204:453-468.

41. N. Malamud (1966). Neuropathology of phenylketonuria. *J. Neuropath. Exp. Neurol.*, 25:254-268.

42. C. M. McKean (1972). The effects of high phenylalanine concentration on serotonin and catecholamine metabolism in the human brain. *Brain Res.*, 47:469-476.

43. J. H. Menkes (1967). The pathogenesis of mental retardation in phenylketonuria and other inborn errors of metabolism. *Pediatrics*, 39:297-308.

44. P. B. Molinoff, L. Landsberg, and J. Axelrod (1969). An enzymatic assay for octopamine and other β-hydroxylated phenylethylamines. *J. Pharmacol. Exp. Ther.*, 179:253-261.

45. A. D. Mosnaim, E. E. Inwang, J. H. Sugerman, and H. C. Sabelli (1973). Identification of 2-phenylethylamine in human urine by infrared and mass spectroscopy and its quantification in normal subjects and cardiovascular patients. *Clin. Chim. Acta*, 40:407-413.

46. T. Nakajima, Y. Kakimoto, and I. Sano (1964). Formation of 2-phenylethylamine in mammalian tissue and its effect on motor activity in the mouse. *J. Pharmacol. Exp. Ther.*, 143:319-325.

47. V. Neuhoff, A. Behbehani, C.-D. Quentin, and H. Prinz (1974). Identification of 5-hydroxytryptophan by dansylation in the cerebrospinal fluid of children with phenylketonuria. *Hoppe Seyler's Z. Physiol. Chem.*, 355:891-894.

48. J. A. Oates, P. Z. Nirenberg, J. B. Jepsin, A. Sjoerdsma, and S. Udenfriend (1963). Conversion of phenylalanine to phenethylamine in patients with phenylketonuria. *Proc. Soc. Exp. Biol. Med.*, 112:1078-1081.

49. C. M. B. Pare, M. Sandler, and R. S. Stacey (1957). 5-Hydroxytryptamine deficiency in phenylketonuria. *Lancet*, i:551-553.

50. C. M. B. Pare, M. Sandler, and R. S. Stacey (1958). Decreased 5-hydroxytryptophan decarboxylase activity in phenylketonuria. *Lancet*, ii:1099-1101

51. W. A. Pedemont, A. D. Mosnaim, and M. Bulat (1976). Penetration of phenylacetic acid across the blood-cerebrospinal fluid barrier. *Research Communications in Chemical Pathology and Pharmacology*. In press.

52. T. L. Perry (1962). Urinary excretion of amines in phenylketonuria and mongolism. *Science*, 136:879-880.

53. E. Roberts and D. G. Simonsen (1963). Some properties of L-glutamic decarboxylase in mouse brain. *Biochem. Pharmacol.*, 12:113-134.

54. J. M. Saavedra (1974). Enzymatic isotopic assay for and presence of 2-phenylethylamine in brain. *J. Neurochem.*, 22:211-216.

55. J. M. Saavedra and J. Axelrod (1973). Demonstration and distribution of phenylethanolamine in brain and other tissues. *Proc. Nat. Acad. Sci. USA*, 70:769-772.

56. J. M. Saavedra, J. T. Coyle, and J. Axelrod (1974). Developmental characteristics of phenylethanolamine and octopamine in the rat brain. *J. Neurochem.*, 23:511-515.

57. R. P. Silkaitis and A. D. Mosnaim (1976). Pathways linking L-phenylalanine and 2-phenylethylamine with p-tyramine in rabbit brain. *Brain Res.* In press.

58. J. W. Schweitzer, A. J. Friedhoff, and R. Schwartz (1975). Phenethylamine in normal urine: Failure to verify high values. *Biol. Psychiat.*, 10:277-285.

59. S. N. Shah, N. A. Peterson, and C. M. McKean (1972). Lipid composition of human cerebral white matter and myelin in phenylketonuria. *J. Neurochem.*, 19:2369-2376.

60. S. N. Shah, N. A. Peterson, and C. M. McKean (1972). Impaired myelin formation in experimental hyperphenylalaninemia. *J. Neurochem.*, 19:479-485.

61. B. Singh and J. de Champlain (1972). Altered ontogenesis of central noradrenergic neurons following neonatal treatment with 6-hydroxydopamine. *Brain Res.*, 48:432-437.

62. S. Udenfriend, S. Stein, P. Böhlen, and W. Dairman (1972). Fluorescamine: A reagent for assay of amino acids, peptides, proteins, and primary amines in the picomole range. *Science*, 178:871-872.

63. J. M. Vavich and R. R. Howell (1971). Rapid identification and quantitation of urinary metabolites of phenylalanine in phenylketonuria by gas chromatography. *J. Lab. Clin. Med.*, 77:159-167.

64. S. K. Wadman, D. Ketting, P. K. DeBree, C. van der Heiden, M. Th. Grimberg, and H. Kruijswijk (1975). Permanent chemical phenylketonuria and a normal phenylalanine tolerance in two sisters with a normal mental development. *Clin. Chim. Acta*, 65:197-204.

65. S. K. Wadman, F. J. van Sprang, C. van der Heiden, and D. Ketting (1971). Quantitation of urinary phenylalanine metabolites in phenylketonuria. In: *Phenylketonuria*, edited by H. Bickel, F. P. Hudson, and L. I. Woolf. G. T. Verlag, Stuttgart, pp. 65-72.

66. J. Willner, H. F. Le Fevre, and E. Costa (1974). Assay by multiple ion detection of phenylethylamine and phenylethanolamine in rat brain. *J. Neurochem.*, 23:857-859.

67. L. I. Woolf (1951). Excretion of conjugated phenylacetic acid in phenylketonuria. *Biochem. J.*, 49:IX-X.

Chapter 20

URINARY AND BRAIN PHENYLETHYLAMINE LEVELS UNDER NORMAL AND PATHOLOGICAL CONDITIONS

H. Spatz and N. Spatz

Laboratory of Psychopharmacology
and Experimental Neuropsychiatry
Hospital Nacional Jose T. Borda
Buenos Aires, Argentina

I. INTRODUCTION

During the last 10 years, several studies have been carried out on urinary and brain phenylethylamine (PEA) based on the assumption that this substance plays an important role as a neurohumoral agent. Phenylethylamine has been found in the urine of normal subjects and patients with phenylketonuria (1,2). In 1964, Nakajima et al. (3) detected PEA in the organs of mammals pretreated with monoamine oxidase inhibitors (MAOI). Approximately at the same time, we initiated studies in our laboratory on the influence of certain monoamines on animal behavior. Fischer et al. demonstrated that catecholamines or its precursors induced a decrease in the psychomotor activity of mice, a phenomenon also observed after the administration of reserpine (4). Thus, it seemed rather unlikely that catecholamines would be involved in the maintenance of a normal "excitatory" state and that in contrast with the stimulatory theory of catecholamine action popular at that time (5-8), it was necessary to consider the possible existence of other neurohumoral agents with a central ergotropic action. In view of the chemical similarity between amphetamine and PEA, comparative pharmacological studies were carried out which indeed showed that both substances had stimulatory effects when injected in mice (9-13). Thus, it was postulated that if PEA played a role as a neurohumoral agent, disturbances in its metabolism would lead to affective disorders. Therefore, we initiated studies on PEA excretion in normal and depressed subjects and on the action of depressant and antidepressant drugs on brain PEA levels in rats (14-17). The integration of the results obtained in these different studies led Fischer et al. (9,11, 12) to hypothesize that PEA is the stimulatory neurohumoral agent of the ergotropic system of Hess (13). Our urinary and brain PEA levels were confirmed by Mosnaim et al. (18,19) but not so by other investigators (20-23). It appears, therefore, that urinary and brain PEA levels are actually a controversial matter. In this chapter, we discuss published and unpublished work on this matter, hoping to add some new light on this issue.

II. URINARY PEA IN NORMAL SUBJECTS

Our first quantitative data were obtained with a spectrophotofluorometric method (4). Interfering substances were eliminated from the urine by coupling with diazotated sulfanilic acid, PEA extracted with chloroform and then allowed to react with dimethylaminocinnamaldehide. The product of this reaction fluoresces at 520 μm when excited at 470 μm. Figure 1 shows excitation spectra of authentic and urinary PEA, both being very similar. Figure 2 shows the corresponding

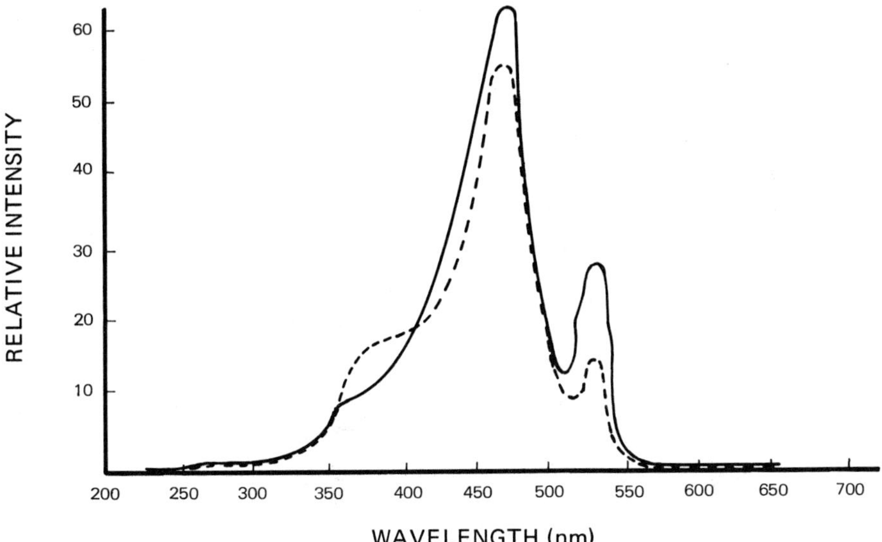

FIG. 1. Excitation spectra of: - - - - - authentic PEA and ——— urinary PEA.

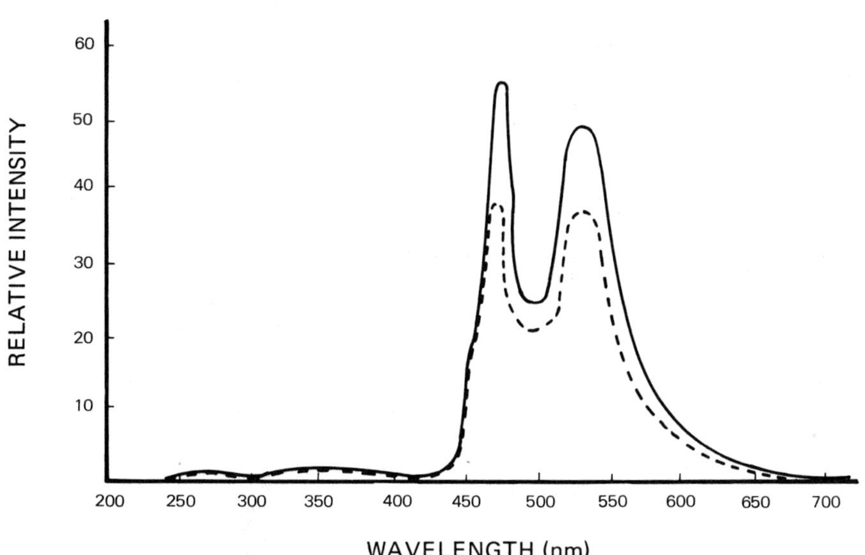

FIG. 2. Emission spectra of: - - - - - authentic PEA and ——— urinary PEA.

emission spectra, which are very similar too. The specificity of this method was demonstrated by thin layer chromatography. The urines were treated with diazotated sulfanilic acid and then extracted with chloroform. The chloroform extracts were concentrated and analyzed chromatographically using five different solvent systems. The plates treated with ninhydrine showed the presence of only one amine, and its R_f was very similar to that of PEA (Table 1).

Later, we developed another quantitative method (17). Interfering substances were eliminated by coupling with diazotated sulfanilic acid, PEA was extracted with chloroform and immediately treated with S_2C. The corresponding dithiocarbamate was then quantified by gas-liquid chromatography. Figures 3 and 4 show a chromatogram of authentic PEA dithiocarbamate and a typical chromatogram of an urinary extract obtained by this method. It can be seen that the chromatogram of the urine shows the presence of a few peaks, one of them with PEA dithiocarbamate's retention time. The specificity of this method was demonstrated by elution of the PEA dithiocarbamate corresponding fraction and infrared analysis. Figures 5 and 6 show the infrared spectra of authentic and urinary PEA, which are almost identical. We then introduced another modification in the method. Instead of using S_2C, we treated the chloroform extracts with a mixture of pyridine and acetic anhydride transforming PEA into its acetamide.

TABLE 1

R_f of Authentic PEA and of Urinary PEA
After Treatment of Urine Samples with Diazotated
Sulfanilic Acid and Subsequent Extraction with Chloroform

Solvent	R_f authentic PEA	R_f PEA of Urine
96% ethanol-25% aqueous ammonia (8 : 2)	0.65	0.64
Phenol-water (8 : 3)	0.58	0.57
Potassium chloride 20% in water	0.46	0.46
Chloroform-acetic acid (95 : 5)	0.05	0.05
Isopropanol-25% aqueous ammonia-water (85 : 5 : 15)	0.92	0.91

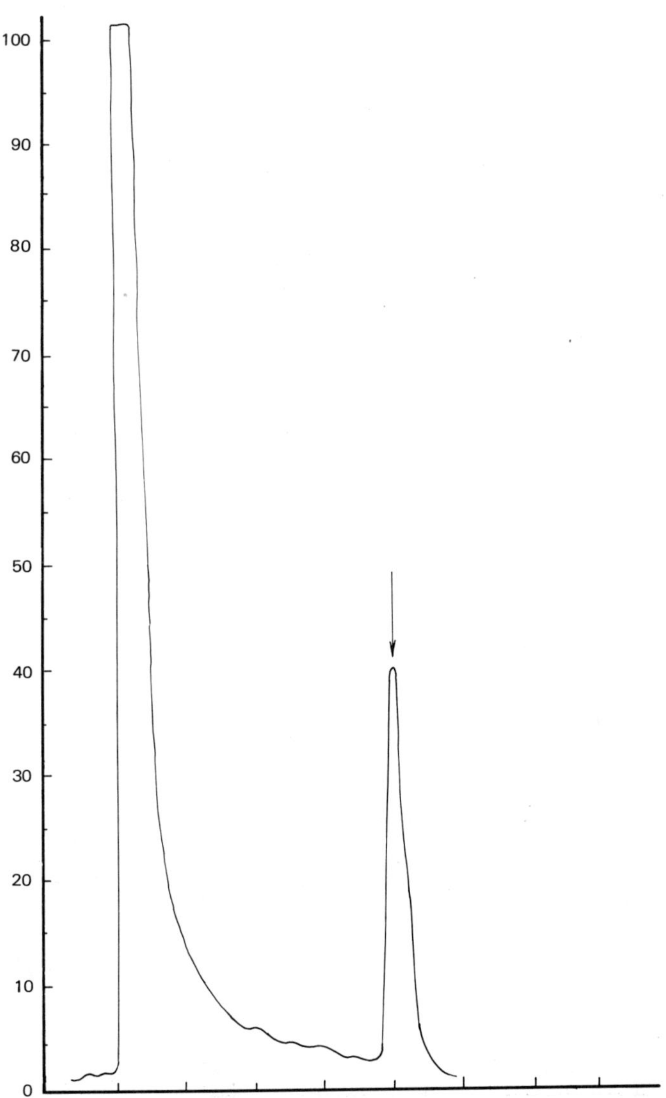

FIG. 3. Gas chromatographic record of authentic PEA dithiocarbamate.

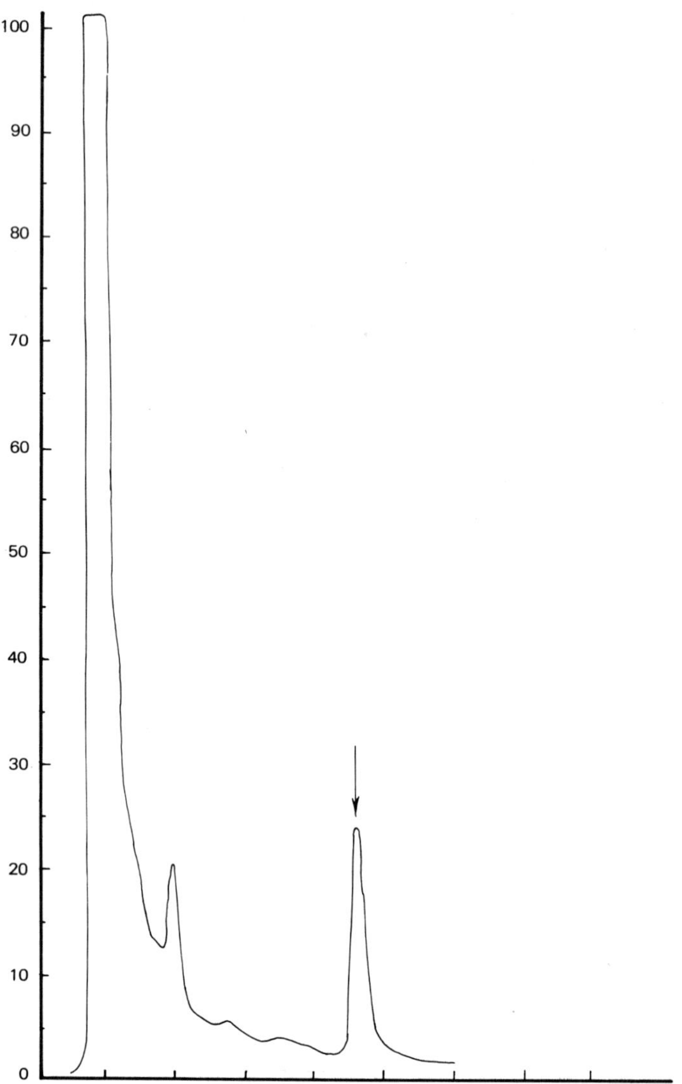

FIG. 4. Gas chromatographic record of a urinary extract.

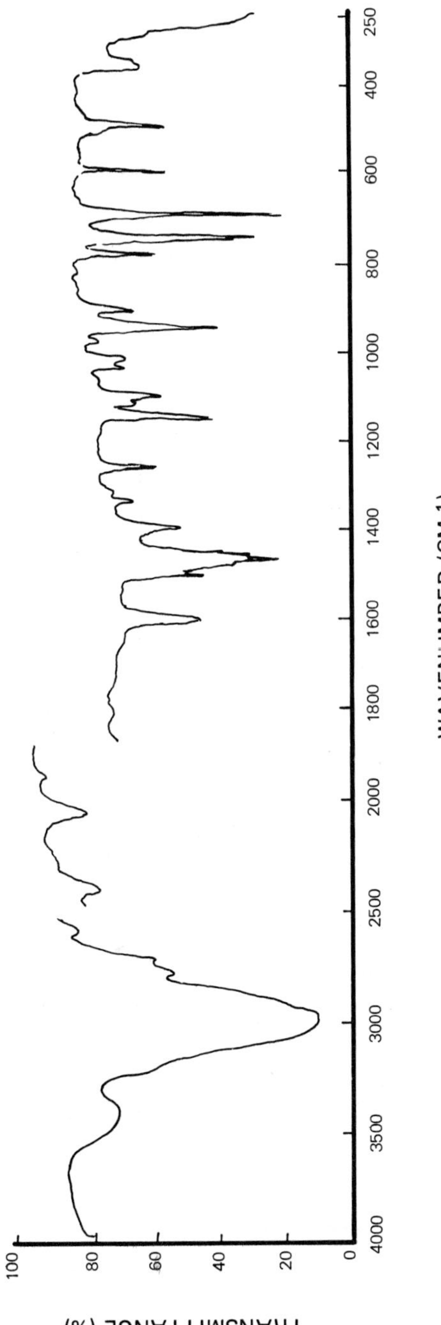

FIG. 5. Infrared spectrogram of urinary PEA dithiocarbamate.

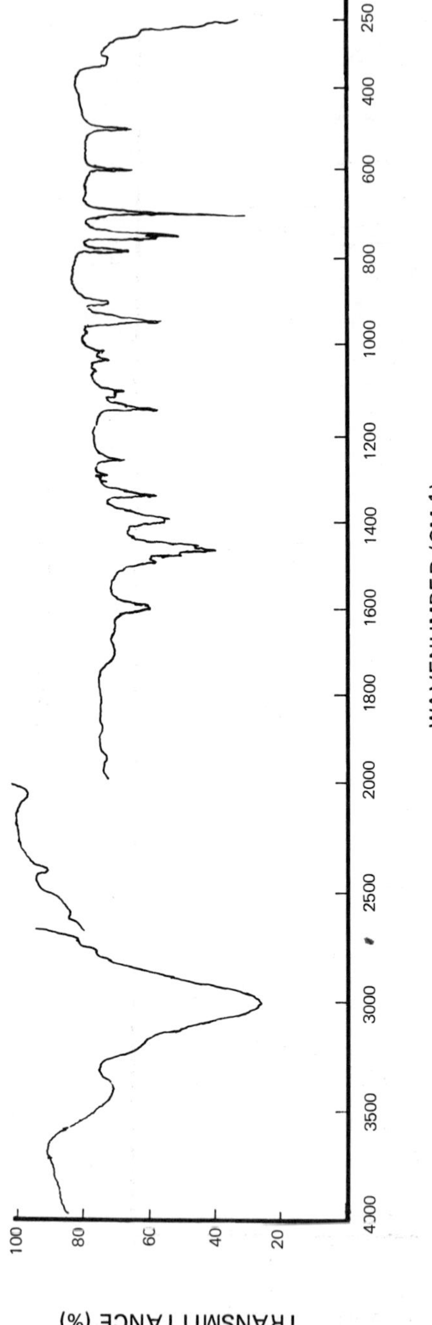

FIG. 6. Infrared spectrogram of authentic PEA dithiocarbamate.

This product was quantified also by gas-liquid chromatography. The specificity of the method was demonstrated by analyzing the eluted fraction by mass spectrometry. Figures 7 and 8 show the mass spectra of authentic and urinary PEA acetamide. Also in this case, both spectra corresponded to the same substance.

Table 2 shows the results obtained after analyzing the urine of 130 normal subjects by the three methods. Table 3 shows the results obtained in a group of five normal subjects whose urines were analyzed by both gas-chromatographic methods. On the basis of the results obtained by the three different methods whose specificity was confirmed by thin layer chromatography, infrared spectrometry, and mass spectrometry, we can state that urinary normal PEA values are over 80 μg/24 hr. Mosnaim et al. (18,19) reported similar values, not so Boulton et al. (20) and Schweitzer et al. (21) who find much lower quantities. It must be emphasized that the conditions of the urine collection are very important. In our experience, urinary PEA decomposes rapidly even when bacterial action is inhibited. This is apparently due to the presence in the urine of oxidative substances of unknown structure. During our first experiments, we could accidentally observe that 24-hr normal frozen stored urines collected in plastic bottles containing 10 ml of chloroform gave significantly lower values than fresh ones. Also, in the chromatograms obtained treating the chloroform extracts with S_2C, a peak was present with a lower retention time than that of PEA dithiocarbamate. This peak was not present when fresh urines were analyzed, or when they were collected in plastic bottles containing 10 ml of chloroform and 100 mg of ascorbic acid. Apparently, the aforementioned peak corresponds to a decomposition product of PEA. It must be noted too that PEA excretion can vary considerably during the day. Also, urinary pH influences the elimination of amines. For these reasons we always collect urine samples during a complete 24 hr period in plastic bottles containing 10 ml of chloroform and 100 mg of ascorbic acid as preservatives and keep it refrigerated at 5°C during the collection period. The urines are then stored at -60°C until

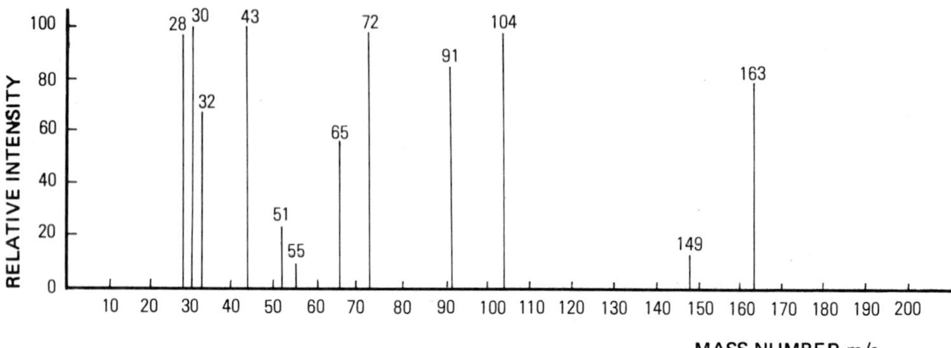

FIG. 7. Mass spectrum of authentic PEA acetamide.

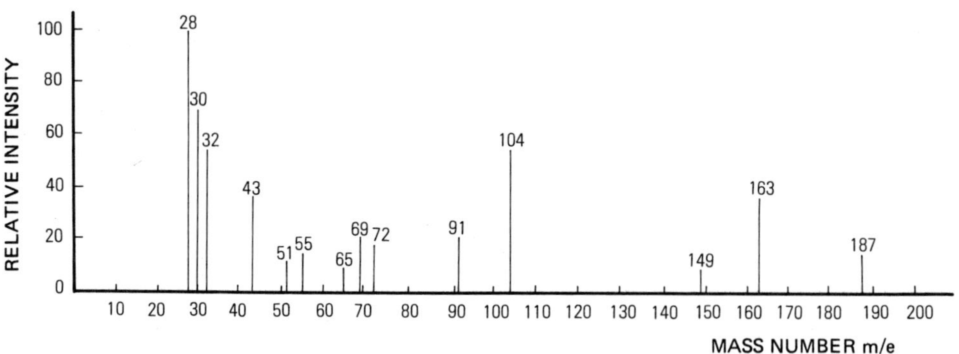

FIG. 8. Mass spectrum of urinary PEA acetamide.

TABLE 2

Amounts of PEA (μg/24 hr) Determined in Normal Urines by Different Methods

Method	Number of Cases	Limits	Mean Value	SD	Number of Reference
Spectrophotofluorometric	10	101-399	239	26	14
	17	105-775	336	180	16
GLC of PEA dithiocarbamate	13	40-1.320	400	105	17
	30	95-1.720	360	121	25
GLC of PEA acetamide	60	80-1.830	372	128	25

TABLE 3

Levels of PEA (μg/24 hr) Determined in Urines from Normal Subjects Using two GLC Methods

Subject	GLC of PEA acetamide PEA (μg/24 hr)	GLC of PEA dithiocarbamate PEA (μg/24 hr)
1	336	328
2	132	137
3	410	426
4	511	498
5	97	92

analyzed. Patients are given the same standard diet during the day of urine collection and the day prior to it (24).

III. URINARY PEA IN DEPRESSIVE STATES

In view of the possibility that PEA could have a stimulatory action playing an important role in the modulation of affective states, urines of depressed patients were analyzed based on the assumption that a disorder in PEA metabolism could be responsible for the symptoms or etiology of this disease. The first determinations using the fluorometric method (4) indicated that urinary PEA values in endogenous depressed patients were much lower than those of normal subjects but not so in the case of reactive and atypical depressions where the values were similar to normal ones.

Table 4 shows that the urinary PEA Excretion using three different analytical methods for its determination is significantly reduced in depressed patients as compared to normal subjects. Tables 5 and 6 describe the results obtained in 60 normal subjects and 98 endogenous depressed patients using the pyridine-acetic anhydride

TABLE 4

Daily Urinary PEA Excretion (μg/24 hr) in Endogenous Depressed Patients as Determined by Three Different Methods

Method	Number of Cases	Limits	Mean Value	SD	Number of Reference
Spectrophotofluorometric	14	30-225	114	49	16
GLC of PEA dithiocarbamate	30	14-146	57	7	17
GLC of PEA acetamide	98	0-78	39	12	25

TABLE 5

Daily Urinary PEA Excretion (μg/24 hr) in Healthy Subjects as Determined by GLC of PEA Acetamide

Daily Elimination (μg/24 hr)	Number of Cases	Percent of Cases
80-100	6	10
101-200	29	48.3
201-400	14	23.3
401-800	5	8.3
801-1,200	2	3.3
1,201-2,000	4	6.7

TABLE 6

Daily Urinary PEA Excretion (μg/24 hr) in Endogenous Depressed Patients as Determined by GLC of PEA Acetamide

Daily Elimination (μg/24 hr)	Number of Cases	Percent of Cases
0-20	25	25.5
21-40	39	39.8
41-60	22	22.4
61-70	8	8.2
71-80	4	4.1

method (24). It can be observed that the majority of normal individuals have urinary PEA values between 120 and 400 μg/24 hr, while in depressed patients the values ranged from 0 to 80 μg/24 hr, but most of them were below 60 μg/24 hr. These results support Fischer's hypothesis of a biochemical disturbance of PEA metabolism as an etiological factor for endogenous depression.

IV. PEA IN OTHER MENTAL DISEASES

Considerable work must yet be done in this field. It was postulated that in other mental disorders in which depression is one of the observed symptoms, an alteration in PEA metabolism could be present. Thus, in seven out of 10 cases of schizophrenia undergoing treatment (see Table 7), we found low urinary PEA values (25) in contrast to the first three cases studied and reported by us (15). Depression is one of the symptoms observed in schizophrenic patients undergoing treatment, and as yet it is not clear whether the low PEA values can

TABLE 7

Daily Urinary PEA Excretion (μg/24 hr) in Schizophrenic Patients

Subject	PEA (μg/24 hr)
1	62
2	43
3	132
4	27
5	177
6	41
7	23
8	16
9	420
10	32

be attributed to the illness per se or to the effect of neuroleptic drugs with which these patients are treated. Considerable difficulties in obtaining 24-hr urine sample from untreated schizophrenics make this study troublesome. For this reason we are investigating at the present time the effect of neuroleptics on brain PEA levels in animals.

Heller et al. (26) found in all cases of Parkinson's disease low urinary PEA values. Also, in this illness depression is a frequently encountered symptom. Heller reported too that PEA antagonizes Parkinson-like states in animals (27) and hypothesized about the neurohumoral mechanism involved in the regulation of extrapyramidal activity, which states that PEA and dopamine on one hand, and acetylcholine, serotonin, and tryptamine on the other, control the activity of the structures involved in its function. Mosnaim et al. found low PEA values after myocardial infarction, and depression is also a common symptom of this condition (28).

V. NORMAL PEA BRAIN LEVELS

As in the case of urinary PEA levels, normal brain PEA levels are at the moment also controversial. Durden et al. (22) reported values in the range of nanograms per gram of rat brain. Saavedra and Axelrod's values are approximatively 10-fold higher. We reported mean values of about 0.49 μg/g in rat brain. Mosnaim et al. (28,29) and Inwang et al. (30) found similar amounts in mammal and human brains. The methods employed by us to quantify brain PEA are very similar to those described in this chapter for urinary determinations. After killing the rats by decapitation, the brains are quickly extracted, weighed, and immediately homogenized in a 20% sodium sulfate solution in 0.5 N HCl. A pool of three to five brain homogenates is shaken with chloroform, allowed to stand 2 hr at 3°C, centrifuged, and the supernatant treated identically as described for the urine PEA methods.

Normal rat brain PEA levels obtained at first by the spectrophotofluorometric method and later derivatizing the amine with S_2C or pyridine-acetic anhidride followed by gas chromatographic analysis,

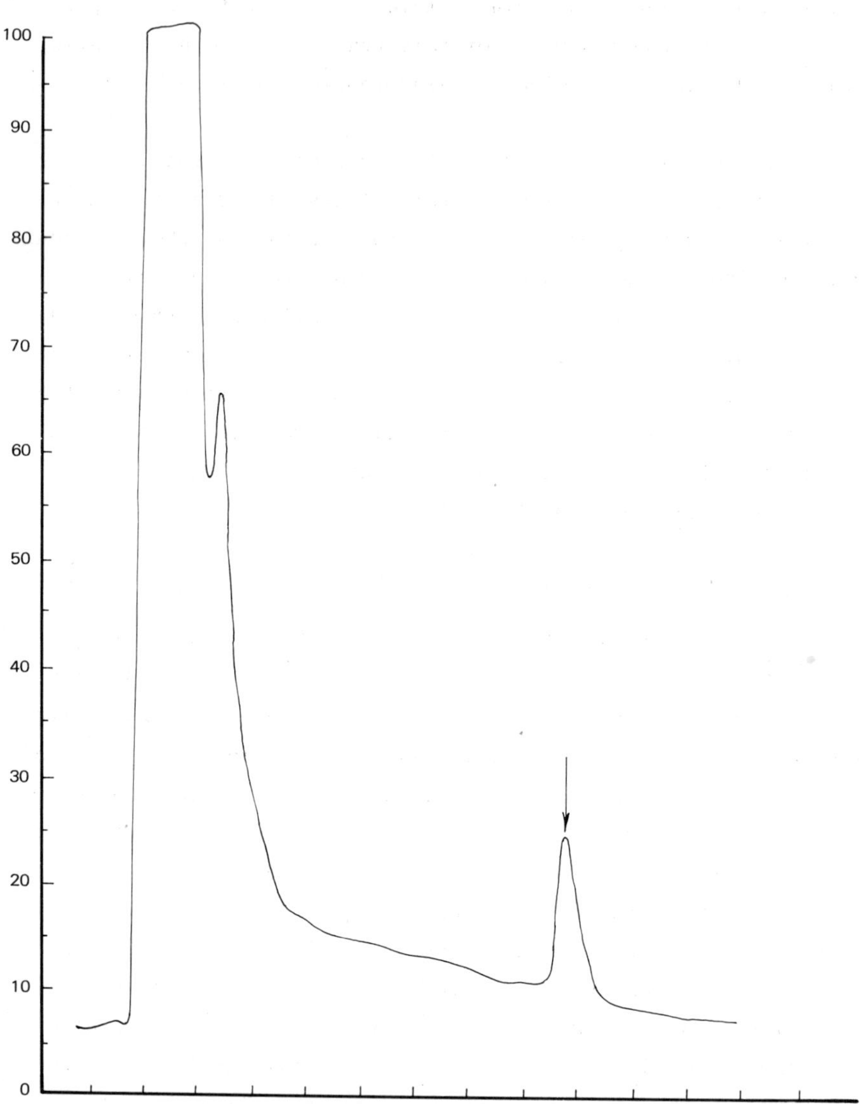

FIG. 9. Gas chromatographic record of authentic PEA acetamide.

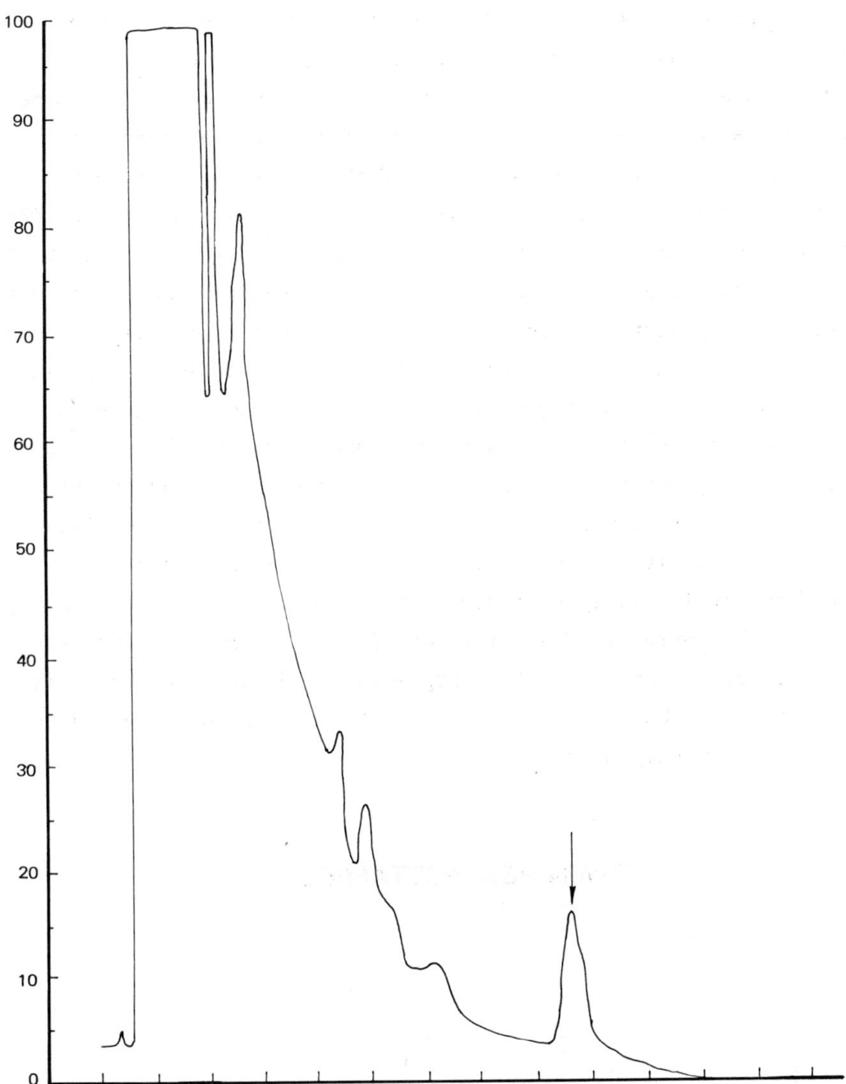

FIG. 10. Gas chromatographic record of brain extract.

were similar. In the three methods, we used the same deproteinazing agent. Phenylethylamine acetamide gives higher peaks than the corresponding dithiocarbamate. For this reason, our last determinations were carried out by this procedure. Figures 9 and 10 show chromatograms of authentic PEA acetamide and of a brain extract. In the latter, a peak with the same retention time as PEA acetamide can be observed. We could demonstrate the specificity of the method by isolation from the gas chromatograph of the derivatized brain PEA followed by mass spectrometry. Figure 11 shows mass spectrum of PEA acetamide obtained from brain PEA. When compared with Fig. 7, which shows spectrum of authentic PEA acetamide eluted from the gas chromatograph, it is evident that the so-called brain PEA is identical to an authentic PEA. This is conclusive evidence that the substance estimated by us is PEA. At the moment it is not possible to give an explanation for the differences between our values and those of other authors. The possibility of PEA being present in a bound and unbound form in the brain must be studied as this could explain the different levels obtained with various procedures. If PEA is present in a bound form, it is likely that according to the physicochemical properties of the deproteinazing media or of the different agents used to homogenize the brains, PEA linkages could be broken or not. In this last case, PEA would precipitate with proteins or other water insoluble brain components.

BRAIN PEA ACETAMIDE

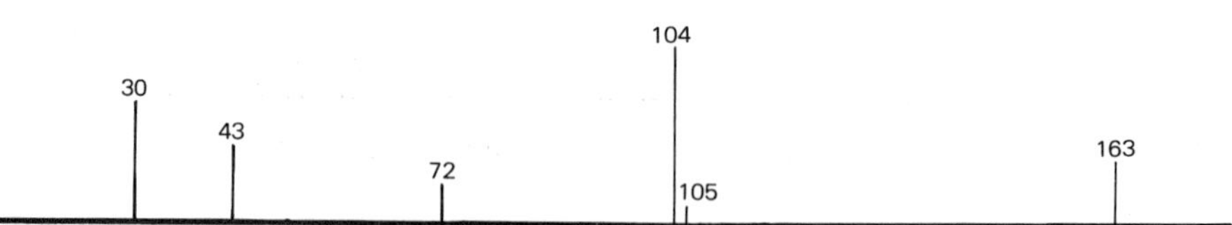

FIG. 11. Mass spectrum of brain PEA acetamide.

VI. URINARY PEA AFTER TREATMENT WITH ANTIDEPRESSANT DRUGS

Urinary PEA levels of endogenous depressed patients increased after treatment with antidepressant drugs such as MAOI, impramine or other tryciclics (15,16,32,33). This increment was observed mostly in the cases with clinical improvements. Cases in which the treatment failed did not show significant changes in PEA elimination. Table 8 summarizes a series of studies on patients treated with different preparations. These results led our group to study the clinical and pharmacological effects of PEA itself and of its possible precursors (33-39). Phenylethylamine induced an increment of the locomotory activity in mice, but has not been shown to have any therapeutic effects in man (35). This is probably due to the fact that PEA does not cross the hematoencephalic barrier in the necessary degree when administered at routine dosages.

In view of the lack of toxicity of D- and L-phenylalanine, clinical assays were carried out with both isomers. D-Phenylalanine has a very marked antidepressant effect in endogenous depressed patients, without any side effects, not so the L-isomer with which the results obtained were not clear cut. Some patients showed anxiety or excitation, possibly due to a placebo effect or as a natural symptom of the illness, which does not disappear under the effect of L-phenylalanine (38). In a double blind study, comparing the effects of imipramine and D-phenylalanine, the amino acid showed a faster and more effective antidepressant action (39). With both drugs, urinary PEA values increased when clinical improvement was observed. When the medication was suspended after 15 days of treatment with D-phenylalanine, urinary PEA values decreased in most cases in 1 week, parallel with a reappearance of the original symptoms (Table 9).

TABLE 8

Daily Urinary PEA Excretion in Depressed Patients Prior to and After Treatment with Antidepressant Drugs[a]

Case	Medication Administered	Therapeutic Result	PEA Treatment (μg/24 hr) Before	After
1	Desmethylimipramine 120 mg/day, 18-21 days	W	83	40
2		R	42	380
3		C.R.	40	900
4		W	87	40
5		I	40	130
6		C.R.	60	220
7		I	42	110
8		C.R.	40	330
9		C.R.	60	180
10		N.C.	40	40
11		I	36	130
12	Merck M.K. 50 mg/day, 7 days	I	20	330
13	Chlorimipramine 50-75 mg/day 11 days	I	90	555
14	10 mg tranylcypromine + 2 mg trifluorperazine 2-3 times/day	I	34	180
15		C.R.	14	678
16		C.R.	60	178
17		C.R.	42	320

[a]W, worsened; R, remission; C.R., complete remission; I, improvement; N.C., no change.

TABLE 9

Daily Urinary PEA Excretion (μg/24 hr)
in Depressed Patients Prior to and
After 15 Day Treatment with D-Phenylalanine

Patient	PEA Before Treatment	PEA 10 days After	PEA 20 days After
1	61	328	363
2	30	441	32
3	35	441	113
4	83	481	244
5	26	730	284
6	67	983	31

VII. PEA BRAIN LEVELS UNDER THE EFFECT OF DIFFERENT DRUGS

Certain drugs with depressant (i.e., reserpine) or antidepressant (i.e., MAOI, tricyclic) properties modify considerably brain PEA levels (40-42). Thus, it is suggested that reserpine not only depletes catecolamines and serotonin, but also PEA. This depletion must be considered when interpreting the depressant and other pharmacological actions of this drug, including pseudo-Parkinsonism, MAOI as expected raised brain PEA levels. In this case too, our results differ considerably from those obtained by Durden et al. (22) and Saavedra et al. (23) who report a 50- to 100-fold increment of brain PEA under the effect of this kind of drug, while we observed brain PEA values only two- to threefold higher than normal ones. In view of the fact that there exists a relative narrow limit, conditioned by an homeostatic equilibrium, in which the concentration of any given substance can increase without having dramatic or deleterious effects, this 50- to 100-fold increment of brain PEA levels cannot be well explained. Tricyclic antidepressants raise brain PEA too, the mechanisms of this action is not yet clear. Gabay et al. (43) postulate the existence of two MAO systems, one of which is

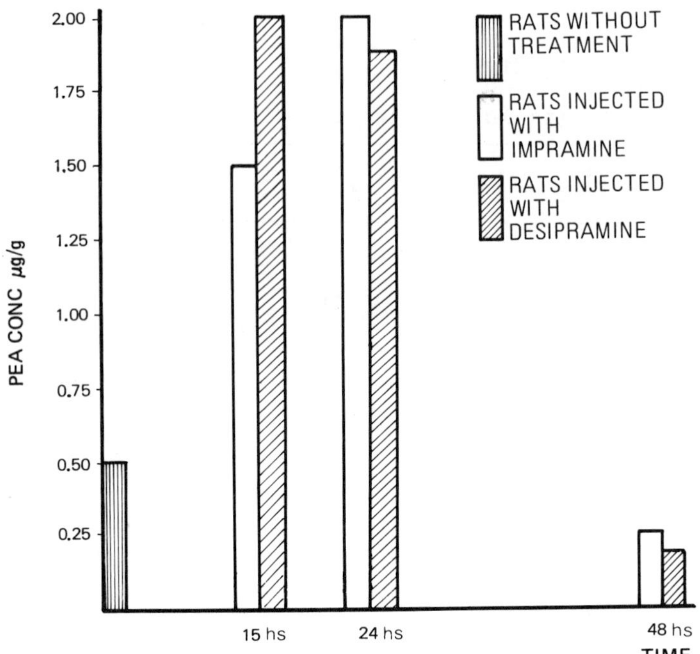

FIG. 12. Action of imipramine and desmethylimipramine on brain PEA levels.

inhibited by these drugs. When studying the action of imipramine and desmethylimipramine on rat brain levels, a strange effect could be observed (44): 24 hr after a 50 mg/kg weight i.p. injection, PEA values are normal again, but 48 hr after the injection, PEA levels were significantly lower (Fig. 12). In another study, we have determined rat brain PEA levels at different times after a single intraperitoneal administration of amphetamine and related substances. The results obtained were summarized in Table 10, which clearly indicated that drugs belonging to the amphetamine group cause significant alterations of brain PEA levels. The increase observed half an hour after the injection may be attributed to a MAOI effect and the partial depletion may be the consequence of a release of PEA and subsequent enzymatic degradation. The releasing effect shows a certain parallelism with the pharmacological potence of the assayed

TABLE 10

Rat Brain PEA Levels After
Intraperitoneal Injection of Different Amphetamines

Substance	Dose (mmol/kg)	PEA (μg/g) 30 min	1 hr	2 hr	24 hr
Water		0.50 ± 0.02	0.48 ± 0.01	0.49 ± 0.03	0.50 ± 0.02
D-Amphetamine	0.025	0.32 ± 0.05	0.26 ± 0.01	0.05 ± 0.01	0.28 ± 0.04
D,L-Amphetamine	0.025	0.80 ± 0.04	0.34 ± 0.02	0.22 ± 0.02	0.44 ± 0.03
Ephedrine	0.025	0.56 ± 0.02	0.28 ± 0.04	0.40 ± 0.02	0.50 ± 0.02
Fenfluramine	0.025	0.54 ± 0.01	0.46 ± 0.03	0.48 ± 0.02	0.51 ± 0.02
Fenfluramine	0.125	0.96 ± 0.04	0.32 ± 0.01	0.22 ± 0.01	0.48 ± 0.02
Diethylpropion	0.025	0.52 ± 0.03	0.48 ± 0.03	0.42 ± 0.02	0.52 ± 0.03
Diethylpropion	0.125	0.78 ± 0.04	0.32 ± 0.01	0.28 ± 0.01	0.49 ± 0.02

drug. Thus, it is observed that D,L-amphetamine, which is pharmacologically less potent than D-amphetamine causes a much lower decrease and faster recovery of brain PEA content. Ephedrine, also a less active drug, causes a substantial decrease of PEA, too, but normal values are reached faster. Other drugs assayed at a dose of 0.025 mmol/kg caused few changes. Nevertheless, when injected at a dose of 0.125 mmol/kg, ephedrine produced a clear decrease of brain PEA levels. This agrees with the fact that these drugs are administered in dosages five- to 10-fold higher than the usual ones for amphetamine. In view of these results, the pharmacological action of amphetamines can be explained by a complex mechanism which include a MAOI action, a PEA release, and a simultaneous or later direct effect on PEA receptors. Phenylethylamine depletion induced by these drugs can account for the depression observed during the withdrawal syndrome in human subjects (45).

Mosnaim et al. (46) studied the action of D- and L-phenylalanine on brain PEA levels in the rabbit finding that at a dose of 200 mg/kg both isomers induce an increment of PEA concentration. We confirmed these results in rats (47); a 100 mg/kg dose of D- and L-phenylalanine

TABLE 11

Rat Brain PEA Levels After
Intraperitoneal Injection of D- and L-Phenylalanine

Time (hr)	PEA Brain Levels After Treatment with D-Phenylalanine		PEA Brain Levels After Treatment with L-Phenylalanine	
	5 mg/kg	100 mg/kg	5 mg/kg	100 mg/kg
0	0.49 ± 0.05	0.49 ± 0.05	0.49 ± 0.03	0.48 ± 0.02
1	1.72 ± 0.12	3.08 ± 0.27	0.60 ± 0.03	3.15 ± 0.08
2	1.13 ± 0.11	3.12 ± 0.35	0.59 ± 0.05	3.10 ± 0.34
4	1.10 ± 0.10	3.42 ± 0.25	0.50 ± 0.06	3.00 ± 0.27
8	0.81 ± 0.06	2.50 ± 0.21	0.49 ± 0.03	1.38 ± 0.15
24	0.79 ± 0.07	2.00 ± 0.24	0.50 ± 0.02	0.60 ± 0.08
48	0.57 ± 0.03	1.70 ± 0.13	0.49 ± 0.03	0.48 ± 0.03

raise brain PEA levels up to 3.0 μg/g, the values remain high for 48 hr and then descend reaching normal values at 72 hr. At lower dosages of D- and L-phenylalanine there exist considerable differences between both aminoacids (Table 11). At a dose of 5 mg/kg, D-phenylalanine induces a considerably greater and long lasting increment of brain PEA. This finding explains the differences in the antidepressant action between D- and L-phenylalanine. When the L-isomer is administered to humans at a daily dosage of 100 to 200 mg, which is approximately the amount available in a normal diet, no pharmacological effects can be expected. Only with an overload of L-phenylalanine might there be an increment of PEA biosynthesis and also an antidepressant or other pharmacological effects.

The case is different with D-phenylalanine. This substance seems to be an artificial or maybe a natural precursor of PEA and relative low dosages induce an increment in brain PEA. This finding explains the antidepressant action of the D-isomer in the treatment of endogenous depression, an illness in which a disease of PEA metabolism seems to be involved.

VIII. CONCLUSIONS

Let us conclude the results reported in this chapter with:

1. Normal human urinary PEA values are higher than 80 μg/24 hr.
2. Normal PEA levels in Wistar rats brains are about 0.49 μg/g.
3. Endogenous depressed patients, not so reactive ones, excrete significantly lower quantities of PEA in the urine than normal subjects.
4. Antidepressant drugs raise brain PEA levels in rats and urinary PEA elimination in human beings.
5. These results support Fischer's hypothesis of a regulating role for PEA on affective states and also of the existence of a metabolic disorder in the formation or elimination of this substance in endogenous depression.

REFERENCES

1. J. A. Oates, P. Z. Niremberg, J. B. Jepson, A. Sjordama, and S. Udenfriend (1963). Conversion of phenylalanine to phenethylamine in patients with phenylketonuria. *Proc. Soc. Exp. Biol. Med.*, 112:1078.

2. J. B. Jepson, W. Lovenberg, and A. Zaltzman (1960). Amine metabolism studies in normal and phenylketonuric patients by monoaminooxidase inhibition. *Biochem. J.*, 74:50.

3. T. Nakajima, Y. Kakimoto, and I. Sano (1964). Formation of beta-phenethylamine in mammalian tissue and its effect on motor activity in the mouse. *J. Pharm. Exp. Ther.*, 143:319.

4. E. Fischer and M. Lopez Amalfará (1962). Effect of catecholamines on the psychomotor activity of mice. *Nature*, 193:590.

5. B. B. Brodie and P. A. Shore (1957). A concept for a role of serotonin and norepinephrine as chemical mediators in the brain. *Ann. N.Y. Acad. Sci.*, 66:631.

6. A. M. Carlson, T. T. Linqvist, T. Magnuso (1962). *Adrenergic Mechanisms*, Ciba Foundation Symposium, edited by J. R. Irané, G. E. W. Wolstenhome, and M. O'Connor. Little, Brown and Co., Boston.

7. G. M. Everett (1961). Some electrophysiological and biochemical correlates of motor activity. *Neuro-Psychopharmacology*, 2:479.

8. A. B. Rothballer (1959). The effects of catecholamines on the central nervous system. *Pharmacol. Rev.*, 11:494.

9. E. Fischer, R. L. Ludmer, and H. C. Sabelli (1967). The antagonism of phenethylamine to catecholamines in mouse motor activity. *Acta Physiol. Lat. Am.*, 17:15.

10. E. Fischer, J. M. Saavedra, and B. Heller (1968). Effects of catecholamines, adrenergic substances and their blocking agents on the searching behavior of mice. *Arzneim.-Forsch.*, 18:780.

11. E. Fischer (1965). Monoaminooxidase inhibitors. *Lancet*, ii:245.

12. E. Fischer and B. Heller (1972). Phenethylamine as a neurohumoral agent in brain. *Behav. Psychiat.*, 4:8.

13. W. R. Hess (1949). *Das Zwischen hirn.* Schwabe, Basel.

14. H. Spatz and N. Spatz (1971). Quantitative determination of beta-phenethylamine in the urine. *Biochem. Med.*, 6:1.

15. E. Fischer, H. Spatz, J. M. Saavedra, H. Reggiani, and B. Heller (1972). *B. Biol. Psychiat.*, 5:139.

16. E. Fischer, H. Spatz, B. Heller, and H. Reggiani (1972). Phenethylamine content of human urine and rat brain: its alterations in pathological conditions and after drug administration. *Experientia*, 28:307.

17. E. Fischer, H. Spatz, R. Fernandez Labriola, E. Rodriguez Casanova, and N. Spatz (1973). Quantitative gas chromatographic determination and infrared spectrographic identification of urinary phenethylamine. *Biol. Psychiat.*, 7:161.

18. A. D. Mosnaim and E. E. Inwant (1973). A spectrophotometric method for the quantification of 2-phenethylamine in biological specimens. *Anal. Biochem.*, 54:561.

19. A. D. Mosnaim, E. E. Inwang, J. H. Sugerman, W. J. DeMartini, and H. C. Sabelli (1973). Ultraviolet spectrophotometric determination of 2-phenethylamine in biological samples and its possible correlation with depression. *Biol. Psychiat.*, 6:235.

20. A. A. Boulton and L. Milward (1971). Separation, detection and quantitative analysis of urinary beta-phenylethylamine. *J. Chromatogr.*, 57:287.

21. J. W. Schweitzer, A. J. Friedhoff, and R. Schwartz (1975). Phenethylamine in normal urine: Failure to verify high values. *Biol. Psychiat.*, 10:277.

22. D. A. Durden, S. R. Philips, and A. A. Boulton (1973). Identification and distribution of beta-phenylethylamine in the rat. *Can. J. Biochem.*, 51:995.

23. J. M. Saavedra (1974). Enzymatic isotopic assay for and presence of beta-phenylethylamine in brain. *J. Neurochem.*, 22:211.

24. H. Spatz and N. Spatz. Submitted to publisher.

25. N. Spatz and H. Spatz. Submitted to publisher.

26. B. Heller and E. Fischer (1973). Diminution of phenethylamine in the urine of Parkinson patients. *Arzneim.-Forsch.*, 23:884.

27. B. Heller (1970). *Fenetilamina y Parkinson experimental Proc. 1er. Simp. suda mericano sobre el estado actual del tratamiento en la enfermedad de Parkinson y parkinsonismos.* Bs. As., Argentina.

28. A. D. Mosnaim and H. C. Sabelli (1971). Quantitative determination of the brain level of beta-phenylethylamine-like substance in control and drug treated mice. *Pharmacologist*, 13:283.

29. A. D. Mosnaim and E. E. Inwang (1973). A spectrophotometric method for the quantification of 2-phenylethylamine in biological specimens. *Anal. Biochem.*, 54:561.

30. E. E. Inwang, A. D. Mosnaim, and H. C. Sabelli (1973). Isolation and characterization of phenethylamine and phenethanolamine from human brain. *J. Neurochem.*, 20:1469.

31. A. D. Mosnaim, E. E. Inwang, J. H. Sugerman, and H. C. Sabelli (1973). Identification of 2-phenylethylamine in human urine by infrared and mass spectrometry and its quantification in normal subjects and cardiovascular patients. *Clin. Chim. Acta*, 46:407.

32. E. Fischer, H. Spatz, B. Heller, T. Fledel, and N. Spatz (1974). Influencia de la medicación antidepresiva sobre la eliminación de fenetilamina por la orina. *Neuropsiquiatría (Argent.)*, 5:1.

33. R. Fernandez Labriola, E. Rodriguez Casanova, H. Spatz, and N. Spatz (1974). Influence of antidepressive therapy on phenethylaminuria. *An. Psiq. Biol.*, 3:238.

34. N. M. Fernandez Pardal, C. Lopez, J. Fernandez Pardal, and E. Fischer (1973). Parkinson experimental y su tratamiento con fenetilamina y fenilalanina. *An. Psiq. Biol.*, 2:103.

35. E. Fischer, B. Heller, M. Nachón, and H. Spatz (1975). Therapy of depression by phenylalanine. *Arzneim.-Forsch.*, 25:132.

36. H. Spatz, E. Fischer, B. Heller, M. Nachón, and M. A. Di Santo (1974). Efectos de la administración de d-fenilalanina sobre el curso de episodios depresivos y la fenetilaminuria. *Therapia*, 1:52.

37. H. Spatz, B. Heller, M. Nachón, and E. Fischer (1975). Effects of d-phenylalanine on clinical picture and phenethylaminuria in depression. *Biol. Psych.*, 10:235.

38. M. Nachón. Personal communication.

39. M. Nachón. Submitted to publisher.

40. H. Spatz and N. Spatz (1971). Fenetilamina cerebral: Valores normales y bajo la acción de distintas anfetaminas. *Proc. 1er. Simp. Argent. sobre depresión.* Bs. As., Argentina.

41. N. Spatz and H. Spatz (1971). Estudios sobre el contenido cerebral de fenetilamina bajo la acción combinada de agentes IMAO y neurolépticos. *Ler. Simp. Argent. sobre depresión.* Bs. As., Argentina.

42. E. Fischer, B. Heller, H. Spatz, and H. Reggiani (1972). Thin layer chromatographic assay of phenethylamine content of rat brain and its changes after imipramine and reserpine administration. *Arzneim.-Forsch.*, 22:1560.

43. S. Gabay and A. J. Valcourt (1968). Studies on monoamino oxidase I. Purification and properties of the rabbit liver mitocondrial enzyme. *Biochim. Biophys. Acta,* 1959:440.

44. H. Spatz and N. Spatz (1973). Concentración de beta-feniletilamina en cerebros de ratas tratadas con desipramina. *A Psiq. Biol.*, 2, suppl. 4:28

45. N. Spatz and H. Spatz. Submitted to publisher.

46. A. D. Mosnaim, E. Inwang, and H. C. Sabelli (1974). The influence of psychotropic drugs on the levels of endogenous 2-phenylethylamine in rabbit brain. *Biol. Psych.*, 227-234.

Chapter 21

A CRITIQUE OF CURRENT METHODS FOR THE ANALYSIS OF PHENETHYLAMINE IN BIOLOGICAL MEDIA

Jack W. Schweitzer and Arnold J. Friedhoff

Department of Psychiatry
Millhauser Laboratories
New York University School of Medicine
New York, New York

I. INTRODUCTION

Phenethylamine (PEA) was first isolated from rotting beef pancreas about 100 years ago (18). Subsequent studies have revealed that it can be formed by the enzymatic decarboxylation of phenylalanine and that its major metabolite is phenylacetic acid.

Rather surprisingly, the first inquiry with regard to the presence of PEA in human urine was made about 25 yr after Folling described the syndrome of phenylketonuria (1). Jepson et al. (21), by means of a semiquantitative procedure, were barely able to distinguish normal and phenylketonuric subjects on the basis of PEA excretion since the maximum level for the latter group was estimated to be about 20 μg/day. However, it was shown that by blocking monoamine oxidase (MAO) (22,28) or by pretreating normal subjects or heterozygote carriers of phenylketonuria with phenylalanine and a monoamine oxidase inhibitor (MAOI), much larger amounts could be excreted. At about the same time, Nakajima et al. (27) were investigating PEA levels in rabbit brain. They reported a level of 10 ng/g after pretreatment with a MAOI.

Phenylethylamine was investigated next with respect to psychiatric illness. Fischer et al. (13), acting on their hypothesis that the antidepressive effect of treatment with MAOI might result from the elevation of brain PEA, investigated urinary levels of this amine in normals and psychotics. The normal level was found to be 33 μg/liter. Seven schizophrenics, reportedly, excreted 25 to 800 μg/liter, but the amine could not be detected in the urine of depressives. This finding stimulated a variety of research on the possible role of PEA in biological psychiatry and in CNS function. Unfortunately, widely divergent levels have been reported by different investigators, both for brain and urinary PEA, and this has tended to confound the interpretation of these findings. For this reason, we have undertaken an assessment of current procedures.

The following has been restricted to a critique of methodologies used for the analysis of endogenous levels of PEA in brain and urine. We have deliberately avoided any discussion of conclusions that may have been drawn on the basis of reported levels, since the prime purpose of this report has been to bring to the attention of investigators the suitability or nonsuitability of particular methods. Some current hypotheses relating to possible CNS roles for PEA can be found in other chapters and in the recent literature (12,31).

II. ANALYSIS OF PHENETHYLAMINE—GENERAL REMARKS

Phenethylamine is surprisingly volatile. When present in organic solvent extracts that are subjected to evaporation, a major portion can be swept out with the solvent vapors. Although losses incurred in this way can be reduced by derivatization or by conversion to a salt, it is better, perhaps, to be able to account for losses than to aim for quantitation at each purification or transfer step in the procedure. One way that this can be accomplished is by the introduction of an internal standard at the beginning of the analysis. Substances such as [1-^{14}C]PEA or a deuterated form of PEA, the latter useful only in conjunction with mass spectrometry, seem most appropriate. If the internal standard can be quantified separately from endogenous PEA when the latter is undergoing analysis, one can obtain quantitative information with regard to losses incurred during purification procedures.

The primary amino group of PEA undergoes reactions that are typical for aliphatic amines, and thus this amine can be derivatized by a wide variety of reagents. Some that have actually been used for quantifying PEA are listed in Table 1. Quantitative and semiquantitative techniques for the detection of PEA in biological media include

TABLE 1

Some Reagents Reported As Suitable for the Detection of Phenethylamine

Reagent	Method of Detection[d]	Reference
Ninhydrin[a]	A	2,28,29
2,4-Dinitrofluorobenzene[b]	A,C	11,12
Ceric sulfate	A	20,24,25,26
p-Dimethylaminocinnamaldehyde[a]	B	17,36
Alloxan[a]	B	5
Dansyl chloride[a,b,c]	B,E	4,5,6,8,10
Acyl anhydrides[b]	D	32
Perfluoroacylanhydrides[b]	C,E	33,38
CS_2	D	16,17
Fluorescamine	B	23
2,4-Dinitrobenzenesulfonic acid[b]	C	7

[a]Spray reagent.

[b]For preparation of stable derivatives.

[c]Dansyl chloride is 5-N,N-dimethylamino-1-naphthylenesulfonyl chloride.

[d]A, visual or ultraviolet spectrophotometry; B, visual or fluorimetry; C, electron capture; D, flame ionization; E, mass spectrometry.

absorption or fluorescence spectrophotometry (2,15,17,19,20,24,25,36), visual estimation from spots on thin layer or paper chromatograms (5, 6,13,14,21,27,29,35,37,39), radiolabel assay (30), flame ionization or electron capture detection following gas chromatography (7,11,12, 16,22,23,28,33,34), and mass spectrometry (3,8,9,10,38).

The direct measurement of PEA in biological media, as we shall soon show, is far more tedious than might first appear. In the event that some investigators should want to analyze for phenylacetic acid levels instead, on the assumption that the presence of this major metabolite of PEA would be a valid indicator of PEA, we would like to point out that this acid is also formed by the further oxidation of phenylpyruvic acid and the latter is both a precursor of and a metabolite of phenylalanine (1).

III. ESTIMATION OF PHENETHYLAMINE IN BRAIN TISSUE

Levels reported for rat brain, as shown in Table 2, have either been below 2 ng/g or above 300 ng/g. An estimate of 180 ng/g has been reported for one human brain (20). As might be expected from this bimodal distribution, the several methodologies can also be divided into two groups. All of the high values have been estimated from solutions that were purified by a maximum of one chromatographic separation, and this solution was then judged to be pure by such qualitative tests as ultraviolet or infrared spectroscopy and mass spectrometry. Mass spectral evidence will be examined in the next section (also see Table 4). Those methodologies by which low levels were found incorporated a number of criteria for enhancing specificity. While it is difficult, if not impossible, to indicate specifically why procedures by which high values were obtained may be unsound, the other methods, inherently more sensitive and specific, appear to be more applicable. For this reason we outline only those procedures by which specificity was assured. These, we must stress once again, are the methods by which the low values were obtained.

Saavedra (30) homogenized brain tissue in acid, centrifuged, and extracted the equivalent of about one-fifth of a whole rat brain,

TABLE 2

Reported Phenethylamine Levels in Brain

Animal	Level (ng/g)	How Detected	Reference
Rabbit	400	Colorimetry	24
Rat	492	Fluorimetry	15
Rat	534	Fluorimetry	14
Rabbit	341	Flame ionization[a]	7
Rat	60 after MAOI	Electron capture[a]	12
Rat	10 after MAOI	Visually	27
Rat	1.7	Mass spectrometry[a]	38
Rat	1.8	Mass spectrometry	9
Rat	1.5	Radioassay	30

[a]Method of detection after separation by gas chromatography.

at pH 11, with toluene. The organic phase was extracted with dilute acid, and this aqueous phase, containing brain PEA, adjusted to the proper pH, was incubated with dopamine-β-hydroxylase (DBH). This enzyme is capable of placing a hydroxyl group on the β-carbon of many β-phenethylamines. Following this, a second enzyme, phenethanolamine-N-methyltransferase (PNMT), was added and, in the presence of [methyl-^{3}H]S-adenosylmethionine, formed N[^{14}C]methylphenethanolamine. This product was extracted from the incubation medium at pH 10 with 97 : 3 toluene-isoamyl alcohol and a portion of this was counted. A parallel recovery with 10 ng of PEA added to another, identical portion of the original aqueous supernatant was also run and from which the overall recovery was estimated to be 50 to 60%. In typical assays, a blank (no DBH) contained 300 cpm, the recovery sample, 41,500 cpm, and the sample itself, 2400 cpm. From these data, one can estimate that the whole brain contained about 2.5 ng. While it is possible that all of the radioactivity identified as product instead may have been due to other substances, it is not reasonable to consider that the brain tissue could have contained much more than was reported by him.

Similar results have been obtained by Durden et al. (9) and by Willner et al. (38) by means of mass spectral analysis. These two independent methodologies included the following basic steps: homogenization of brain tissue in an acidic medium, addition of a few nanograms of deuterated PEA (see Table 2) to function as an internal standard, partial purification by extraction and/or chromatography, derivatization, further purification by thin layer or gas chromatography and finally, quantitative analysis by mass spectrometry. Quantitation was achieved by comparing the ratio of the molecular ion of derivatized PEA and derivatized deuterated PEA detected in the sample with similar ratios obtained from known mixtures of the authentic compounds. The detection of derivatized deuterated PEA in the sample established that some PEA was carried through the procedure. It is unreasonable to argue that deuterated PEA and endogenous PEA would, in some manner, separate during workup. Durden et al. (9) reported a mean level of 1.8 ng/g, and Willner et al. (38), 1.7 ng/g. What is surprising about these two procedures, which were very likely to have been arrived at independently, is the similar number and kind of steps that were incorporated. During the development of our procedure for urinary PEA, which is described later, we found it necessary to introduce about the same number of purification steps. In the absence of any one step, high and nonreproducible levels were invariably obtained.

IV. ESTIMATION OF PHENETHYLAMINE IN URINE

Phenethylamine levels in urine, as were found for brain, again fall into two separate, widely different categories. As shown in Table 3, reported levels are either below 50 μg or above 300 μg/24 hr urine. Those investigators who have reported high values have again had to rely largely on demonstrations that their urine extracts were pure and have then analyzed such extracts by relatively nonspecific techniques. The apparent impurity of such extracts can be demonstrated

TABLE 3

Reported Phenethylamine Levels in Urine of Normal Subjects

Laboratory	PEA Level	Number of Subjects
Jepson et al. (21)	< 20 μg/day	> 1
Perry (29)	0 μg/day	6
Oates et al. (28)	< 3 μg/hr	6
Fischer et al. (13)	33.5 μg/liter urine	11
Boulton and Milward (5)	47 ± 44 μg/day	19
Fischer et al. (17)	336 ± 180 μg/day	17
Fischer et al. (16)	400 ± 105 μg/day	13
Mosnaim and Inwang (24)	453 ± 50 μg/day	27
Schweitzer et al. (33)	10.3 ± 18.2 μg/day	18
Reynolds and Gray (41)	6.8 ± 2.9 μg/day	12

by the authors' own data. Inwang et al. (20) and Mosnaim et al. (26), respectively, have provided mass spectra of presumed PEA as isolated from brain and urine. These spectra were obtained from extracts that were purified beyond the point where quantitative analysis would normally be carried out. Since neither report included a companion spectrum of authentic PEA, we compared the spectra recorded by them with that of authentic PEA generated on two different types of mass spectrometers. One of these, Perkin-Elmer model MS 270, is a double focusing instrument in which the ions are separated by a magnetic sector. The other, Finnigan model 3000, is of the quadrupole type. As can be seen from the data in Table 4, the instruments provided nearly identical fragmentation patterns. In contrast, the spectra reported by the above cited investigators (Table 4) would indicate that their material was contaminated. Our purpose here is not to challenge the conclusion that PEA was present in these samples, but to point out that the spectra (which they have referred to as "simplified" spectra) appear to indicate the presence of other substances as well. In particular, the abundant fragments at m/e 55, 56, and 57 in these spectra are totally absent from our

TABLE 4

Mass Spectra of Authentic PEA and Material Isolated from Brain and Urine[a]

	Spectra of Authentic PEA		Spectra of Isolated Material	
	A	B	Brain	Urine
m/e	Rel. Ab.	Rel. Ab.	Rel. Ab.	Rel. Ab.
30	100	100	100	100
36	17	17	38	6
37	1	1	23	
38	6	5		
39	4	2	38	
44	1	0	75	51
50	2	0		
51	3	2	38	4
52	1	0		
55	0	0	63	88
56	0	0	50	79
57	0	0	63	90
63	3	1		
65	6	7	38	4
67	0	0		10
68	0	0		8
71	0	0	50	7
73	0	0		
77	3	2	38	5
91	13	10	45	11
92	6	4	26	5
103	2	1		
104	1	0		
117	1	0		
118	1	0		
119	1	0		
120	1	0		
121	6	4	25	3
122	1	0		

[a]A and B refer to mass spectral data obtained on Perkin-Elmer MS 270 and Finnigan 3000 mass spectrometers, respectively, and are rounded to the nearest whole number. Brain and urine data were taken from Inwang et al. (20) and Mosnaim et al. (26), respectively. All relative abundances (Rel. Ab.) and mass to charge ratios (m/e) reported by those authors are reproduced in this table. See text for explanation.

spectra of authentic material. Parenthetically, Mosnaim et al. (25) and Inwang et al. (20) have reported that m/e 55 and 57 also predominate in "simplified" spectra for presumed phenethanolamine which was isolated from brain and urine. A reference spectrum was again not provided and one that we obtained on the P-E MS 270 did not reveal significant abundances for those masses.

As part of their analysis, the above investigators used ceric sulfate to generate a chromophore with PEA. In demonstrating (25) that, of some 20 to 30 selected compounds that also produced absorbence when reacted with ceric sulfate, none were carried through their purification steps prior to analysis, these authors may have erred in assuming that numerous other compounds present in urine might likewise also be eliminated. In fact, as their own mass spectra would indicate, some of these endogenous substances must have been present, and from a comparison of their reported levels and ours (Table 3), we must conclude that some of these endogenous substances were measured as PEA.

Our experience with urinary PEA began some years ago in connection with the development of a gas chromatographic method for the analysis of amphetamine (32). At that time we chose PEA to be the internal standard because, in addition to certain favorable chemical characteristics, we were unable to detect it in significant quantities (less than 30 ug/liter urine) on gas chromatographic records. In developing our procedure (33) for PEA analysis, we used the same principle: the use of an internal standard to provide an estimate of losses. For our present purpose, $[1-^{14}C]$PEA was found to be convenient and in addition was useful for locating PEA on chromatograms. In common with the mass spectrometric procedures for the analysis of brain PEA, we found it necessary to carry out several chromatographic purification steps prior to quantitative analysis. After the addition of about 0.6 μg of $[1-^{14}C]$PEA to a urine sample containing 100 mg of creatinine, bases were extracted at high pH. The concentrated extract was subjected to thin layer chromatography and the eluted labeled area was derivatized with heptafluorobutyric anhydride. The product was rechromatographed and the labeled area was eluted. One portion was counted, and another portion was injected into a gas chromatograph fitted with a Poly A-103 column. Electron capture (^{63}Ni) permitted the detection of subnanogram quantities. The peak appearing at the correct retention time for authentic N-heptafluorobutyryl-PEA was matched against a curve prepared from external standards. The value obtained, corrected for the contribution of the internal standard to the peak height and adjusted up to the original quantity of label added to the urine, was the endogenous PEA level for that urine sample. We would again like to emphasize that techniques such as these yield maximum levels. It is entirely possible that the peak seen at the correct retention time for the derivative may have, in fact, been due almost entirely to other substances. The quantitative recovery of one microgram of unlabeled PEA added to parallel urine samples (33) indicated to us that the method was suitably sensitive and, therefore, of use for determining whether previously reported low or high values were, in the main, correct. Our analyses of urines from 18 normal subjects yielded a mean level of 10.3 ± 18.2 μg/24 hr urine (Table 3). The same method has been found useful for the estimation of PEA in 1- to 2-g samples of chocolate (34). Levels of 1 to 15 μg/g were obtained, depending on brand and type, in agreement with a current estimate (40).

V. SUMMARY AND CONCLUSIONS

Average levels of PEA vary according to the investigator. Some laboratories report consistently high values and others, uniformly low (Tables 2 and 3). Methodologies by which low values were obtained have incorporated an internal standard and numerous purification steps. The failure to add an internal standard and the known volatility of PEA probably prevented those investigators who obtained high values from carrying out apparently necessary additional purifications. Instead, they placed too great a stress on the use of other quantitative and qualitative tests to verify that their extracts were pure. We would characterize these two basically different concepts as being serial and parallel approaches toward establishing validity. By the former approach, additional purification steps are introduced until the method is sufficiently sensitive and specific. By the latter, the same partially pure extract is analyzed by several different techniques. Unfortunately, for the specific analysis of PEA, all tests of such extracts yielded similar high values and forced the investigators to conclude that endogenous levels of PEA were high.

Of the acceptable methods, the double enzyme-isotope transfer method perfected by Saavedra (30) appears to be the most convenient procedure, considering the requirements of apparatus, time and skill, for the determination of PEA in brain. If it is used for the analysis of PEA in biological media other than brain, its specificity should be re-evaluated.

ACKNOWLEDGMENTS

This work was supported in part by Research Scientist Award No. 14024 to AJF and Grant No. MH08618, both from the United States Public Health Service.

REFERENCES

1. M. D. Armstrong (1963). Biochemistry, Chapter 4. In: *Phenylketonuria*, edited by F. L. Lyman. Charles C Thomas, Springfield, Ill.
2. U. Bachrach, M. Segal, and R. Rosansky (1958). Effect of tetracyclines on formation of amines by bacteria. *Proc. Soc. Exp. Biol. Med.*, 97:874-876.
3. G. B. Baker and A. A. Boulton (1974). Identification and quantitative analysis of tryptamine, p-tyramine and β-phenylethylamine in subcellular fractions isolated from rat brain both in the presence and absence of pargyline. *Proc. Amer. Soc. Neurochem.*, New Orleans.
4. A. A. Boulton and G. B. Baker (1975). The subcellular distribution of β-phenylethylamine, p-tyramine and tryptamine in rat brain. *J. Neurochem.*, 25:477-481.
5. A. A. Boulton and L. Milward (1971). Separation, detection and quantitative analysis of urinary β-phenylethylamine. *J. Chromatogr.*, 57:287-296.

6. A. A. Boulton, S. R. Philips, and D. A. Durden (1973). The analysis of certain amines in tissues and body fluids as their dansyl derivatives. *J. Chromatogr.*, 82:137-142.

7. R. L. Borison, A. D. Mosnaim, and H. C. Sabelli (1974). Biosynthesis of brain 2-phenylethylamine: Influence of decarboxylase inhibitors and D-amphetamine. *Life Sci.*, 15:1837-1848.

8. D. A. Durden, B. A. Davis, and A. A. Boulton (1974). Qualitative and quantitative mass spectrometry of some noncatecholic biogenic amines and related compounds as their dansyl derivatives. *Biomed. Mass Spectromet.*, 1:83-95.

9. D. A. Durden, S. R. Philips, and A. A. Boulton (1973). Identification and distribution of β-phenylethylamine in the rat. *Can. J. Biochem.*, 51:995-1002.

10. L. E. Dyck and A. A. Boulton (1975). The effect of some drugs on the urinary excretion of some aryl alkyl amines in the rat. *Res. Commun. Chem. Pathol. Pharmacol.*, 11:73-77.

11. D. J. Edwards and K. Blau (1972). Analysis of phenylethylamines in biological tissues by gas-liquid chromatography with electron capture detection. *Anal. Biochem.*, 45:387-402.

12. D. J. Edwards and K. Blau (1973). Phenethylamines in brain and liver of rats with experimentally induced phenylketonuria-like characteristics. *Biochem. J.*, 132:95-100.

13. E. Fischer, B. Heller, and A. H. Miro (1968). β-Phenylethylamine in human urine. *Arzneim.-Forsch.*, 18:1486.

14. E. Fischer, B. Heller, H. Spatz, and H. Reggiani (1972). Thin layer chromatographic assay of phenethylamine content of the rat brain and its changes after reserpine and imipramine administration. *Arzneim.-Forsch.*, 22:1560.

15. E. Fischer, H. Spatz, B. Heller, and H. Reggiani (1972). Phenethylamine content of human urine and rat brain, its alterations in pathological conditions and after drug administration. *Experientia*, 28:307-308.

16. E. Fischer, H. Spatz, R. S. Fernandez Labriola, E. M. Rodriguez Casanova, and N. Spatz (1973). Quantitative gas-chromatographic determination and infrared spectrographic identification of urinary phenethylamine. *Biol. Psychiat.*, 7:161-165.

17. E. Fischer, H. Spatz, J. M. Saavedra, H. Reggiani, A. H. Miro, and B. Heller (1972). Urinary elimination of phenethylamine. *Biol. Psychiat.*, 5:139-147.

18. M. Guggenheim (1940). *Die Biogenen Amine*. S. Karger, A. G. Verlag, Basel, pp. 420-421.

19. E. E. Inwang, P. U. Madubuike, and A. D. Mosnaim (1973). Evidence for the excretion of 2-phenethylamine glucuronide in human urine. *Experientia*, 29:1080-1081.

20. E. E. Inwang, A. D. Mosnaim, and H. C. Sabelli (1973). Isolation and characterization of phenylethylamine and phenylethanolamine from human brain. *J. Neurochem.*, 20:1469-1473.

21. J. B. Jepson, W. Lovenberg, P. Zaltzman, J. A. Oates, A. Sjoerdsma, and S. Udenfriend (1960). Amine metabolism, studied in normal and phenylketonuric humans by monoamine oxidase inhibition. *Biochem. J.*, 74:5P.

22. R. J. Levine, P. Z. Nirenberg, S. Udenfriend, and A. Sjoerdsma (1964). Urinary excretion of phenethylamine and tyramine in normal subjects and heterozygous carriers of phenylketonuria. *Life Sci.*, 3:651-656.

23. A. D. Mosnaim, R. L. Borison, P. U. Madubuike, and H. C. Sabelli (1974). Synthesis and metabolism of brain 2-phenylethylamine. *Proc. Amer. Soc. Neurochem.*, New Orleans.

24. A. D. Mosnaim and E. E. Inwang (1973). A spectrophotometric method for the quantification of 2-phenylethylamine in biological specimens. *Anal. Biochem.*, 54:561-577.

25. A. D. Mosnaim, E. E. Inwang, J. H. Sugerman, W. J. DeMartini, and H. C. Sabelli (1973). Ultraviolet spectrophotometric determination of 2-phenylethylamine in biological samples and its possible correlation with depression. *Biol. Psychiat.*, 6:235-257.

26. A. D. Mosnaim, E. E. Inwang, J. H. Sugerman, and H. C. Sabelli (1973). Identification of 2-phenylethylamine in human urine by infrared and mass spectroscopy and its quantification in normal subjects and cardiovascular patients. *Clin. Chim. Acta*, 46:407-413.

27. T. Nakajima, Y. Kakimoto, and I. Sano (1964). Formation of β-phenylethylamine in mammalian tissue and its effect on motor activity in the mouse. *J. Pharmacol. Exp. Ther.*, 143:319-325.

28. J. A. Oates, P. Z. Nirenberg, J. B. Jepson, A. Sjoerdsma, and S. Udenfriend (1963). Conversion of phenylalanine to phenethylamine in patients with phenylketonuria. *Proc. Soc. Exp. Biol. Med.*, 112:1078-1081.

29. T. L. Perry (1962). Urinary excretion of amines in phenylketonuria and mongolism. *Science*, 136:879-880.

30. J. M. Saavedra (1974). Enzymatic isotopic assay for and presence of β-phenylethylamine in brain. *J. Neurochem.*, 22:211-216.

31. H. C. Sabelli and W. J. Giardina (1973). Amine modulation of affective behavior. In: *Chemical Modulation of Brain Function*, edited by H. C. Sabelli. Raven Press, New York, pp. 225-259.

32. J. W. Schweitzer and A. J. Friedhoff (1970). Amphetamines in human urine: Rapid estimation by gas-liquid chromatography. *Clin. Chem.*, 16:786-788.

33. J. W. Schweitzer, A. J. Friedhoff, and R. Schwartz (1975). Phenethylamine in normal urine: Failure to verify high values. *Biol. Psychiat.*, 10:277-285.

34. J. W. Schweitzer, A. J. Friedhoff, and R. Schwartz (1975). Chocolate, β-phenethylamine and migraine reexamined. *Nature*, 257:256.

35. J. W. T. Seakins, R. S. Ersser, and I. S. E. Gibbins (1970). Studies on the origin of faecal amino acids in cystic fibrosis. *Gut*, 11:600-609.

36. H. Spatz and N. Spatz (1972). Spectrophotophlorometric determination of Beta-phenylethylamine in blood and urine. *Biochem. Med.*, 6:1-6.

37. H. M. Stevens and P. D. Evans (1973). Identification tests for bases formed during the putrefaction of visceral material. *Acta Pharmacol. Toxicol.*, 32:525-552.

38. J. Willner, H. F. LeFevre, and E. Costa (1974). Assay by multiple ion detection of phenylethylamine and phenylethanolamine in rat brain. *J. Neurochem.*, 23:857-859.

39. P. H. Wu and A. A. Boulton (1975). Metabolism, distribution and disappearance of injected β-phenylethylamine in the rat. *Can. J. Biochem.*, 53:42-50.

40. M. Saxby. Personal communication.

41. G. P. Reynolds and D. O. Gray (1976). A method for the estimation of 2-phenylethylamine in human urine by gas chromatography. *Clin. Chim. Acta*, 70:213-217.

Chapter 22

USE OF D-PHENYLALANINE IN THE TREATMENT OF DEPRESSION: THEORETICAL AND CLINICAL IMPLICATIONS

J. A. Yaryura-Tobias

Research Division
North Nassau Mental Health Center
Manhasset, New York

I. INTRODUCTION

Historically, depression has been described and reported in numerous instances throughout the centuries in secular and religious books, usually referred to as melancholia (from the greek melas: black, and chole: bile). In today's world, where man is questioned at a material and cosmic level, the input of stimuli overwhelms the capacity to process the incoming data. It is then that emotional disturbances take place, as the ability of man to cope with them fails.

Depression and schizophrenia are the two most important forms of mental illness, not only for the lack of knowledge that surrounds them, but also for the great number of people affected. World-wide distribution is the same, if the diagnostic elements are agreed upon. Of these two entities, our interest is centered on depression.

A. Definition

The word "depression" can be defined as a mood disorder, characterized by lowered spirits.

B. Classification

Depression can be classified in three main groups: a. organic, b. neurotic, and c. reactive. These groups may be subdivided into subgroups (see Table 1), for a better understanding of depression, which will help to diagnose, treat and give a prognosis.

TABLE 1

Classification of Depression

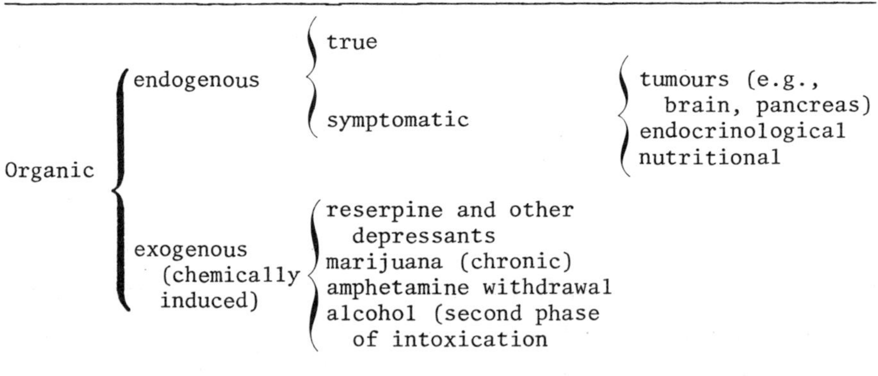

Organic	endogenous	true	
		symptomatic	tumours (e.g., brain, pancreas) endocrinological nutritional
	exogenous (chemically induced)	reserpine and other depressants marijuana (chronic) amphetamine withdrawal alcohol (second phase of intoxication	

Neurotic (personality)

Reactive	existential socioeconomic emotional

C. Symptomatology

In the onset of depression, precipitant factors are important and can be defined as any psychological or organic stimulus which will trigger the mechanism of any form of depression. Subjective symptoms are low self-esteem, self-deprecation, guilt feelings, loss of interest, decreased libido, loss of "joie de vivre," feelings of hopelessness and helplessness, suicidal ideation, and lack of drive. There is also a slowness of the stream of thought and loss of concentration. At times, anxiety and agitation can be present. In depression, reality is never impaired; therefore, psychotic depression may be in itself a completely different entity, because the disorder of the mood is secondary to an altered thought process.

Physical symptoms are mostly present in organic forms of depression and are helpful in distinguishing it from neurotic or reactive forms (see Table 2). To establish this distinction from the beginning is important, because a proper diagnosis will be the only way to apply the right therapeutic approach and arrive at a prognosis.

As in many other illnesses that affect humans, depression, in its true form, is a generalized illness. We may say that all organic systems are affected by depression. Although the main symptomatology reflects brain pathology such as mood, thought, and behavior changes, careful clinical examinations will reveal that other organs are directly or indirectly involved. For instance, cerebellar implications may be suspected if the patient complains of loss of equilibrium, or if his coordination is impaired. Some times the illness can be monosymptomatic, or masked by anxiety. The correction of any physical symptom may not only provide information that may eventually be helpful for the programming of the treatment, but also might control unwanted variables which may jeopardize the overall treatment. Example: a patient who is depressed and has also a low arterial blood pressure reading of 80/50 mmHg will complain of dizziness, weakness, and fatigue. If untreated, these symptoms may overlap and consequently mask similar symptoms of true depression.

TABLE 2

Physical symptoms of True Depression

CNS	disturbed sleeping pattern cephalea dizziness loss of equilibrium motor incoordination slow speech tremor (voluntary) hypomotoricity
ENT	loss of ocular gloss rhinitis
CV	tachycardia arterial hypotension
GI	anorexia loss of weight saburra constipation bloating dyspepsia vesicular dysfunction hepatatic insufficiency
Lung	dyspnea acrocyanosis
GU	impotence pollakiuria dysmenorrhea
Skin	dry warm perspiration

D. Diagnosis

Diagnosis is usually based upon subjective symptoms verbally manifested by the patient and objective symptoms observed by patient, family, and physician. These physical symptoms are diagnosed, not uncommonly, as hypochondriasis by the psychiatrist or internist. Therefore, the need of depressed patients to believe in a psychological cause even if the depression is endogenous, is reinforced by the clinician and the patient's search for a psychological reason. Sadness and boredom, two states of mind which may be confused with depression, may be ruled out if they are remembered when establishing a diagnosis.

The interest of this chapter is primarily dedicated to true depression. Is its clinical diagnosis possible? The answer is yes, if other organic factors are screened.

There is a group of symptoms which can help to separate true from reactive or neurotic depression. In the latter, symptoms usually are dynamic in nature and closely related to mind mechanisms. In the former, physical characteristics are predominant. The patient presents a sleeping pattern of frequent awakenings during the night with a final awakening in early morning where sleep cannot be reconciled again. One of the main symptoms that differentiates true from neurotic and reactive depressions is that the patient wakes up depressed. In neurotic depression, the patient is depressed through the day; and in reactive, he is at his worst by the evening where daily events, as depressant factors, reach their peak.

Other elements which can be helpful are seasonal rhythm of recurrence and the therapeutic response. In most instances, non-organic depressions respond more favorably to psychotherapy than does true depression. Contrariwise, biological therapies are more efficacious in the latter.

In summary, the clinical diagnosis of true depression can be established by utilizing the proper knowledge of the symptomatology.

II. BIOLOGICAL THEORIES OF TRUE DEPRESSION

At the beginning of the 1960s, two main hypotheses on the biochemistry of depression were advanced: 1) the norepinephrine (NE) theory and 2) the serotonin (5-HT) theory. As it is known in the diseases of the affect, biogenic amines have been the subject of speculation for these many years. However, almost 15 yr have elapsed without arriving at conclusive results. A good review has been published by Mendels (33).

Later, a third theory of depression was advanced by Fischer et al. (14) and, independently, by Dewhurst et al. (9) postulating phenylethylamine (PEA) as a physiologic stimulant of ergotropic properties.

The basic research in PEA is described elsewhere in this book. However, for the better understanding of this chapter, and in order to make clinical correlations, we shall summarize some of the findings.

According to Fischer et al. (17), DOPA and catecholamines in inhibit the exploratory behavior in mice, while PEA stimulates it. Also, PEA antagonizes the inhibitor effect of DOPA and catecholamines in the exploratory behavior of mice.

The chemical similarity of PEA and amphetamine is known. Also, they are related to hallucinogenic compounds. To help prove it as a neurohumoral agent, its presence was demonstrated in animal and human urine (36,38,41). Its action can be modified by depressor agents such as reserpine and antidepressant substances such as monoamine oxidase inhibitors (MAOI) and tricyclicals (10-12,20,23,39,49). In addition, tetrahydrocannabinol increases its concentration in the brain (50).

The main concentration of PEA occurs in the cerebellum (38); thus, its brain distribution may be indicative of one of the neurochemical pathways in the mechanism of depression. Excess administration of phenylalanine to neonatal rats produces neuronal lesions in the cerebellum (1). Experimental work of Heath et al. (26) in human subjects would indicate a mediatory action of the cerebellum in the neurophysiological aspects of emotion. Thereby, the cerebellum not only regulates synergysm, symmetry, and equilibrium, but it also may participate in the regulation of emotions as well. Moreover, in true depression, some physical signs could be of cerebellar nature (e.g., loss of equilibrium and tremors) (64, personal communication).

It was hypothesized by Sandler (51) that a subgroup of migraneous patients presented an abnormal PEA metabolism, and that this form of migraine is a pulmonary disease (51). The access of free monoamines to the lungs brought about by defective inactivation (52) may account for the release of vasoactive substances from the lungs into the systemic circulation, where a PEA-oxidizing defect is present (53).

Moreover, reserpine, which has depressant and vasoactive properties, may cause headache as a side effect. Experimental migraine has been induced by reserpine (29). Headaches could also be triggered by monoamine-containing foods (24), such as chocolate or wine (25), which are known to contain PEA (52).

Pulmonary tuberculosis, asthma, and manic depressive illness includes depression, among other symptoms (46,55). Furthermore, depressed patients may complain of dyspnea and acrocyanosis. Whether the pathophysiology of dyspnea alters the pulmonary metabolism of tyramine and PEA (substances that are related to depression) is yet to be determined. For us, dyspnea appears as an interesting factor in true depression. The biochemical treatment of tuberculosis with the introduction of iproniazid caused euphoric and/or manic symptoms. These findings started the era of antidepressants, which are iproniazid derivatives (8,70). Theoretically, a pulmonary model of depression could be entertained. However, PEA was not detected in cat and rabbit lungs and has not yet been measured in human lungs (37).

III. PHENYLETHYLAMINE'S ROLE

A. Phenylethylamine and Pyridoxine

A vitamin that may be correlated to the chemistry of depression is pyridoxine (Py). This vitamin is an important co-enzyme in the metabolism of cerebral monoamines; substances not only indispensable in brain, but in heart, gastro-intestinal, and pulmonary chemistry, as well.

Loo (30) has established a relationship between PEA and Py; it seems that Py becomes depleted when PEA forms a complex with it. The same author has identified a new phenylalanine metabolite derived from exogenous phenylalanine and Py in phenylketonuria.

Empirically, Py has an energetic and an antidepressant activity if given in high doses with vitamin B_1. It was shown later on that females under anovulatory drug treatment, who manifest symptoms of depression as an untoward effect, are low in Py. Subsequently, these symptoms are reversed by pyridoxine administration (61). It also acts as an anticonvulsant (5) and indirectly, as an antiaggressor factor (66). It has been claimed that it has antipsychotic properties (2) and that it increases the elimination of urinary DA in schizophrenic patients (65).

In the conversion of DOPA to DA, tryptophan to 5-HT or to anthranillic acid, Py is utilized.

If we briefly look into phenylalanine, a precursor of PEA, we find that phenylpyruvate is a potent inhibitor of 5-hydroxytryptophan (5-HTP) carboxylase in pig kidney homogenate (44) and an inhibitor of DOPA decarboxylase in vitro (10). Furthermore, PEA depletes DA (23).

B. Phenylethylamine and Alcohol

In alcoholic intoxication, there is first a phase of excitation: increased motor activity, verborrhagia and happiness, followed by a second phase: hypomotoractivity with unsteady gait, slow thought process, and depression.

The first phase is characterized by hyperglycemia, as glucose is required to metabolize alcohol, and the second phase is characterized by hypoglycemia. In acute alcoholism, glucose disturbance may play a role in symptoms of depression, aggressive behavior, and cephalea. Whether PEA is increased in alcoholic intoxication has yet to be determined. Is the typical "hangover headache" related to a PEA disturbance?

In 1953, it was reported that the i.v. administration of 50 mg of pyridoxine to patients with acute alcoholic intoxication resulted in a sudden disappearance of the acute symptoms (57). Arterial hypotension was the only side effect observed.

In 1958, I treated about 50 acute alcoholic patients by the administration of 50 to 100 mg of pyridoxine with similar results (63). In addition, in hyperphenylalaninemic conditions, glycolysis, and energy generation may be impaired (60).

C. Phenylethylamine and Melanin

Phenylalanine is a tyrosine precursor and, in normal individuals, a melanin precursor. It is known that in phenylketonuria, a relative decrease of melanin formation is present, and clinically manifested by fair skin, blond hair, and blue eyes.

Disturbance of pigmentation is seen in chlorpromazine (CPZ)-treated patients, with CPA known to be a mood depressant. Changes of pigmentation are reported in schizophrenic patients and also in Addisonian patients, where adrenal physiology is hypoactive and depression one of the psychiatric symptoms manifested.

The incidence of endogenous depression in populations of fair skins against dark skins living in the original habitat has not been established. Yet, in areas where cold temperatures and less sun exposure are present, a trend to depression, suicide, slow movements, and so on is sociologically and culturally known and outstanding, if compared to regions close to the Equator. Interestingly enough, these countries' inhabitants drink alcoholic beverages in excess. Wine, known to have PEA (M. J. Saxby, 1974, personal communication), would exert its stimulant action, not only through its alcoholic but also through its PEA content.

D. Phenylethylamine in Human Clinical Research

First report on the presence of PEA in urine of normal individuals was made by Jepson et al. (28). Others confirmed their findings (4,15,18,19,37,42,45,56) (see Table 3).

Fischer et al. (19) demonstrated that urinary PEA is decreased in true depression and increased in schizophrenia. Further research has confirmed low urinary PEA values in depressed individuals (4,35, 47) (see Table 4).

Subsequently, oral PEA was given to relieve clinical symptoms of true depression. This study, done by Fischer's group, had to be discontinued because PEA caused severe cephaleas. One way to obviate the problem was to administer phenylalanine, hoping that enough of it would be converted into PEA. Therefore, PEA was replaced by the oral

TABLE 3

Urinary PEA in Normal Subjects[a]

Laboratory	PEA (mg/24 hr)	Number of Subjects
Jepson et al. (28)	< 20	> 1
Oates et al. (42)	< 3	6
Fischer et al. (17)	33.5 mg/liter	11
Boulton & Milward (4)	47 ± 44	19
Fischer et al. (18)	400 ± 105	13
Mosnaim et al. (36)	453 ± 50	27
Schweitzer et al. (56)	10.3 ± 18.2[a]	18

[a]Per 100 mg of creatinine.

TABLE 4

Urinary PEA in True Depression

Laboratory	PEA (μg/24 hr)	Number of Subjects
Fischer et al. (20)	114 ± 49	13
Yaryura-Tobias et al. (64)	34 ± 14	6
Mosnaim and Sabelli (39)	352 ± 85	24
Boulton and Milward (4)	0	20
Rodriguez Casanova and Fernandez Labriola (48)	44 ± 14	30

administration of D,L- and D-phenylalanine 100 to 300 mg/day. Under D,L-phenylalanine, three cases improved out of six and under the D-form, seven cases improved out of nine. Improvement was observed within the first week and, in a few cases, agitation. In psychotic patients with concomittant symptoms of depression, D,L-phenylalanine seems to aggravate their mental symptoms.

Our own experience has shown that of all forms, the D-form is the best to relieve symptoms of depression. Untoward effects are usually mild and easily controllable. Improvement appears within the first 7 days, treatment usually lasting 2 weeks. In most instances, there was no need to continue treatment (64). This is a puzzling observation, for antidepressant therapy requires no less than 6 months to avoid a recurrence of symptoms.

Later, these findings were confirmed by other investigators (16,27,40,59). Furthermore, in those cases where phenylalanine was partly ineffective per se, it enhanced the action of tricylical antidepressants.

In another study, we decided to examine the relationship between PEA values in urines of true depression patients and 5 hr oral glucose tolerance tests (5OGTT), mainly because of the presence of glucose disturbance in psychiatric illnesses (68) (see Table 5). A sample of 12 patients (nine females and three males), ages ranging from 28 to 58 ($\overline{X}$ = 41.66), was selected. Diagnosis was true depression with a duration range from first to last episode of 1 to 25 yr ($\overline{X}$ = 7.33). Phenylethylaminuria values ranged from 12 to 76 μg/24 hr ($\overline{X}$ = 38.08) and normals ranging from 130 to 350 μg/24 hr. Twenty-four urines were collected. Out of the 12 patients, two had normal 5OGTT, one showed a prediabetic curve, and nine functional hypoglycemic responses. All patients received oral D-phenylalanine ($\overline{X}$ = 200 mg) per day in divided doses for 2 weeks. Results were as follows: five complete remission, two partial improvement, one mild improvement, three negative results, one discontinued treatment for personal reasons. In two cases, there was a moderate relapse within 2 weeks. Another trial of one month of D-phenylalanine was positive. In

TABLE 5

Pretreatment Phenylethylaminuria and 60GTT Values in True Depression

NO. = 12	AGE	DURATION/YEARS	URINARY PEA μg/24 HR NORMAL VAL. (130-350)	60GTT FAST	30 MIN	1 HOUR	2 HOUR	3 HOUR	4 HOUR	5 HOUR	6 HOUR	TYPE	IMPROVEMENT
M.L.	58	3	31	80	80	80	100	80	80	100	100	F	VERY GOOD
M.S.	36	6	41	80	117	77	72	69	95	92	90	F	NEGATIVE
N.T.	41	12	46	80	119	110	79	69	70	81	85	F	NEGATIVE
CH.T.	38	4	40	90	130	120	80	100	70	—	—	N	GOOD
M.S.	35	10	31	67	107	—	90	86	73	91	72	F	GOOD
M.D.	53	25	21	90	100	90	90	90	100	90	—	F	GOOD
C.V.	41	2	23	86	165	210	270	170	—	—	—	P_D	GOOD
L.M.	37	7	66	75	80	80	75	60	60	60	70	H	VERY GOOD
M.P.	28	3	76	84	84	70	56	72	72	70	70	H	MILD
Z.C.	58	1	12	80	140	130	110	100	80	70	—	F	NEGATIVE
G.D.	44	10	28	49	84	42	48	49	36	35	—	H	GOOD
H.A.	31	5	42	100	160	120	110	105	80	70	60	N	GOOD

N = NORMAL CURVE
F = FLAT CURVE
H = FUNCTIONAL HYPOGLYCEMIA
P_D = PREDIABETIC

psychotic patients with depressive symptomatology, D-phenylalanine aggravated the psychosis (67).

IV. DISCUSSION

We have reviewed physical symptoms and the role of PEA in true depression. Theoretically, there are two main organic systems involved: a) cerebral and b) pulmonary. The body as a whole cannot ignore one system if both systems share, in importance, the same substance, in this case PEA. Therefore, psychiatric, cerebellar, and pulmonary symptoms should be looked for.

Pigmentation may also be an important indicator of a PEA metabolic error. For instance, light hair and skin color associated to phenylketonuria has suggested an impairment of tyrosin metabolism. Furthermore, repigmentation has been observed where these patients were placed on a low phenylalanine diet (10). In phenylketonuria, cutaneous findings are not rare (21).

The incidence of functional hypoglycemia in hypophenylethylaminuric patients may be an index of phenylalanine changes, as usually seen in phenylketonuric patients with low levels of phenylalanine (6). Phenylalanine intake modifies behavioral variables in phenylketonuric children (22). Additional evidence can be provided by the fact that cerebral monoamines interact and keep a molecular balance. In 1967, Woolley pointed out that in phenylketonuria, the tissues are deficient in 5-HT and catecholamines. Moreover, the administration of melatonin or 5-HTP prevented a defect in maze-learning ability in infant mice fed with large quantities of phenylalanine (62). Also, there is a decrease of brain serotonin in phenylalanine fed rats (69). In humans suffering from phenylketonuria, as early as 1957, Pare et al. found low blood 5-HT levels (43), which was corroborated later on in human brain autopsies (31). Therefore, phenylketonuria could serve as an oppositional model to study some of the problems encountered in the physiopathology of PEA disturbances.

Phenylalanine appears to act on the basal ganglia, as it aggravates Parkinson's disease when given in doses varying from 1.6 to 12.6 g/day (7).

Hallucinogenic PEA also interact with 5-HT turnover (3). Most neurotransmitters have been implied in the etiology of schizophrenia and depression. Yet, when administered, only 5-HT, tryptophan, and levoDOPA caused psychiatric disturbances. Phenylethylamine seems elevated in urine of schizophrenic patients (13). D,L- and D-Phenylalanine aggravates psychosis as previously signaled in this chapter. Moreover, in 1975, Shulgin and Carter (58) introduced two new PEA; 4-methyl-2,5-dimethoxy-phenethylamine and 4-bromo-2,5-dimethoxyphenethylamine. These agents are active in man at oral levels of 0.1 to 0.2 mg/kg, causing euphoria, body awareness, and increased perceptual receptiveness. Recently, Sandler (54) postulated PEA as a schizophrenic agent.

All this experimental and clinical data has taken us from molecular biology to the medical treatment of the patient, indicating the relationship of PEA and true depression. Some comments should be made in this regard. From the various laboratories measuring urinary PEA, uniformity in their given values is not found. This could be explained based on the different methodologies used or regional dietary differences (56) or different ways of classifying true depression. However, despite the lack of uniformity in PEA values, group differences (controls, true depressives, schizophrenics) have been reported by all investigators with the exception of Schweitzer et al.

Further clinical supportive evidence of the antidepressant action of PEA is that its urinary excretion in true depressed patients is increased after the administration of tranylcipromine, imipramine, and desipramine (48), or D-phenylalanine (59) (see Table 6).

From the clinical viewpoint, it seems that D-phenylalanine is most effective in true depressed patients where previous biological therapies failed to bring about improvement. These observations

TABLE 6

Urinary PEA (μg/24 hr) Before and After Treatment[a]

Laboratory	Medication	Pre-treatment	Post-treatment
Rodriguez Casanova and Fernandez Labriola (48)	MAOI/Tricyclical	44 ± 14 (1)	265 ± 218
Spatz et al. (59)	D-Phenylalanine	$\overline{X}$ = 43.7 (2)	$\overline{X}$ = 187 (3)

[a](1), 30 pts; (2), 7 pts; (3), 4 pts.

indicate that true depressed patients refractory to other forms of treatment may constitute a subgroup which may benefit from a trial of D-phenylalanine. The efficacy of this treatment is usually observed within 5 to 14 days, similar to desipramine therapy.

The present evidence, although highly stimulating for academic exercise, is far from being definitive. If NE and 5-HT occupies a place in the biochemical theories of true depression (32), the same can be said of PEA, for they all share the patient. However, brain autopsies in suicidal patients failed to show changes in NE and DA concentrations (34). If the reader reviews the mechanism of actions of the monoamines, a biochemical equilibrium is always required for the proper physiology to operate. The only substantial hypothesis is that monoamines are ultimately related to the physiopathology of psychoses and affective disorders.

New concepts in brain neuroanatomy and recent investigations in the cholinergic system would indicate caution, as a first step, before forwarding final conclusions in the organic theories of depression.

Finally, I wonder what will happen 20 years from now.

REFERENCES

1. L. S. Adelman, J. D. Mann, D. W. Caley, and N. H. Bass (1973). Neuronal lesions in the cerebellum following the administration of excess phenylalanine to neonatal rats. *J. Neuropathol. Exp. Neurol.*, 3:380-393.

2. J. V. Ananth, T. A. Ban, and H. E. Lehman (1973). Potentiation of therapeutic effects of nicotinic acid by pyridoxine in chronic schizophrenics. *Can. Psychol. Ass. J.*, 18:377-384.

3. N. E. Anden, H. Corrodi, K. Fuxe, and J. L. Meek (1974). Hallucinogenic phenylethylamines: Interactions with serotonin turnover and receptors. *Eur. J. Pharmacol.*, 25:176-184.

4. A. A. Boulton and L. Milward (1971). Separation, detection and quantitative analysis of urinary β-phenylethylamine. *J. Chromatogr.*, 57:287-296.

5. B. D. Bower (1961). The trypophan load test in the syndrome of infantile spasms with oligophrenia. *Proc. Roy. Soc. Med.*, 54: 540-544.

6. F. Brambilla, M. Giardini, and R. Russo (1975). Prospects for a pharmacological treatment of phenylketonuria. *Dis. Ner. Syst.*, 36:257-260.

7. G. C. Cotzias, M. H. Van Woert, and L. M. Schiffer (1967). Aromatic amino acids and modification of Parkinsonism. *N. Eng. J. Med.*, 276:374-379.

8. J. Delay, P. Deniker, J. Buisson, and A. Haim (1959). Le traitement des ētats dèpressifs par les derives de lacide kso-nicotinique, isoniazide et iproniazide. *Ann. Med. Psychol.*, 117:125-132.

9. W. G. Dewhurst (1968). New theory of cerebral amine function and its clinical application. *Nature*, 218:1130-1133.

10. J. H. Fellman (1956). Inhibition of dopa decarboxylase by aromatic acids associated with phenylpyruvic oligophrenia. *Proc. Soc. Exp. Biol. Med.*, 93:413-414.

11. R. Fernandez Labriola and E. Rodriguez Casanova (1973). Modification de las aminas biogenas en depresivos tratados con desipramina. *An. Psiq. Bio., Supl. Al*, 4:34-36.

12. R. Fernandez Labriola, E. Rodriguez Casonova, H. Spatz, and N. Spatz (1974). Influence of anti-depressive therapy on phenethylaminuria. *An. Psiq. Bio.*, 3:237-238.

13. E. Fischer and B. Heller (1972). Fenetilamina como agente neurohumoral, sus efectos psicofisiologicos, su papel en enfermedades mentales y en el mecanismo de accion de drogas. *Minerva Psiq.*, 1:18-25.

14. E. Fischer, P. L. Ludmer, and H. C. Sabelli (1967). The antagonism of phenylethylamine to catecholamines on mouse activity. *Acta Phisiol. Lat. Am.*, 17:15-21.

15. E. Fischer, B. Heller, and A. A. Miro (1968). β-Phenylethylamine in human urine. *Arzneim.-Forsch.*, 18:1486.

16. E. Fischer, B. Heller, M. Nachon, and H. Spatz (1975). Therapy of depression by phenylalanine. *Arzneim.-Forsch.*, 25:132.

17. E. Fischer, J. M. Saavedra, and B. Heller (1968). Effects of catecholamines, adrenergic substances and their blocking agents on the searching behavior of mice. *Arzneim.-Forsch.*, 18:780-786.

18. E. Fischer, H. Spatz, R. Fernandez Labriola, E. Rodriguez Casanova, and N. Spatz (1973). Quantitative gas-chromatographic determination and infrared spectrographic identification of urinary phenethylamine. *Biol. Psychiat.*, 7:161-165.

19. E. Fischer, H. Spatz, J. M. Saavedra, H. Reggiani, A. H. Miro, and B. Heller (1972). Urinary elimination of phenethylamine. *Biol. Psychiat.*, 5:139-147.

20. E. Fischer, H. Spatz, B. Heller, and H. Reggiani (1972). Phenethylamine content of human urine and rat brain; Its alterations in pathological conditions and after drug administration. *Experientia*, 28:307-308.

21. T. L. Fleisher and I. Zeligman (1960). Cutaneous findings in phenylketonuria. *Arch. Dermatol.*, 81:898-903.

22. W. K. Frankenburg, A. D. Goldstein, and C. O. Olson (1973). Behavioral consequences of increased phenylalanine intake by phenylketonuric children: A pilot study describing a methodology. *Am. J. Ment. Defic.*, 77:524-532.

23. K. Fuxe, H. Grobecker, and J. Jonsson (1967). The effect of β-phenylethylamine on central and peripheral monoamine-containing neurons. *Eur. J. Pharmacol.*, 2:202-207.

24. E. Hanington (1967). Preliminary report on tyramine headache. *Br. Med. J.*, ii:550-551.

25. E. Hanington, M. Horn, and J. Wilkinson (1970). Further observations on the effects of tyramine. In: *Background to Migraine. Third Migraine Symposium*, edited by A. L. Cochran. Heinemann, London. 113-119.

26. R. G. Heath, A. W. Cox, and L. S. Lustick (1974). Brain activity during emotional states. *Am J. Psychiat.*, 131:8,858-862.

27. O. J. Ipar, E. Fischer, B. Heller, and M. N. Nachon (1974). Terapia de la depresion con D-fenilalanina. *Prensa. Med. Arg.*, 61:498.

28. J. B. Jepson, W. Lovenberg, P. Zaltzman, J. A. Oates, A. Sjoerdsma, and S. Udenfriend (1960). Amine metabolism, studied in normal and phenylketonuric humans by monoamine oxidase inhibition. *Biochem. J.*, 74:5P (abstr.).

29. R. W. Kimball, A. P. Friedman, and E. Vallejo (1960). Effect of serotonin in migraine patients. *Neurology*, 10:107-111.

30. Y. H. Loo (1967). Characterization of a new phenylalanine metabolite in phenylketonuria. *J. Neurochem.*, 14:813-821.

31. C. M. McKean (1972). The effects of high phenylalanine concentrations on serotonin and catecholamine metabolism in the human brain. *Brain Res.*, 47:469-476.

32. J. W. Maas (1975). Biogenic amines and depression. *Arch. Gen. Psychiat.*, 32:1357-1361.

33. J. Mendels (Ed.) (1975). *The Psychobiology of Depression.* Halsted Press, New York.

34. S. G. Moses and E. Robins (1975). Regional distribution of norepinephrinc and dopamine in brains of depressive suicides and alcoholic suicides. *Psychopharmacol. Comm.*, 1(3):327-337.

35. A. D. Mosnaim, E. E. Inwang, J. H. Sugarman, W. J. DeMartini, and H. C. Sabelli (1973). Ultraviolet spectrophotometric determination of 2-phenylethylamine in biological samples and its possible correlation with depression. *Biol. Psychiat.*, 6:235-257.

36. A. D. Mosnaim, E. E. Inwang, and H. C. Sabelli (1973). Isolation and characterization of phenylethylamine and phenylethanolamine from human brain. *J. Neurochem.*, 20:1469-1473.

37. A. D. Mosnaim and E. E. Inwang (1973). A spectrophotometric method for the quantification of 2-phenylethylamine in biological specimens. *Anal. Biochem.*, 54:561-577.

38. A. D. Mosnaim, H. C. Sabelli, and E. E. Inwang (1973). The influence of psychotropic drugs on the levels of endogenous 2-phenylethylamine in rabbit brain. *Biol. Psychiat.*, 8:227-234.

39. A. D. Mosnaim and H. C. Sabelli (1971). Quantitative determination of the brain levels of a b-phenylethylamine-like substance in control and drug treated mice. *Pharmacologist*, 13:2831.

40. M. N. Nachon and M. A. Di Santo (1974). Tratamiento de episodios depresivos con el aminoacido fenilalanina. In: *Proc. 1st World Congr. Biol. Psychiat.*, Buenos Aires.

41. T. Nakajima, Y. Kakimoto, and I. Sano (1964). Formation of phenylethylamine in mammalian tissues and its effect on motor activity in the mouse. *J. Pharmacol. Exp. Ther.*, 143:319-325.

42. J. A. Oates, P. Z. Nirenburg, J. B. Jepson, A. Sjoerdsma, and S. Udenfriend (1963). Conversion of phenylalanine to phenethylamine in patients with phenylketonuria. *Proc. Soc. Exp. Biol. Med.*, 112:1078-1081.

43. C. M. Pare, M. Sandler, and R. S. Stacey (1972). 5-Hydroxytryptamine deficiency in phenylketonuria. *Lancet*, i:551-553.

44. M. S. Patel and I. J. Arinze (1975). Phenylketonuria: Metabolic alterations induced by phenylalanine and phenylpyruvate. *Am. J. Clin. Nutrit.*, 28:183-188.

45. T. L. Perry (1962). Urinary excretion of amines in phenylketonuria and mongolism. *Science,* 136:879-880.

46. V. Ramachandran, K. V. Thiruvengadam, M. G. Zakria, and A. Kathiresan (1974). Psychiatric symptoms and personality traits in asthma patients. *Indian J. Psychiat.,* 16(2):170-174.

47. E. Rodriguez Casanova and R. Fernandez Labriola (1972). Estudios sobre la fenetilaminuria en depresiones. *An. Psiq. Biol.,* 1:6-9.

48. E. Rodriguez Casanova and R. Fernandez Labriola (1973). Eliminacion de fenetilamina antes y despues del tratamiento antidepresivo de la depresion endogena. *An. Psiq. Biol.,* 2:69-73.

49. H. C. Sabelli, W. J. Giardina, A. D. Mosnaim, and E. E. Inwang (1972). *Abst. Commun. Int. Congr. Pharmacol.,* San Francisco, p. 198.

50. H. C. Sabelli, A. D. Mosnaim, A. J. Vasquez, and L. M. Pedemonte (1973). (-)-Trans-Δ^9-Tetrahydrocannabinol-induced increase in brain 2-phenylethylamine: Its possible role in the behavioral effects of marihuana. In: *Drug Addiction: Behavioral Aspects,* edited by J. Singh. Futura Publishing, New York.

51. M. Sandler (1972). Migraine: A pulmonary disease? *Lancet,* i:618-619.

52. M. Sandler, M. B. H. Youdim, and E. Hanington (1974). A phenylethylamine-oxidising effect in migraine. *Nature,* 250: 335-337.

53. M. Sandler, M. B. H. Youdim, E. Hanington, S. J. Corne, and R. J. Stephens (1975). Phenylethylamine-oxidising defect in migraine: Circulatory implications. *Clin. Exp. Pharmacol. Physiol.,* 2:79-81.

54. M. Sandler and G. P. Reynolds (1976). Does phenylethylamine cause schizophrenia? *Lancet,* i:70-71.

55. S. Saxl (1933). Ein Fall von Asthma bronchiale bei einem Manisch-Depressiven. *Wiener Klinische Wochenschrift,* 50: 1515-1516.

56. J. W. Schweitzer, A. J. Friedhoff, and R. Schwartz (1975). Phenethylamine in normal urine: Failure to verify high values. *Biol. Psychiat.,* 10:277-285.

57. Scientific Service (Roche Labs.) (1962). *Treatment of alcoholics with benadon.*

58. A. T. Shulgin and M. F. Carter (1975). Centrally active phenethylamines. *Psychopharmacol. Comm.,* 1(1):93-98.

59. H. Spatz, B. Heller, M. Nachon, and E. Fischer (1975). Effects of D-phenylalanine on clinical picture and phenethylaminuria in depression. *Biol. Psychiat.,* 10(2):235-239.

60. G. Weber, R. I. Glazer, and R. A. Ross (1970). Regulation of human and rat brain metabolism: Inhibitory action of phenylalanine and phenylpyruvate on glycolysis, protein, lipid, DNA and RNA metabolism. In: *Advances in Enzyme Regulation*, edited by G. Weber. Pergamon, New York, p. 13.

61. N. F. Winston (1973). Oral contraceptives, pyridoxine and depression. *Am. J. Psychiat.*, 130:1217-1222.

62. D. Woolley (1967). Involvement of the hormone serotonin in emotion and mind. In: *Neurophysiology and Emotion*. Rockefeller Press, New York, pp. 108-116.

63. J. A. Yaryura-Tobias (1959). Pyridoxine in acute alcoholism. *Mtg. Asoc. de Medicos del Hosp. Fernandez*, Buenos Aires.

64. J. A. Yaryura-Tobias, B. Heller, H. Spatz, and E. Fischer (1974). Phenylalanine for endogenous depression. *J. Ortho. Psychiat.*, 3(2):80-81.

65. J. A. Yaryura-Tobias, T. Fledel, and F. A. Neziroglu (1975). Vitamins, neuroleptics and anti-parkinsonian drugs in catecholamine excretion. *Biol. Psychiat.*, 11:363-366.

66. J. A. Yaryura-Tobias and F. A. Neziroglu (1975). Violent behavior, brain dysrhythmia, and glucose dysfunction: A new syndrome. *J. Ortho. Psychiat.*, 4:182-188.

67. J. A. Yaryura-Tobias and F. A. Neziroglu (1976). Phenylethylamine and glucose in true depression. *J. Ortho. Psychiat.*, 5:199-202.

68. J. A. Yaryura-Tobias and F. Neziroglu (1976). Psychosis and disturbance of glucose metabolism. *J. Prev. Med.*, 2:38-45.

69. A. Yuwiler and E. Geller (1969). Brain serotonin changes in phenylalanine-fed rats: Synthesis storage and degradation. *J. Neurochem.*, 16:999-1005.

70. E. A. Zeller and B. Barsky (1952). In vivo inhibition of liver and brain monoamine oxidase by 1-isonicotinly-z-isopropyl hydrazine (19910). *Proc. Soc. Exp. Biol. Med.*, 81:459-461.

AUTHOR INDEX

Numbers in parentheses are reference numbers and indicate that an author's work is referred to although his or her name is not cited in text. Underlined numbers give the page on which the complete reference is listed.

SUBJECT INDEX

PROPERTY OF BRISTOL-MYERS CO.
MEAD JOHNSON
RESEARCH CENTER
JAN 22 1979
LIBRARY